Organometallic Chemistry: A Practical Approach

Organometallic Chemistry: A Practical Approach

Andreas Hobbs

MURPHY & MOORE
www.murphy-moorepublishing.com

Published by Murphy & Moore Publishing,
1 Rockefeller Plaza,
New York City, NY 10020, USA

ISBN: 978-1-63987-420-0

Cataloging-in-Publication Data

Organometallic chemistry : a practical approach / Andreas Hobbs.
p. cm.
Includes bibliographical references and index.
ISBN 978-1-63987-420-0
1. Organometallic chemistry. 2. Chemistry, Organic. I. Hobbs, Andreas.
QD411 .O74 2022
547.05--dc23

For information on all Murphy & Moore Publications
visit our website at www.murphy-moorepublishing.com

MURPHY & MOORE

Table of Contents

Preface

This book aims to help a broader range of students by exploring a wide variety of significant topics related to this discipline. It will help students in achieving a higher level of understanding of the subject and excel in their respective fields. This book would not have been possible without the unwavered support of my senior professors who took out the time to provide me feedback and help me with the process. I would also like to thank my family for their patience and support.

Organometallic chemistry refers to the study of organometallic compounds that have at least one chemical bond between a metal and a carbon atom of an organic compound. This metal can be alkaline earth, alkaline, transition metals and sometimes metalloids like tin, boron and silicon. Carbon monoxide, carbide and cyanide are some of the common organometallic compounds. These compounds are used as catalysts in a wide range of industrial chemical reactions. Some of the important reactions that organometallic compounds undergo are electron transfer, cyclometalation, nucleophilic abstraction, transmetalation and organometallic substitution reaction. This field of study draws from the areas of organic and inorganic chemistry. The various sub-specialties of this discipline are organolithium chemistry, organoaluminium chemistry, organotitanium chemistry, organosilver chemistry, organoplatinum chemistry and organouranium chemistry. This book attempts to understand the multiple branches that fall under the discipline of organometallic chemistry. Different approaches, evaluations, methodologies and advanced studies related to this discipline have been included herein. Constant effort has been made to make the understanding of the difficult concepts of this field as easy and informative as possible, for the readers.

A brief overview of the book contents is provided below:

Chapter - Introduction

Organometallic compounds are the compounds which contain at least one metal-to-carbon bond where carbon is part of organic group. Organometallic chemistry is the branch of chemistry that deals with the study of organometallic compounds including alkaline, alkaline earth, metalloids and transition metals, etc. This chapter discusses organometallic chemistry in detail.

Chapter - Concepts of Organometallic Chemistry

There are many concepts that are studied under the domain of organometallic chemistry. Some of them are 18- electron rule, pi backbonding, ligand field theory, frontier molecular orbital theory, etc. This chapter has been carefully written to provide an easy understanding of the concepts associated with organometallic chemistry.

Chapter - Organometallic Compounds of Main Group Elements

Organometallic compounds of main group elements belong to the s-block and p-block elements. It includes organomagnesium compounds, organoaluminium compounds, organoboron compounds, organolithium reagent, Ziese's salt, etc. The topics elaborated in this chapter will help in gaining a better perspective of organometallic compounds of main group elements.

Chapter – Organometallic Compounds of Transition Metals

Organometallic compounds of transition metals belong to the d-block and f-block and follow similar rules as the main-group organometallics. Organotin compounds, organoiron compounds, organocobalt compounds, organotitanium compounds, etc. come under it. All the organometallic compounds of transition metals have been carefully analyzed in this chapter.

Chapter – Organometallic Reactions

Organometallic reactions form a new bond through reduction in a formal oxidation state of metals. They show the basic and nucleophilic character of the carbon atoms in organometallic compounds. Some of the reactions are hydrogenation reaction, carbometalation, transmetalation, oxidative addition reaction, etc. This chapter delves into the aspects of organometallic reactions for a thorough understanding of the subject.

Andreas Hobbs

1

Introduction

Organometallic compounds are the compounds which contain at least one metal-to-carbon bond where carbon is part of organic group. Organometallic chemistry is the branch of chemistry that deals with the study of organometallic compounds including alkaline, alkaline earth, metalloids and transition metals, etc. This chapter discusses organometallic chemistry in detail.

Organometallic chemistry is the study of chemical compounds containing bonds between carbon and metal atoms. In more general terms, it is the study of compounds containing metal-element bonds that are largely covalent in character. Organometallic chemistry combines aspects of inorganic chemistry and organic chemistry. Living systems contain a variety of organometallic compounds, prominent examples being hemoglobin and chlorophyll. The specialized field focused on the study of such compounds is known as bioinorganic chemistry.

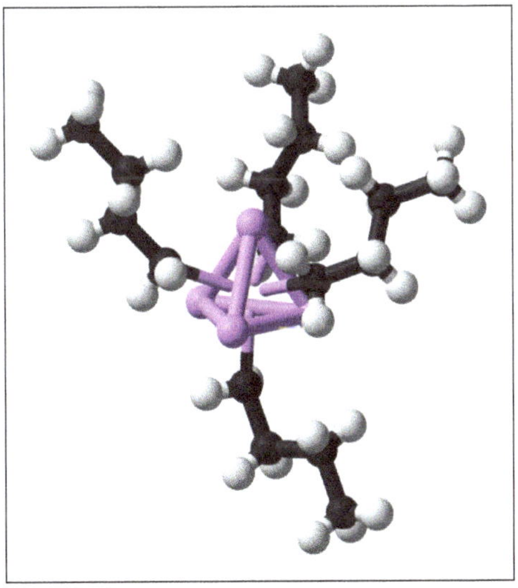

A model of a molecule of n-butyllithium, an organometallic compound.

Organometallic compounds have a variety of practical uses. For example, ferrocene and MMT (methylcyclopentadienyl manganese tricarbonyl) are added to gasoline as antiknock agents. The industrial manufacture of acetic acid from methanol and carbon monoxide (by the Monsanto process) involves the use of a rhodium-carbonyl complex. A titanium-based organometallic compound, called the Ziegler-Natta catalyst, is used in the production of polyethylene and other polymers. Also, the ruthenium-BINAP complex is used in the production of fine chemicals and pharmaceuticals.

A number of researchers have been awarded the Nobel Prize in Chemistry for their work in the area of organometallic chemistry. For example, the 1973 Nobel Prize was awarded to Ernst Fischer and Geoffrey Wilkinson for their work on metallocenes. In 2005, Yves Chauvin, Robert H. Grubbs, and Richard R. Schrock shared the Nobel Prize for their work on metal-catalyzed olefin metathesis.

ORGANOMETALLIC COMPOUNDS

Organometallic Compounds are chemical compounds which contain at least one bond between a metallic element and a carbon atom belonging to an organic molecule. Even metalloid elements such as silicon, tin, and boron are known to form organometallic compounds which are used in some industrial chemical reactions.

The catalysis of reactions wherein the target molecules are polymers or pharmaceuticals can be done with the help of organometallic compounds, resulting in an increase in the rate of the reactions.

Generally, the bond between the metal atom and the carbon belonging to the organic compound is covalent in nature. When metals with relatively high electropositivity (such as sodium and lithium) form these compounds, a carbanionic nature is exhibited by the carbon which is bound to the central metal atom.

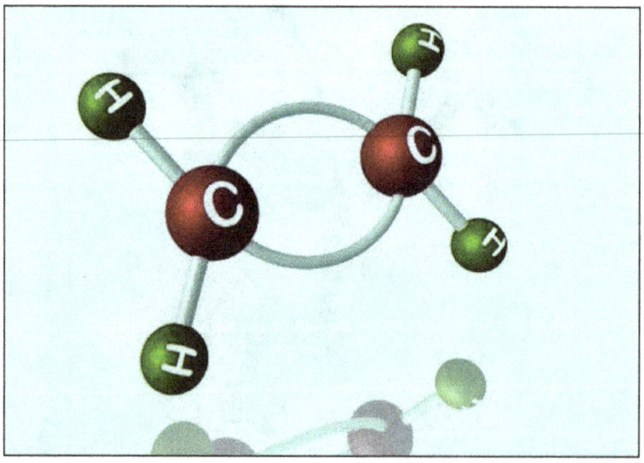

An example of an organometallic compound wherein carbons belonging to a benzene molecule bond with chromium is illustrated above. A few more examples of these types of compounds are Grignard reagents, tetracarbonyl nickel, and dimethyl magnesium.

Properties of Organometallic Compounds

A few properties of organometallic compounds are listed below as short points:

- The bond between the metal and the carbon atom is often highly covalent in nature.
- Most of the organometallic compounds exist in solid states, especially the compounds in which the hydrocarbon groups are aromatic or have a ring structure.
- The compounds consisting of highly electropositive metals such as sodium or lithium are very volatile and can undergo spontaneous combustion.
- In many cases, organometallic compounds are found to be toxic to humans (especially the compounds that are volatile in nature).
- These compounds can act as reducing agents, especially the compounds formed by highly electropositive metals.

From the points given above, it can be observed that the properties of organometallic compounds differ amongst each other based on the properties of the metals that constitute them.

Applications

Organometallic Compounds have a broad range of applications in the field of chemistry. Some of them are given below:

- In some of the commercial chemical reactions, organometallic compounds are used as homogeneous catalysts.
- These compounds are used as stoichiometric reagents in both industrial and research-oriented chemical reactions.
- These compounds are also used in the manufacture of some of the semiconductors, which require the use of compounds such as trimethylgallium, trimethylaluminum, trimethylindium, and trimethyl antimony.
- They are also used in the production of light emitting diodes (or LEDs).
- These compounds are employed in bulk hydrogenation processes such as the production of margarine.
- These compounds are used as catalysts and reagents during the synthesis of some organic compounds.
- The complexes formed from organometallic compounds are useful in the facilitation of the synthesis of many organic compounds.

The points given above emphasize the importance of organometallic compounds. However, they are also the cause of many environmental concerns due to the highly toxic nature of some of these types of compounds.

Elements that form Organometallics

A wide variety of elements of the periodic table can form organometallic compounds. Examples are given below:

- Period 2 elements: Organolithium chemistry, organoberyllium chemistry, organoborane chemistry.
- Period 3 elements: Organomagnesium chemistry, organoaluminum chemistry, organosilicon chemistry.
- Period 4 elements: Organotitanium chemistry, organochromium chemistry, organomanganese chemistry, organoiron chemistry, organocobalt chemistry organonickel chemistry, organocopper chemistry, organozinc chemistry, organogallium chemistry, organogermanium chemistry.
- Period 5 elements: Organopalladium chemistry, organosilver chemistry, organocadmium chemistry, organoindium chemistry, organotin chemistry.
- Period 6 elements: Organoplatinum chemistry, organogold chemistry, organomercury chemistry, organothallium chemistry, organolead chemistry.

Organometallic Complexes

Organometallic complexes are compounds that have dative covalent bonds between a metal and organic ligands. The organic ligands often bind the metal through a hetero – atom such as oxygen or nitrogen; such compounds are called "coordination compounds". Many organic coordination compounds occurred naturally e.g. myoglobin, haemoglobin and chlorophyll. Organometallic complexes are highly basic, reducing agents which serve as catalyst by catalyzing many polymerization reactions and find practical applications in stoichiometric and catalytic processes. Well characterized organometallic complexes can sometimes be detected and isolated from "Syn-gas" and "Water-gas" shift reactions which serve as the principal carbon source form coal.

Identification and Analysis of Organometallic Compounds and Complexes

Technology is now available such as Nuclear Magnetic Resonance (NMR) and infrared spectroscopy are common techniques used to determine structure because biomedical spectroscopy is a multidisciplinary research field involving spectroscopic tools for applications in the field of biomedical science. Vibration spectroscopy such as Raman or infrared spectroscopy is used to determine the chemical composition of a material based on detection of vibration modes of constituent molecules. Some spectroscopic methods are routinely used in clinical settings for diagnosis of disease; an example is Magnetic resonance imaging (MRI). Fourier transform infrared (FTIR) spectroscopic imaging is a form of chemical imaging for which the contrast is provided by composition of the material. Nuclear magnetic resonance spectroscopy, most commonly known as NMR spectroscopy, is a research technique that exploits the magnetic properties of certain atomic nuclei. It determines the physical and chemical properties of atoms or the molecules in which they are contained. It relies on the phenomenon of nuclear magnetic resonance and can provide detailed information about the structure, dynamics, reaction state, and chemical environment of molecules. The intramolecular magnetic field around an atom in a molecule changes the resonance frequency, thus giving access to details of the electronic structure of a molecule similarly. The polyelectrolyte nature of the absorbent studied and the hypothesis that each site occupied by a proton may be available for the absorption of a metal atom, allows the application of a potentiometric titration method, generally used in the study of organometallic complexes in solution.

Most frequently, NMR spectroscopy is used by chemists and biochemists to investigate the properties of organic molecules, although it is applicable to any kind of sample that contains nuclei possessing spin. Suitable samples range from small compounds analyzed with 1-dimensional proton or carbon-13 NMR spectroscopy to large proteins or nucleic acids using 3 or 4-dimensional techniques. The impact of NMR spectroscopy on the sciences has been substantial because of the range of information and the diversity of samples, including solutions and solids. NMR spectra are unique, well-resolved, analytically tractable and often highly predictable for small molecules. Thus, in organic chemistry practice, NMR analysis is used to confirm the identity of a substance. Different functional groups are obviously distinguishable, and identical functional groups with differing neighbouring substituent still give distinguishable signals. NMR has largely replaced traditional wet chemistry tests such as colour reagents for identification. A disadvantage is that a relatively large amount, 2–50 mg, of a purified substance is required, although it may be recovered. Preferably, the sample should be dissolved in a solvent, because NMR analysis of solids requires a dedicated MAS machine and may not give equally well-resolved spectra. The timescale of NMR is relatively long, and thus it is not suitable for observing fast phenomena, producing only an averaged spectrum. Although large amounts of impurities do show on an NMR spectrum, better methods exist for detecting impurities, as NMR is inherently not very sensitive. NMR spectrometers are relatively expensive; universities usually have them, but they are less common in private companies. Modern NMR spectrometers have a very strong, large and expensive liquid helium-cooled superconducting magnet, because resolution directly depends on magnetic field strength. Less expensive machines using permanent magnets and lower resolution are also available, which still give sufficient performance for certain application such as reaction monitoring and quick checking of samples.

Most drugs, catalysts, reagents and inhibitors used today are purely organic compounds. Especially after the enormous success of the cisplatin in tumor treatment, interest in metal complexes has grown. Synthetic organometallic compounds are generally considered to be toxic or non-compatible with biological systems. Despite this perception, the medicinal properties of organometallic compounds, in particular organo-transition metal compounds have been probed for a long time and in the last few years the area has grown considerably.

Catalysts are substances that increase the rate of a reaction but are not consumed in the reaction. Catalysts are widely encountered in nature, industry, and in the laboratory. Many of the catalysts utilized in the chemical industry and the laboratory are organometallic compounds which are shown in the reactions.

Reaction with Metal and Transmetallation

The net reaction of an electropositive metal M and a halogen-substituted hydrocarbon is; if, one metal atom takes the place of another, it is called transmetallation. Transmetallation is favorable when the displacing metal is higher in the electrochemical series than the displaced metal.

Metathesis

The metathesis of an organometallic compound (MR) and a binary halide (EX) is a widely used synthetic route in organometallic chemistry. Metathesis reaction can frequently be predicted from either electronegativity or acid-base considerations. Hydrocarbon groups tends to bond to the more electronegative element; the halogen favour the formation of ionic compounds with the more electropositive metal. In brief, the alkyl and aryl group tends to migrate from the less to the more electronegative element [X= electronegativity]. When the electronegativities are similar, the correct outcome may be predicted, with care, by considering the combination of the softer element with organic group and harder element with fluoride or chloride. An insoluble product or reactant may change the outcome, e.g.; HgPhBr turns out to be insoluble in THF. Metathesis reactions involving the same central element are often referred to as redistribution reactions. Aluminum (Al) is more electropositive than Germanium (Ge), this reaction occurs as it is thermodynamically favorable.

Hydrometallation

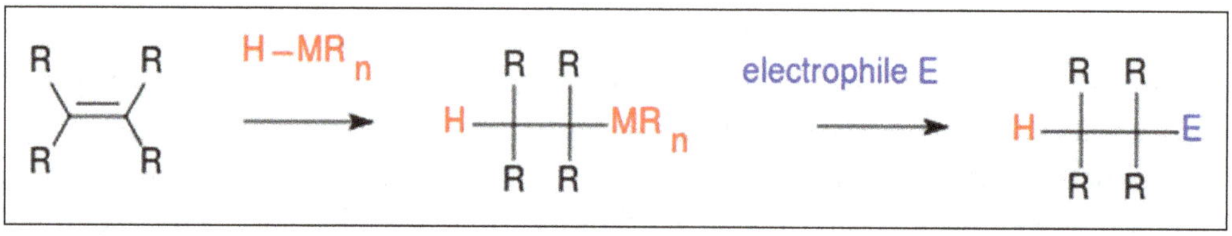

Shows the reaction of alkene forming a new compound with carbon and metal bond.

This is a type of chemical reaction in organometallic chemistry in which a chemical compound with a hydrogen to metal bond (M-H, metal hydride) adds to compounds with an unsaturated bond like an alkene (RC=CR) forming a new compound with a carbon to metal bond (RHC-CRM). The metal (M) is less electronegative than hydrogen (H); the reverse reaction is beta-hydride elimination. The reaction is structurally related to carbometalation. When the substrate is an alkyne the reaction product is a vinylorganometallic.

Hydrosilylation

Hydrosilylation also called catalytic hydrosilation describes the addition of Si-H bonds across bonds. Ordinarily the reaction is conducted catalytically and usually the substrates are unsaturated organic compounds. Alkenes and alkynes give alkyl and vinyl silanes, aldehydes and ketones give silyl ethers. Hydrosilylation has been called the "most important application of platinum in homogeneous catalysis. The catalytic transformation represents an important method for preparing organosilicon compounds. An idealized transformation is illustrated by the addition of triethylsilane to diphenylacetylene shown below:

$$Et_3SiH + (Ph)C \equiv C(Ph) \rightarrow Et_3Si(Ph)C = CH(Ph)$$

Hydrocarbons

In commercial preparation of hydrocarbon, methyllithium (MeLi) exists as a tetrahedral cluster in the solid state and in the solution which serve as are very strong base. Although not usually done with Grignard reagents, organolithium reagents can be used as strong bases. Both Grignard reagents and organolithium reagents react with water to form the corresponding hydrocarbon. This is why so much care is needed to insure dry glassware and solvents when working with organometallic reagents as shown below:

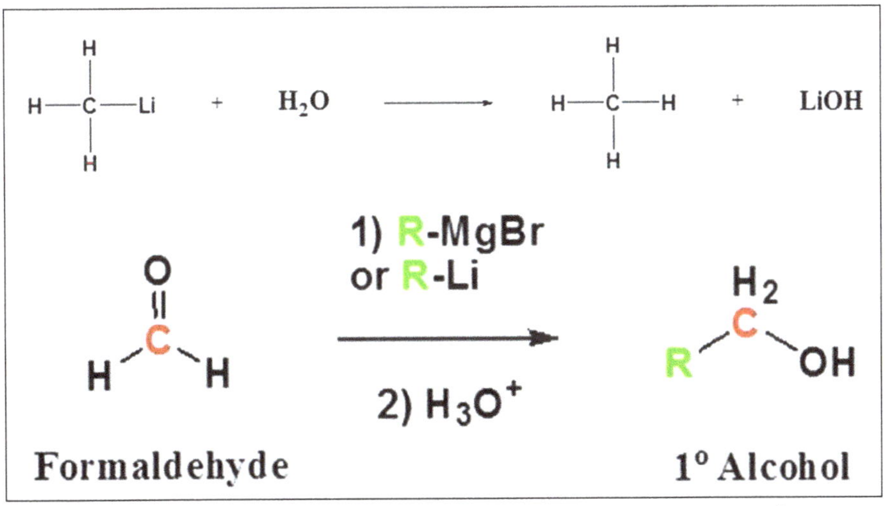

Shows how both Grignard and organolithium reagent serve as strong base to produce corresponding hydrocarbons.

Many of its higher homologs exist in solution as hexamers or equilibrium mixture of aggregates ranging up to haxamers. The larger aggregates can be broken down by Lewis bases, such as Tetramethylethylenediamine (TMEDA); TMEDA is widely employed as a ligand for metal ions. It forms stable complexes with many metal halides, e.g. zinc chloride and copper (I) iodide, and gives complexes that are soluble in organic solvents. In such complexes, TMEDA serves as a bidentate ligand which has an affinity for lithium ions and the structure is shown below:

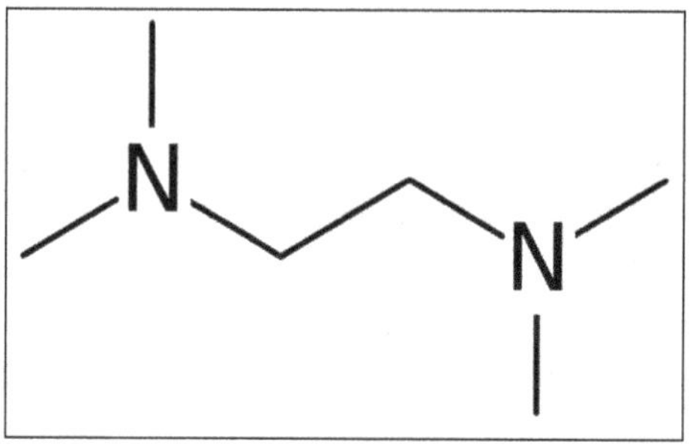

Shows the chemical structure of TMEDA.

Common organolithium compounds have only Lithium per organic group, several polylithiated organic molecules containing several lithium atoms per molecule are known. The simplest example is Li_2CH_2, which can be prepared by the pyrolysis of MeLi which crystallizes in a distorted antifulorite structure. However, the finer details of the orientation of the CH_2 groups are yet to be established. The antifluorite structure is the inverse of the fluorite structure in which the locations of cations and anions are reversed.

Naturally occurring CaF_2 is the principal source of hydrogen fluoride, a commodity chemical used to produce a wide range of materials. Calcium fluoride in the fluorite state is of significant commercial importance as a fluoride source. Hydrogen fluoride is liberated from the mineral by the action of concentrated sulphuric acid as shown below:

$$CaF_2 + H_2SO_4 \rightarrow CaSO_4\,(\text{solid}) + 2HF$$

Look into the structures of CaF_2 (fluorite structure) and K_2O, Na_2o and Li_2o (antifluorite structure). An array of cations and all the tetrahedral holes are filled with anions.

Organometallic Complexes Serve as Highly Basic and Reducing Agents

Sodium naphthalide is an example of an organometallic salt with a delocalized radical anion, $C_{10}H_8$. Such compounds are readily prepared by reacting an aromatic compound with an alkali metal in a polar aprotic solvent. Naphthalene dissolved in THF reacts with Na metal to produce a dark green solution of sodium naphthalide. EPR spectra show that the odd electron is delocalized in an anti-bonding orbital of $C_{10}H_8$.

Formation of radical anion is more favorable when the π of lowest unoccupied molecular orbital (LUMO) of the arene is low in energy. Simple molecular orbital theory (MOT) predicts that the energy of LUMO decreases steadily on going from benzene to more extensively conjugated hydrocarbons. Sodium naphthalide and similar compounds are highly reactive reducing agents. They are preferred to sodium because unlike sodium, they are readily soluble in ethers. The resulting homogeneous reaction is generally faster and easier to control than a heterogeneous reaction between one reagent in solution and pieces of sodium metal, which are often coated with unreactive sodium oxide or with insoluble reaction products. The additional advantage is that by proper choice of the aromatic group the reduction potential of the reagent can be chosen to match the requirements of a particular synthetic task. Alternative route to delocalized anion is the reductive cleavage of acidic C-H bonds by an alkali metal or alkylmetallic compound. The reaction below shows the radicality and basicity of organometallic complexes.

Hydrogenation Reaction

Hydrogenation reactions are extensively used to create commercial goods; hydrogenation is used in the food industry to make a large variety of manufactured goods, like spreads and shortenings, from liquid oils. This process also increases the chemical stability of products and yields semisolid products like margarine. Hydrogenation is also used in coal processing. Solid coal is converted to a liquid through the addition of hydrogen. Liquefying coal makes it available to be used as fuel. An example of an alkenes addition reaction is a process called hydrogenation. In a hydrogenation reaction, two hydrogen atoms are added across the double bond of alkenes, resulting in saturated alkenes. Hydrogenation of a double bond is a thermodynamically favorable reaction because it forms a more stable (lower energy) product in the presence of catalyst such as palladium, Nickel and platinum.

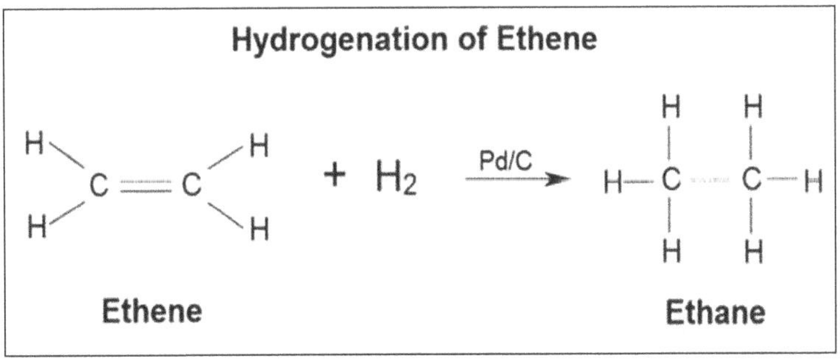

Example of hydrogenation of Ethene using palladium as catalyst.

The Rhodium catalyst has been shown to be suitable promoter for the activation of C-H bonds, which has emerged as a challenging and attractive tool for catalysis. Rhodium catalysis finds increasing interest in the catalytic dehydrogenative cross-coupling, allowing elegant C-C bond construction. Although palladium has been the metal of choice for most examples, Rh catalysts can be suitable promoters for this activation. By using rhodium catalysts from Sigma-Aldrich, access is gained to important couplings such as aryl-aryl, aryl-alkene, and alkene-alkene, viable routes to valuable organic frameworks. Rhodium complexes in solution are the catalysts and are not used up in the reaction. Rhodium, which is more expensive than platinum, can be used even in catalytic processes where the products are inexpensive, because the Rhodium is not consumed. Modifications of this type of catalyst are also employed in the production of pharmaceuticals such as levodopa, which is used to treat Parkinson's disease.

Hydroformylation Reaction

Hydroformylation involves the addition of carbon monoxide and hydrogen to an alkene to form an aldehyde containing one or more carbon atoms than the original alkenes. This catalytic reaction is employed in the petrochemical industry, in the 1960, Rhodium serve as highly active catalysts used. Since the 1970, most hydroformylation relies on catalysts based on Rhodium. Subsequent research led to the development of water-soluble catalysts that facilitate the separation of the products from the catalyst. The catalytic cycle proceeds through a series of organometallic intermediates. The aldehydes produced by hydroformylation are normally reduced to alcohols that are used as solvents, as plasticizers, and in the synthesis of detergents.

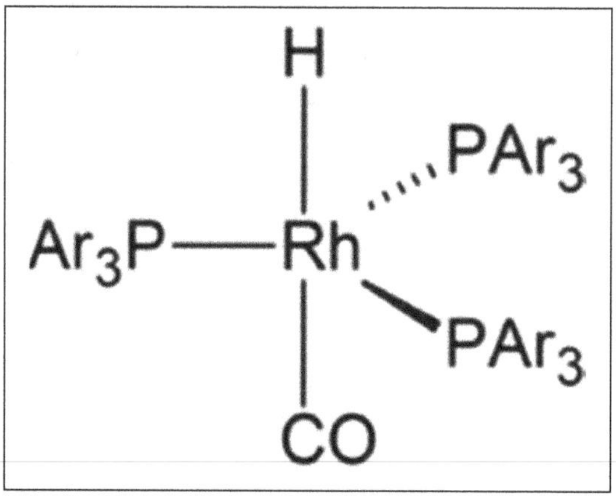

Shown the structure of organorhodium complex which is used as a catalyst in chemical industries.

Alkenes Polymerization Reaction

Polyalkenes, the most common and useful class of synthetic polymers, are often prepared by use of organometallic catalysts, either in solution or supported in a solid surface. In the 1950, the German chemist Karl Ziegler developed a catalyst for ethene polymerization based on a catalyst formed by the reaction of $TiCl_4$ with $Al(C_2H_5)_3$. Soon thereafter, Giulio Natta made use of this type of catalyst for the polymerization of propene to produce polymers with highly regular structures. The intimate details of the reactions of these commercial catalytic processes are not entirely understood, but there are strong indications from more easily studied soluble organometallic catalysts that alkenes coordinate to a metal centre and then insert into a hydrocarbon chain, producing a longer-chain hydrocarbon attached to the metal centre. Repetition of this process leads to extremely-long-chain hydrocarbon polymers, which include many of the most familiar plastics, such as polyethylene and polypropylene. These plastics are used in consumer items ranging from milk containers and plastic bags to artificial limbs and car bumpers. In catalysis silver is active as silver oxide in the Wolff rearrangement. Silver nitrate is used to separate out alkenes as the η^2 -alkene complex. Silver is also present in other carbon-carbon bond skeletal rearrangements such as the quadricyclane to norbornadiene rearrangement, the cubane to cuneane rearrangement and the rearrangement of the cyclobutadiene dimer to cyclooctatetraene.

Applications of Organometallic Compounds and Complexes

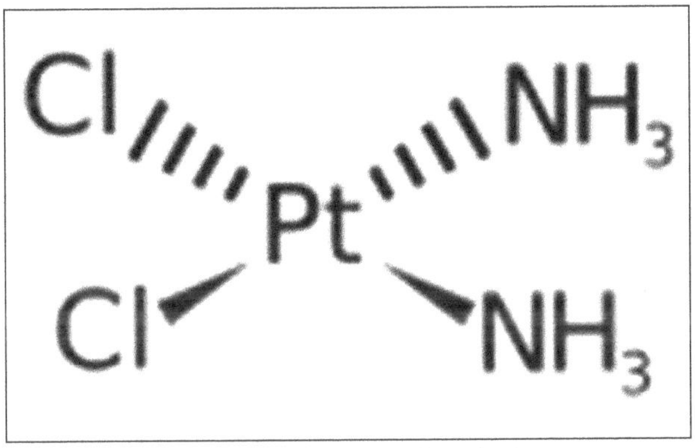

Molecular structure of cisplatin.

A new approach involves modeling of both pharmacological and catalysis properties of established drugs with organometallic fragments for clinical and industrial purposes, several metal complexes have been tested in anticancer therapy, the development of metal complexes with platinum central atoms such as cisplatin, carboplatin and oxaliplatin had an enormous impact on current cancer chemotherapy. In particular, cisplatin has become one of the most widely used drugs and is highly effective in treating several cancers such as ovarian and testicular cancers.

Thousands of other platinum complexes have been synthesized and biologically evaluated for their antitumor properties, from which about forty entered clinical phase I trials but only two carboplatin and oxaliplatin have received worldwide approval. Carboplatin exhibits a tumor inhibiting profile identical to that of cisplatin, however with fewer side effects, whereas oxaliplatin is used in a combination therapy against metastatic colorectal cancer. Preclinical and clinical investigations confirmed that the development of new metal agents with modes of action different from cisplatin is possible. Thus, complexes with iron, cobalt, or gold central atoms have shown promising results in preclinical studies and compounds with titanium, ruthenium, or gallium central atoms have already been evaluated trials.

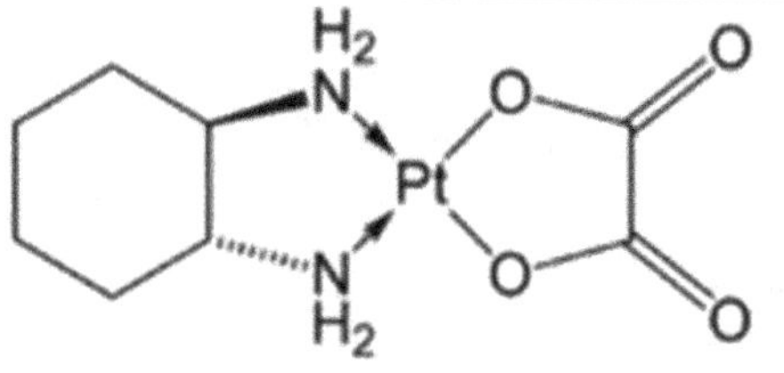

(1)

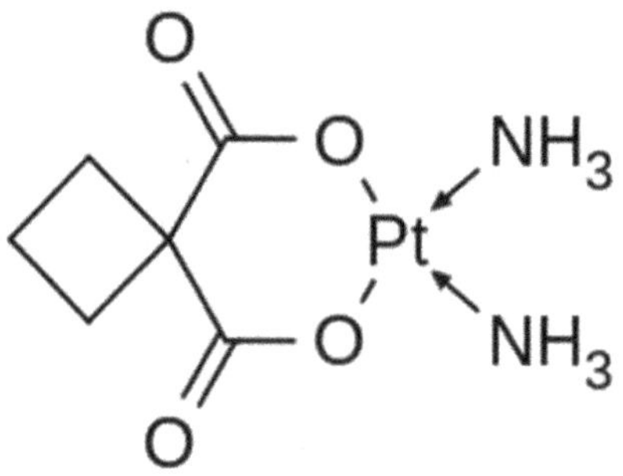

(2)

Molecular structure of Oxaliplatin (1) and Carboplatin (2).

DNA has often been proposed as the target of these organometallic antineoplastic agents, there is a particular emphasis on those that can interact with nucleic acids. Nevertheless, heavy metals are generally toxic by binding to sulfur and nitrogen sites on proteins and, thus, can interfere with a number of modes of metabolism. Several metals also exhibit action through redox activity. Gallium appears to operate through the displacement of metal ions in iron metabolism or bone. In large part, action of Gallium complexes seems to be a consequence of the similarity of Gallium (III) to Iron (III): Gallium interferes with the cellular transport of iron by binding to transferrin, and also interferes with the action of ribonucleotide reductase, which then results in inhibition of DNA synthesis.

Gold compounds are used for treating arthritis and cancer; they have potential for treating HIV/AIDS, malaria and Chagas disease. A property of Gold (III) that greatly complicates its chemistry is that many of its simple complexes can easily be reduced to Gold (I) by a variety of ligands, including thiols and thioethers found on cysteine and methionine residues of peptides and proteins.

The level of anti-HIV activity expressed by these metal complexes were Zn>Ni>Cu>Co>Pd in decreasing order. Scientists are looking for alternative approaches for the treatment of diabetes. Several trace elements, such as zinc and vanadium, exert insulin mimetic in in vitro and in vivo systems because complexes of these metal elements and small organic compounds (ligands) improved glucose utilization in both diabetic model animals and human diabetic patients. Another applications of organometallic complexes is anti-inflammatory therapy due its inherent physic – chemical properties of the complexes itself rather than just that of its constituents and in stoichiometric and catalytic processes, especially processes involving carbon monoxide and alkene-derived polymers. All the world's polyethylene and polypropylene are produced via organometallic catalysts, usually heterogeneously via Ziegler-Natta catalysis. Acetic acid is produced via metal carbonyl catalysts in the Monsanto process and Cativa process. Most synthetic aldehydes are produced via hydroformylation. The bulk of the synthetic alcohols, at least those larger than ethanol, are produced by hydrogenation of hydroformylated derived aldehydes. Similarly, the Wacker process is used in the oxidation of ethylene to acetaldehyde. Organolithium, organomagnesium and organoaluminium compounds are highly basic and highly reducing agents, they catalyze many polymerization reactions, but are also useful stoichiometrically. Organometallic compounds may be found in the environment and some, such as organolead and organomercury compounds are a toxic hazard.

References

- Organometallic-chemistry: newworldencyclopedia.org, Retrieved 28 February, 2020
- organometallic-compounds, chemistry: byjus.com, Retrieved 18 May, 2020
- Organometallic-chemistry-Applications: newworldencyclopedia.org, Retrieved 28 May, 2020
- Clinical-and-Industrial-Application-of-Organometallic-Compounds-and-Complexes-A-Review-330688960: researchgate.net, Retrieved 25 June, 2020

2

Concepts of Organometallic Chemistry

There are many concepts that are studied under the domain of organometallic chemistry. Some of them are 18- electron rule, pi backbonding, ligand field theory, frontier molecular orbital theory, etc. This chapter has been carefully written to provide an easy understanding of the concepts associated with organometallic chemistry.

HYDROCARBON

Hydrocarbon is any of a class of organic chemical compounds composed only of the elements carbon (C) and hydrogen (H). The carbon atoms join together to form the framework of the compound, and the hydrogen atoms attach to them in many different configurations. Hydrocarbons are the principal constituents of petroleum and natural gas. They serve as fuels and lubricants as well as raw materials for the production of plastics, fibres, rubbers, solvents, explosives, and industrial chemicals.

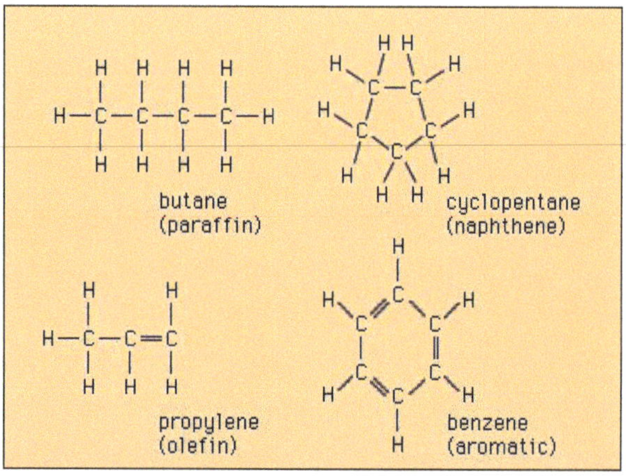

Structures assumed by hydrogen (H) and carbon (C) molecules in four common hydrocarbon compounds.

Many hydrocarbons occur in nature. In addition to making up fossil fuels, they are present in trees and plants, as, for example, in the form of pigments called carotenes that occur in carrots and green leaves. More than 98 percent of natural crude rubber is a hydrocarbon polymer, a chainlike molecule consisting of many units linked together. The structures and chemistry of individual hydrocarbons depend in large part on the types of chemical bonds that link together the atoms of their constituent molecules.

Nineteenth-century chemists classified hydrocarbons as either aliphatic or aromatic on the basis of their sources and properties. Aliphatic described hydrocarbons derived by chemical degradation of fats or oils. Aromatic hydrocarbons constituted a group of related substances obtained by chemical degradation of certain pleasant-smelling plant extracts. The terms aliphatic and aromatic are retained in modern terminology, but the compounds they describe are distinguished on the basis of structure rather than origin.

Aliphatic hydrocarbons are divided into three main groups according to the types of bonds they contain: alkanes, alkenes, and alkynes. Alkanes have only single bonds, alkenes contain a carbon-carbon double bond, and alkynes contain a carbon-carbon triple bond. Aromatic hydrocarbons are those that are significantly more stable than their Lewis structures would suggest; i.e., they possess "special stability". They are classified as either arenes, which contain a benzene ring as a structural unit, or nonbenzenoid aromatic hydrocarbons, which possess special stability but lack a benzene ring as a structural unit.

Structures of representative hydrocarbons

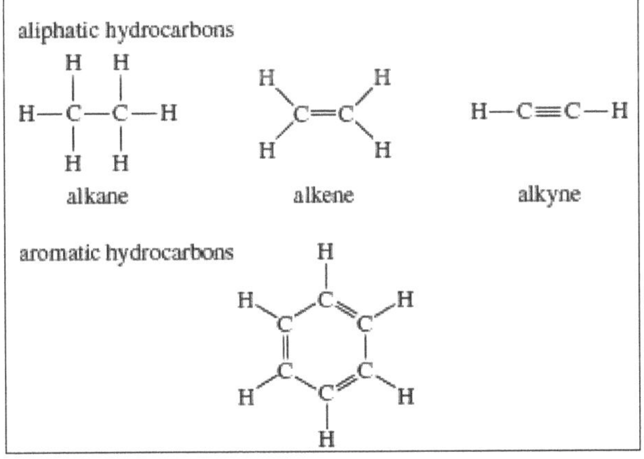

This classification of hydrocarbons serves as an aid in associating structural features with properties but does not require that a particular substance be assigned to a single class. Indeed, it is common for a molecule to incorporate structural unit's characteristic of two or more hydrocarbon families. A molecule that contains both a carbon-carbon triple bond and a benzene ring, for example, would exhibit some properties that are characteristic of alkynes and others that are characteristic of arenes.

Alkanes are described as saturated hydrocarbons, while alkenes, alkynes, and aromatic hydrocarbons are said to be unsaturated.

Aliphatic Hydrocarbons

Alkanes

Alkanes, hydrocarbons in which all the bonds are single, have molecular formulas that satisfy the general expression C_nH_{2n+2} (where n is an integer). Carbon is sp³ hybridized (three electron pairs are involved in bonding, forming a tetrahedral complex), and each C—C and C—H bond is a sigma (σ) bond. In order of increasing number of carbon atoms, methane (CH_4), ethane (C_2H_6), and propane (C_3H_8) are the first three members of the series.

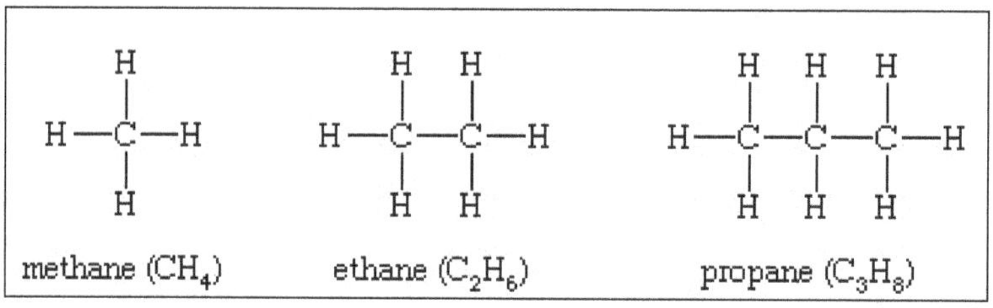

Methane, ethane, and propane are the only alkanes uniquely defined by their molecular formula. For C_4H_{10} two different alkanes satisfy the rules of chemical bonding (namely, that carbon has four bonds and hydrogen has one in neutral molecules). One compound, called n-butane, where the prefix n- represents normal, has its four carbon atoms bonded in a continuous chain. The other, called isobutene has a branched chain.

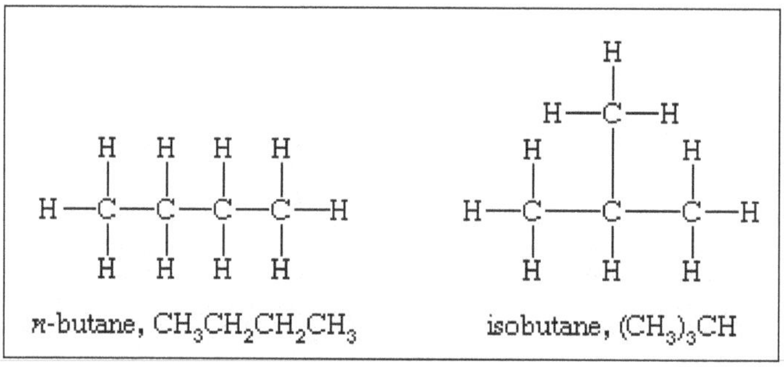

Different compounds that have the same molecular formula are called isomers. Isomers that differ in the order in which the atoms are connected are said to have different constitutions and are referred to as constitutional isomers. (An older name is structural isomers.) The compounds n-butane and isobutane are constitutional isomers and are the only ones possible for the formula C_4H_{10}. Because isomers are different compounds, they can have different physical and chemical properties. For example, n-butane has a higher boiling point (−0.5 °C [31.1 °F]) than isobutane (−11.7 °C [10.9 °F]).

There is no simple arithmetic relationship between the number of carbon atoms in a formula and the number of isomers. Graph theory has been used to calculate the number of constitutionally isomeric alkanes possible for values of n in C_nH_{2n+2} from 1 through 400. The number of constitutional isomers increases sharply as the number of carbon atoms increases. There is probably no upper limit to the number of carbon atoms possible in hydrocarbons. The alkane $CH_3(CH_2)_{388}CH_3$, in which 390 carbon atoms are bonded in a continuous chain, has been synthesized as an example of a so-called superlong alkane. Several thousand carbon atoms are joined together in molecules of hydrocarbon polymers such as polyethylene, polypropylene, and polystyrene.

Table: Number of possible alkane isomers.

Molecular formula	Number of constitutional isomers
C_3H_8	1
C_4H_{10}	2
C_5H_{12}	3
C_6H_{14}	5
C_7H_{16}	9
C_8H_{18}	18
C_9H_{20}	35
$C_{10}H_{22}$	75
$C_{15}H_{32}$	4,347
$C_{20}H_{42}$	366,319
$C_{30}H_{62}$	4,111,846,763

Nomenclature

The need to give each compound a unique name requires a richer variety of terms than is available with descriptive prefixes such as n- and iso-. The naming of organic compounds is facilitated through the use of formal systems of nomenclature.

Nomenclature in organic chemistry is of two types: common and systematic. Common names originate in many different ways but share the feature that there is no necessary connection between name and structure. The name that corresponds to a specific structure must simply be memorized, much like learning the name of a person. Systematic names, on the other hand, are keyed directly to molecular structure according to a generally agreed upon set of rules. The most widely used standards for organic nomenclature evolved from suggestions made by a group of chemists assembled for that purpose in Geneva in 1892 and have been revised on a regular basis by the International Union of Pure and Applied Chemistry (IUPAC). The IUPAC rules govern all classes of organic compounds but are ultimately based on alkane names. Compounds in other families are viewed as derived from alkanes by appending functional groups to, or otherwise modifying, the carbon skeleton.

The IUPAC rules assign names to unbranched alkanes according to the number of their carbon atoms. Methane, ethane, and propane are retained for CH_4, CH_3CH_3, and $CH_3CH_2CH_3$, respectively. The n- prefix is not used for unbranched alkanes in systematic IUPAC nomenclature; therefore, $CH_3CH_2CH_2CH_3$ is defined as butane, not n-butane. Beginning with five-carbon chains, the names of unbranched alkanes consist of a Latin or Greek stem corresponding to the number of carbons in the chain followed by the suffix -ane. A group of compounds such as the unbranched alkanes that differ from one another by successive introduction of CH_2 groups constitute a homologous series.

Table: IUPAC names of unbranched alkanes.

Alkane formula	Name	Alkane formula	Name
CH_4	Methane	$CH_3(CH_2)_6CH_3$	Octane
CH_3CH_3	Ethane	$CH_3(CH_2)_7CH_3$	Nonane
$CH_3CH_2CH_3$	Propane	$CH_3(CH_2)_8CH_3$	Decane
$CH_3CH_2CH_2CH_3$	Butane	$CH_3(CH_2)_{13}CH_3$	Pentadecane
$CH_3(CH_2)_3CH_3$	Pentane	$CH_3(CH_2)_{18}CH_3$	Icosane
$CH_3(CH_2)_4CH_3$	Hexane	$CH_3(CH_2)_{28}CH_3$	Triacontane
$CH_3(CH_2)_5CH_3$	Heptane	$CH_3(CH_2)_{98}CH_3$	Hectane

Alkanes with branched chains are named on the basis of the name of the longest chain of carbon atoms in the molecule, called the parent. The alkane shown has seven carbons in its longest chain and is therefore named as a derivative of heptane, the unbranched alkane that contains seven carbon atoms. The position of the CH_3 (methyl) substituent on the seven-carbon chain is specified by a number (3-), called a locant, obtained by successively numbering the carbons in the parent chain starting at the end nearer the branch. The compound is therefore called 3-methylheptane.

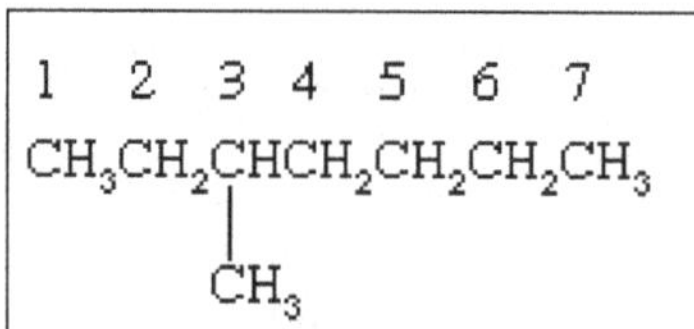

3-methylheptane

When there are two or more identical substituents, replicating prefixes (di-, tri-, tetra-, etc.) are used, along with a separate locant for each substituent. Different substituents, such as ethyl ($-CH_2CH_3$) and methyl ($-CH_3$) groups are cited in alphabetical order. Replicating prefixes are ignored when alphabetizing. In alkanes, numbering begins at the end nearest the substituent that appears first on the chain so that the carbon to which it is attached has as low a number as possible.

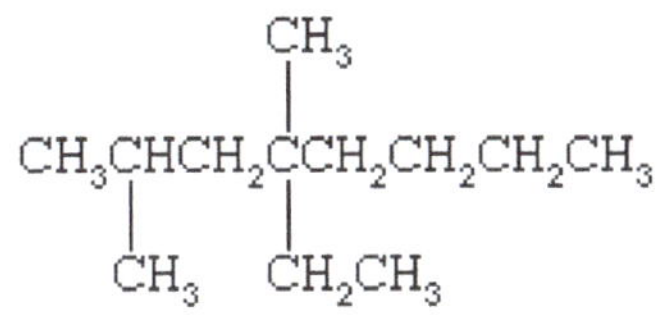

4-ethyl-2,4-dimethyloctane

Methyl and ethyl are examples of alkyl groups. An alkyl group is derived from an alkane by deleting one of its hydrogens, thereby leaving a potential point of attachment. Methyl is the only alkyl group derivable from methane and ethyl the only one from ethane. There are two C_3H_7 and four C_4H_9 alkyl groups. The IUPAC rules for naming alkanes and alkyl groups cover even very complex structures and are regularly updated. They are unambiguous in the sense that, although a single compound may have more than one correct IUPAC name, there is no possibility that two different compounds will have the same name.

Three-dimensional Structures

Most organic molecules, including all alkanes, are not planar but are instead characterized by three-dimensional structures. Methane, for example, has the shape of a regular tetrahedron with carbon at the centre and a hydrogen atom at each corner. Each H–C–H angle in methane is 109.5°, and each C–H bond distance is 1.09 angstroms (Å; 1Å = 1×10^{-10} metre). Higher alkanes such as butane have bonds that are tetrahedrally disposed on each carbon except that the resulting C–C–C and H–C–H angles are slightly larger and smaller, respectively, than the ideal value of 109.5° characteristic of a perfectly symmetrical tetrahedron. Carbon-carbon bond distances in alkanes are normally close to 1.53 angstroms.

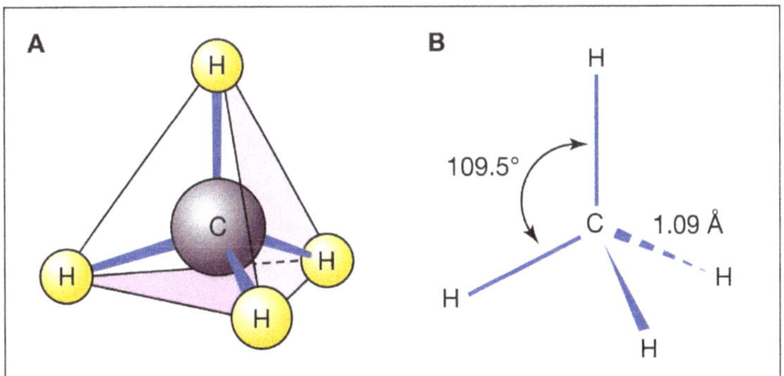

The tetrahedral geometry of methane: (A) stick-and-ball model and (B) showing bond angles and distances. (Plain bonds represent bonds in the plane of the image; wedge and dashed bonds represent those directed toward and away from the viewer, respectively).

An important aspect of the three-dimensional shape of alkanes and other organic molecules is their conformations, the nonidentical arrangements of atoms that are generated by rotation about single bonds. Of the infinite number of conformations possible for ethane—which are related by tiny increments of rotation of one CH_3 group with respect to the other—the eclipsed conformation is the least stable, and the staggered conformation is the most stable. The eclipsed conformation is said to suffer torsional strain because of repulsive forces between electron pairs in the C—H bonds of adjacent carbons. These repulsive forces are minimized in the staggered conformation since all C—H bonds are as far from one another as possible. Although rotation about the C—C bond of ethane is exceedingly rapid (millions of times per second at room temperature), at any instant most of the molecules exist in the staggered conformation.

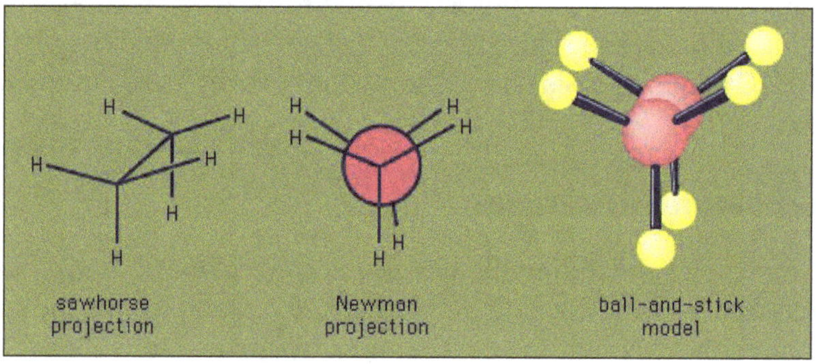

Eclipsed conformation is the least stable of all ethane conformations because the repulsive forces between electron pairs in the C—H bonds of adjacent carbons are maximized.

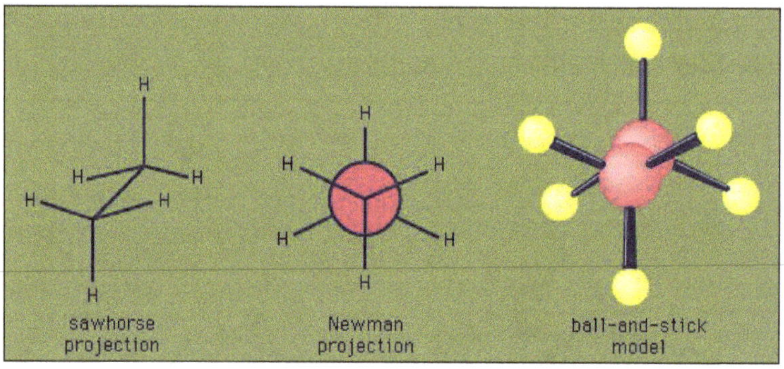

Staggered conformation is the most stable of all ethane conformations because the repulsive forces between electron pairs in the C—H bonds of adjacent carbons are minimized.

For butane, two different staggered conformations, called anti and gauche, are possible. Methyl is a larger substituent than hydrogen, and the greater separation between methyl groups in the anti-conformation makes it slightly more stable than the gauche.

The three-dimensional structures of higher alkanes are governed by the tetrahedral disposition of the four bonds to each carbon atom, by the preference for staggered conformations, and by the greater stability of anti C–C–C–C arrangements over gauche.

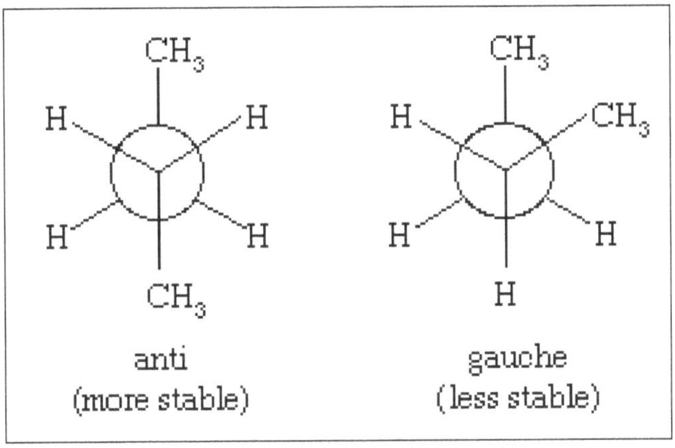

Cycloalkanes

Countless organic compounds are known in which a sequence of carbon atoms, rather than being connected in a chain, closes to form a ring. Saturated hydrocarbons that contain one ring are referred to as cycloalkanes. With a general formula of C_nH_{2n} (n is an integer greater than 2), they have two fewer hydrogen atoms than an alkane with the same number of carbon atoms. Cyclopropane (C_3H_6) is the smallest cycloalkane, whereas cyclohexane (C_6H_{12}) is the most studied, best understood, and most important. It is customary to represent cycloalkane rings as polygons, with the understanding that each corner corresponds to a carbon atom to which is attached the requisite number of hydrogen atoms to bring its total number of bonds to four.

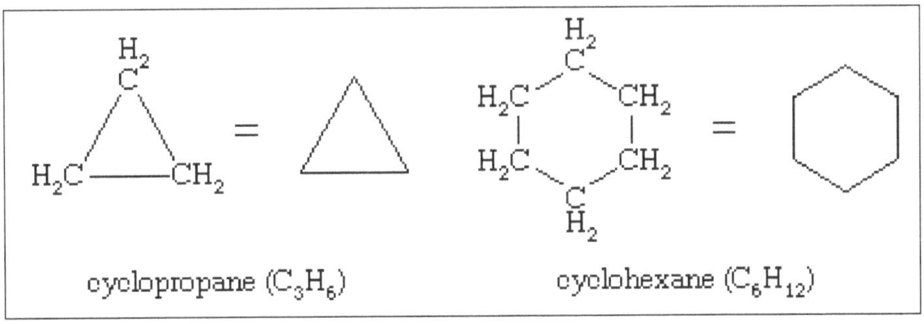

In naming cycloalkanes, alkyl groups attached to the ring are indicated explicitly and listed in alphabetical order, and the ring is numbered so as to give the lowest locant to the first-appearing substituent. If two different directions yield equivalent locants, the direction is chosen that gives the lower number to the substituent appearing first in the name.

The three carbon atoms of cyclopropane define the corners of an equilateral triangle, a geometry that requires the C–C–C angles to be 60°. This 60° angle is much smaller than the normal tetrahedral bond angle of 109.5° and imposes considerable strain (called angle strain) on cyclopropane. Cyclopropane is further destabilized by the torsional strain that results from having three eclipsed C–H bonds above the plane of the ring and three below.

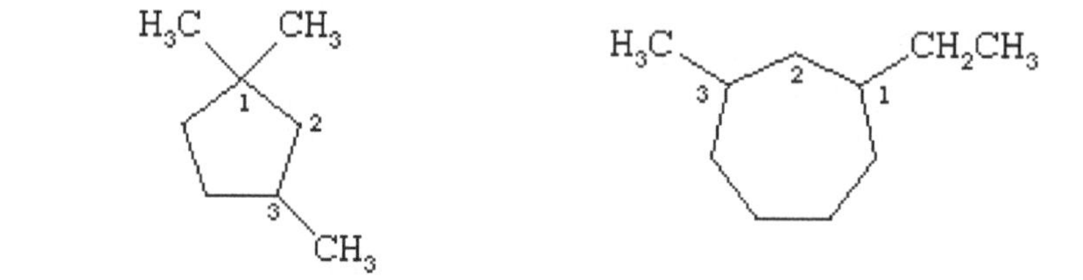

1,1,3-trimethylcyclopentane (not 1,3,3-trimethylcyclopentane) 1-ethyl-3-methylcycloheptane (not 1-methyl-3-ethylcycloheptane)

Cyclopropane is the only cycloalkane that is planar. Cyclobutane (C_4H_8) and higher cycloalkanes adopt nonplanar conformations in order to minimize the eclipsing of bonds on adjacent atoms. The angle strain in cyclobutane is less than in cyclopropane, whereas cyclopentane and higher cycloalkanes are virtually free of angle strain. With the exception of cyclopropane, all cycloalkanes undergo rapid internal motion involving interconversion of nonplanar "puckered" conformations.

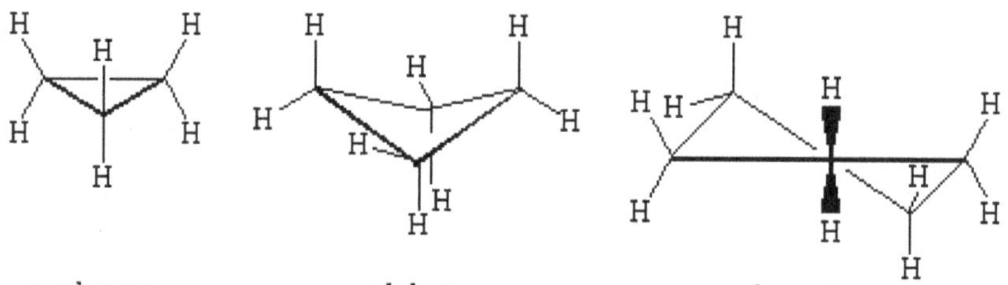

cyclopropane cyclobutane cyclopentane

Many of the most important principles of conformational analysis have been developed by examining cyclohexane. Three conformations of cyclohexane, designated as chair, boat, and skew (or twist), are essentially free of angle strain. Of these three the chair is the most stable, mainly because it has a staggered arrangement of all its bonds. The boat and skew conformations lack perfect staggering of bonds and are destabilized by torsional strain. The boat conformation is further destabilized by the mutual crowding of hydrogen atoms at carbons one and four. The shape of the boat brings its two "flagpole" hydrogen atoms to within 1.80 angstroms of each other, far closer than the 2.20-angstrom distance at which repulsive forces between hydrogen atoms become significant. At room temperature, 999 of every 1,000 cyclohexane molecules exist in the chair form (the other being skew).

There are two orientations of carbon-hydrogen bonds in the chair conformation of cyclohexane. Six bonds are parallel to a vertical axis passing through the centre of the ring and are called axial (a) bonds. The directions of these six axial bonds alternate up and down from one carbon to the next around the ring; thus, the axial hydrogens at carbons one, three, and five lie on one side of the ring and those at carbons two, four, and six on the other. The remaining six bonds are referred to as equatorial (e) because they lie in a region corresponding to the approximate "equator" of the molecule. The shortest distances between nonbonded atoms are those involving axial hydrogens on the same side of the molecule.

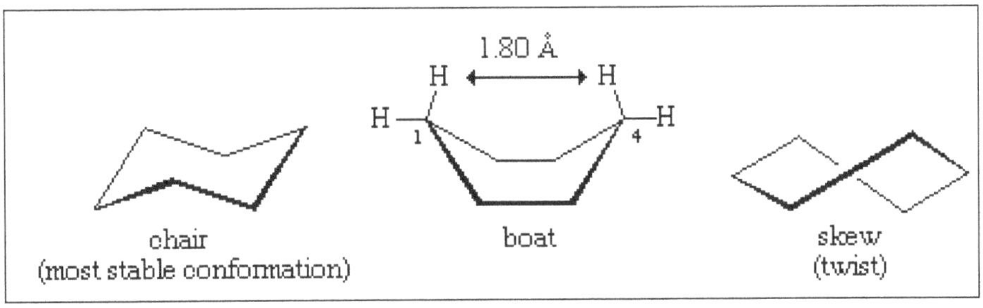

A rapid process of chair-chair interconversion (called ring-flipping) interconverts the six axial and six equatorial hydrogen atoms in cyclohexane. Chair-chair interconversion is a complicated process brought about by successive conformational changes within the molecule. It is different from simple whole-molecule motions, such as spinning and tumbling, and because it is a conformational change only, it does not require any bonds to be broken.

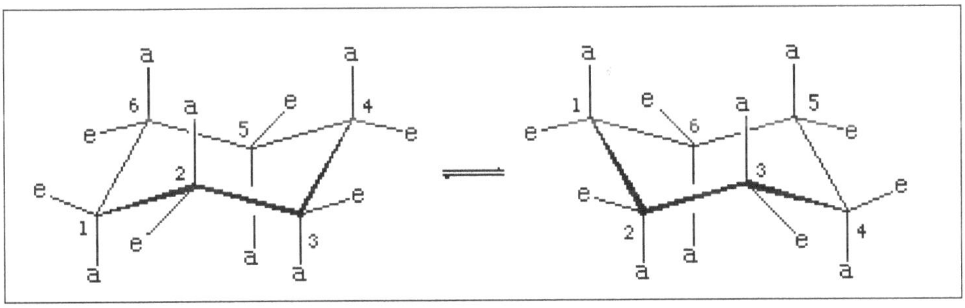

Chair-chair interconversion is especially important in substituted derivatives of cyclohexane. Any substituent is more stable when it occupies an equatorial rather than an axial site on the ring, since equatorial substituents are less crowded than axial ones. In methylcyclohexane, the chair conformation in which the large methyl group is equatorial is the most stable and, therefore, the most populated of all possible conformations. At any instant, almost all the methylcyclohexane molecules in a given sample exist in chair conformations, and about 95 percent of these have the methyl group in an equatorial orientation.

The highly branched tert-butyl group $(CH_3)_3C-$ (tert-butyl) is even more spatially demanding than the methyl group, and more than 99.99 percent of tert-butylcyclohexane molecules adopt chair conformations in which the $(CH_3)_3C-$ group is equatorial.

Conformational analysis of six-membered rings, especially the greater stability of chair conformations with equatorial substituents, not only is important in the area of hydrocarbons but also is essential to an understanding of the properties of biologically important molecules, especially steroids and carbohydrates. Odd Hassel of Norway and Derek H.R. Barton of England shared the Nobel Prize for Chemistry in 1969 for their important discoveries in this area. Hassel's studies dealt with structure, while Barton showed how conformational effects influence chemical reactivity.

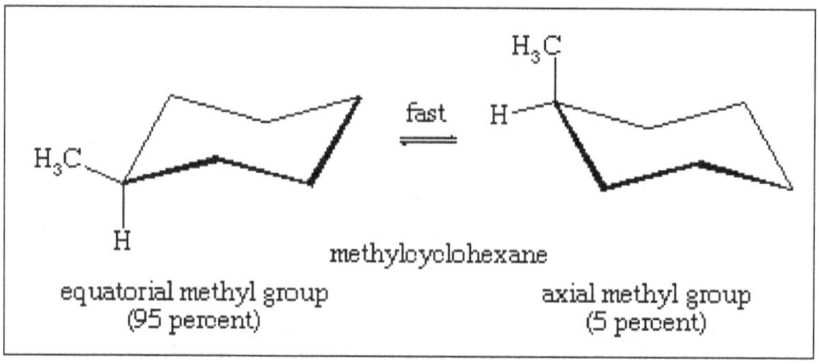

The most stable structures of cycloalkanes and compounds based on them have been determined by a number of experimental techniques, including X-ray diffraction and electron diffraction analyses and infrared, nuclear magnetic resonance, and microwave spectroscopies. These experimental techniques have been joined by advances in computational methods such as molecular mechanics, whereby the total strain energies of various conformations are calculated and compared (chemical bonding: Computational approaches to molecular structure). The structure with the lowest total energy is the most stable and corresponds to the best combination of bond distances, bond angles, and conformation. One benefit of such calculations is that unstable conformations, which are difficult to study experimentally, can be examined. The quality of molecular mechanics calculations is such that it is claimed that many structural features of hydrocarbons can be computed more accurately than they can be measured.

The conformations of rings with 7–12 carbons have been special targets for study by molecular mechanics. Unlike cyclohexane, in which one conformation (the chair) is much more stable than any other, cycloalkanes with 7–12 carbons are generally populated by several conformations of similar energy. Rings with more than 12 carbons are sufficiently flexible to adopt conformations that are essentially strain-free.

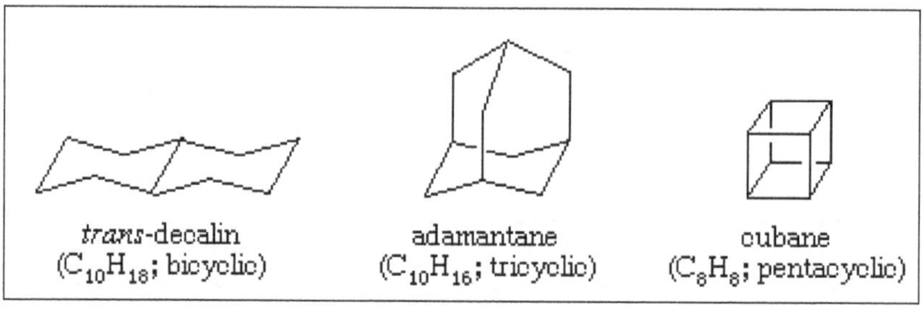

Polycyclic hydrocarbons are hydrocarbons that contain more than one ring. They are classified as bicyclic, tricyclic, tetracyclic, and so forth, according to the number of formal bond disconnections necessary to produce a noncyclic carbon chain. Examples include trans-decalin and adamantane—both of which are present in small amounts in petroleum—and cubane, a compound synthesized for the purpose of studying the effects of strain on chemical reactivity.

Stereoisomerism

Certain substituted derivatives of cycloalkanes exhibit a type of isomerism called stereoisomerism in which two substances have the same molecular formula and the same constitution but differ in the arrangement of their atoms in space. Methyl groups in 1,2-dimethylcyclopropane, for example, may be on the same (cis) or opposite (trans) sides of the plane defined by the ring. The resulting two substances are different compounds, each having its own properties such as boiling point:

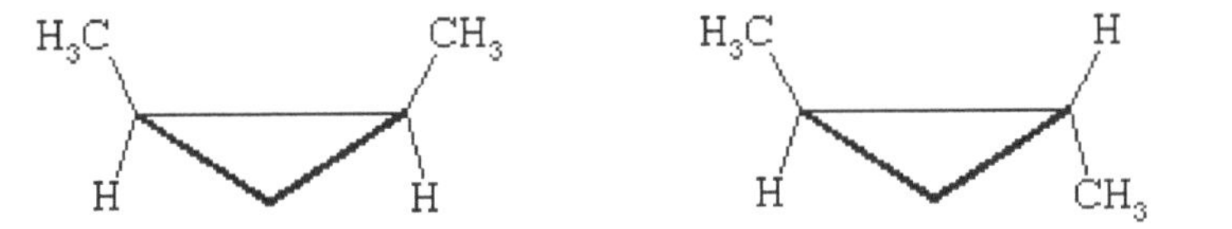

cis-1,2-dimethylcyclopropane (bp 37° C)   *trans*-1,2-dimethylcyclopropane (bp 29° C)

Cis-trans isomers belong to a class of stereoisomers known as diastereomers and are often referred to as geometric isomers, although this is an obsolete term. Cis-trans stereoisomers normally cannot be interconverted at room temperature, because to do so requires the breaking and reforming of chemical bonds.

Physical Properties

Alkanes and cycloalkanes are nonpolar substances. Attractive forces between alkane molecules are dictated by London forces (or dispersion forces, arising from electron fluctuations in molecules; see chemical bonding: Intermolecular forces) and are weak. Thus, alkanes have relatively low boiling points compared with polar molecules of comparable molecular weight. The boiling points of alkanes increase with increasing number of carbons. This is because the intermolecular attractive forces, although individually weak, become cumulatively more significant as the number of atoms and electrons in the molecule increases.

Table: Physical properties of unbranched alkanes.

Name	Formula	Boiling point (°C)	Melting point (°C)
Methane	CH_4	−164	−182.5
Ethane	CH_3CH_3	−88.6	−183.3

Propane	$CH_3CH_2CH_3$	−42	−189.7
Butane	$CH_3(CH_2)_2CH_3$	−0.5	−138.35
Pentane	$CH_3(CH_2)_3CH_3$	+36.1	−129.7
Hexane	$CH_3(CH_2)_4CH_3$	+68.9	−95.0
Heptane	$CH_3(CH_2)_5CH_3$	+98.4	−90.6
Octane	$CH_3(CH_2)_6CH_3$	+125.6	−56.8
Nonane	$CH_3(CH_2)_7CH_3$	+150.8	−51.0
Decane	$CH_3(CH_2)_8CH_3$	+174.1	−29.7
Pentadecane	$CH_3(CH_2)_{13}CH_3$	+270	+10
Octadecane	$CH_3(CH_2)_{16}CH_3$	+316.1	+28.2
Icosane	$CH_3(CH_2)_{18}CH_3$	+343	+36.8
Triacontane	$CH_3(CH_2)_{28}CH_3$	+449.7	+65.8
Tetracontane	$CH_3(CH_2)_{38}CH_3$	—	+81
Pentacontane	$CH_3(CH_2)_{48}CH_3$	—	+92

For a given number of carbon atoms, an unbranched alkane has a higher boiling point than any of its branched-chain isomers. This effect is evident upon comparing the boiling points (bp) of selected C_8H_{18} isomers. An unbranched alkane has a more extended shape, thereby increasing the number of intermolecular attractive forces that must be broken in order to go from the liquid state to the gaseous state. On the other hand, the relatively compact ellipsoidal shape of 2,2,3,3-tetramethylbutane permits it to pack into a crystal lattice more effectively than octane and so raises its melting point (mp).

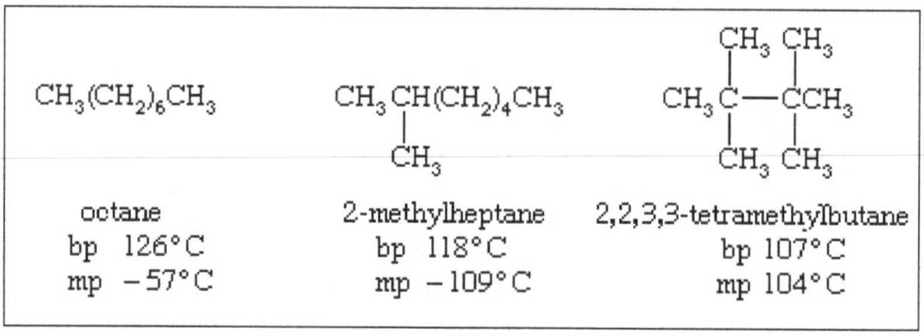

In general, solid alkanes do not often have high melting points. Unbranched alkanes tend toward a maximum in that the melting point of $CH_3(CH_2)_{98}CH_3$ (115 °C [239 °F]) is not much different from that of $CH_3(CH_2)_{148}CH_3$ (123 °C [253 °F]).

The viscosity of liquid alkanes increases with the number of carbons. Increased intermolecular attractive forces, as well as an increase in the extent to which nearby molecules become entangled when they have an extended shape, cause unbranched alkanes to be more viscous than their branched-chain isomers.

The densities of liquid hydrocarbons are all less than that of water, which is quite polar and possesses strong intermolecular attractive forces. All hydrocarbons are insoluble in water and, being less dense than water, float on its surface. Hydrocarbons are, however, usually soluble in one another as well as in organic solvents such as diethyl ether ($CH_3CH_2OCH_2CH_3$).

Sources and Occurrence

The most abundant sources of alkanes are natural gas and petroleum deposits, formed over a period of millions of years by the decay of organic matter in the absence of oxygen. Natural gas contains 60–80 percent methane, 5–9 percent ethane, 3–18 percent propane, and 2–14 percent higher hydrocarbons. Petroleum is a complex liquid mixture of hundreds of substances—including 150 or more hydrocarbons, approximately half of which are saturated.

Approximately two billion tons of methane is produced annually by the bacteria that live in termites and in the digestive systems of plant-eating animals. Smaller quantities of alkanes also can be found in a variety of natural materials. The so-called aggregation pheromone whereby Blaberus craniifer cockroaches attract others of the same species is a 1:1 mixture of the volatile but relatively high-boiling liquid alkanes undecane, $CH_3(CH_2)_9CH_3$, and tetradecane, $CH_3(CH_2)_{12}CH_3$. Hentriacontane, $CH_3(CH_2)_{29}CH_3$, is a solid alkane present to the extent of 8–9 percent in beeswax, where its stability and impermeability to water contribute to the role it plays as a structural component.

With the exception of the alkanes that are readily available from petroleum, alkanes are synthesized in the laboratory and in industry by the hydrogenation of alkenes. Only a few methods are available in which a carbon-carbon bond-forming operation gives an alkane directly, and these tend to be suitable only for syntheses carried out on a small scale.

Chemical Reactions

As is true for all hydrocarbons, alkane's burn in air to produce carbon dioxide (CO_2) and water (H_2O) and release heat. The combustion of 2,2,4-trimethylpentane is expressed by the following chemical equation:

$$\underset{\text{2,2,4-trimethylpentane}}{(CH_3)_3CCH_2CH(CH_3)_2} + \underset{\text{oxygen}}{{}^{25}\!/\!_2\,O_2} \longrightarrow \underset{\text{carbon dioxide}}{8CO_2} + \underset{\text{water}}{9H_2O}$$

The fact that all hydrocarbon combustions are exothermic is responsible for their widespread use as fuels. Grades of gasoline are rated by comparing their tendency toward preignition or knocking to reference blends of heptane and 2,2,4-trimethylpentane and assigning octane numbers. Pure heptane (assigned an octane number of 0) has poor ignition characteristics, whereas 2,2,4-trimethylpentane (assigned an octane number of 100) resists knocking even in high-compression engines.

As a class, alkanes are relatively unreactive substances and undergo only a few reactions. An industrial process known as isomerization employs an aluminum chloride ($AlCl_3$) catalyst to convert unbranched alkanes to their branched-chain isomers. In one such application, butane is isomerized to 2-methylpropane for use as a starting material in the preparation of 2,2,4-trimethylpentane (isooctane), which is a component of high-octane gasoline.

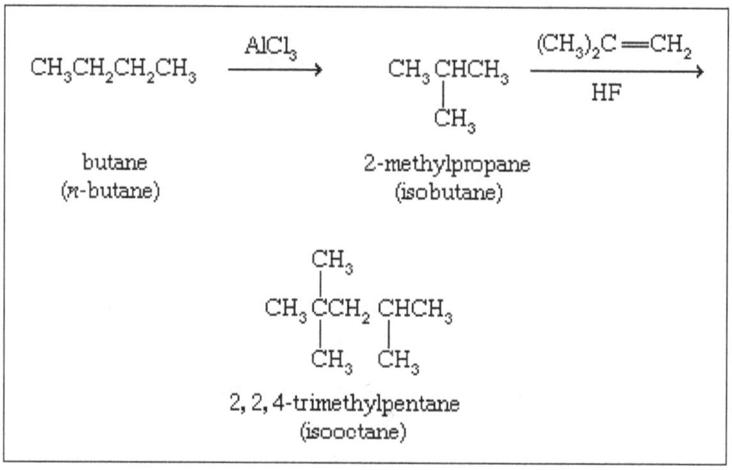

The halogens chlorine (Cl_2) and bromine (Br_2) react with alkanes and cycloalkanes by replacing one or more hydrogens with a halogen. Although the reactions are exothermic, a source of energy such as ultraviolet light or high temperature is required to initiate the reaction, as, for example, in the chlorination of cyclobutane.

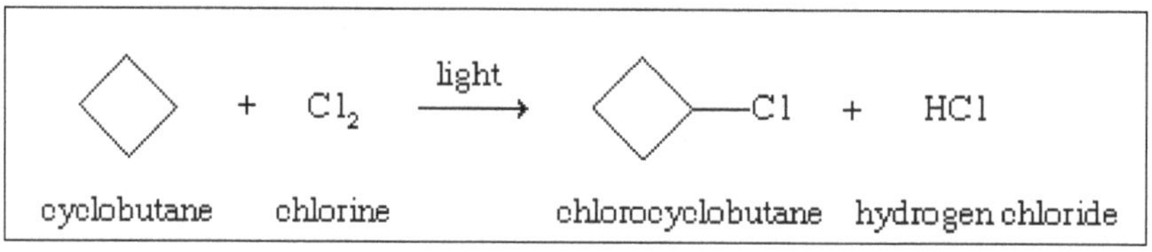

The chlorinated derivatives of methane (CH_3Cl, CH_2Cl_2, $CHCl_3$, and CCl_4) are useful industrially and are prepared by various methods, including the reaction of methane with chlorine at temperatures on the order of 450 °C (840 °F).

The most important industrial organic chemical reaction in terms of its scale and economic impact is the dehydrogenation of ethane (obtained from natural gas) to form ethylene and hydrogen. The hydrogen produced is employed in the Haber-Bosch process for the preparation of ammonia from nitrogen.

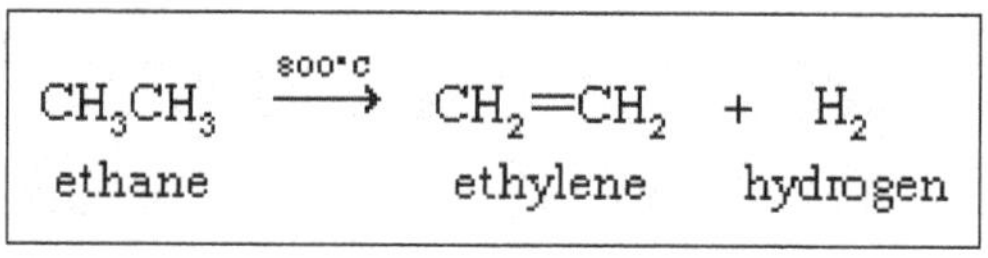

The higher alkanes present in petroleum also yield ethylene under similar conditions by reactions that involve both dehydrogenation and the breaking of carbon-carbon bonds. The conversion of high-molecular-weight alkanes to lower ones is called cracking.

Alkenes and Alkynes

Alkenes (also called olefins) and alkynes (also called acetylenes) belong to the class of unsaturated aliphatic hydrocarbons. Alkenes are hydrocarbons that contain a carbon-carbon double bond, whereas alkynes have a carbon-carbon triple bond. Alkenes are characterized by the general molecular formula CnH2n, alkynes by C_nH_{2n-2}. Ethene (C_2H_4) is the simplest alkene and ethyne (C_2H_2) the simplest alkyne.

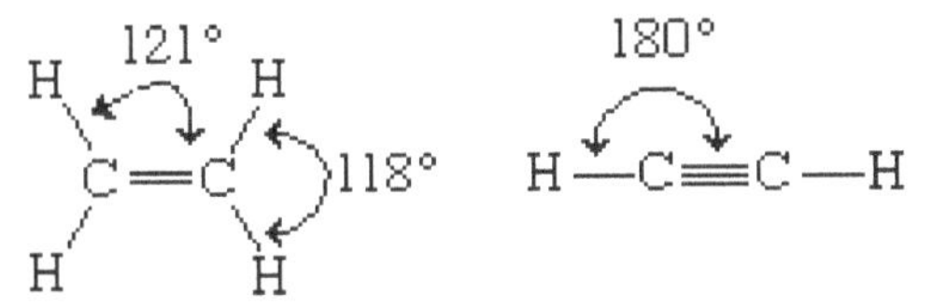

ethene (ethylene) ethyne (acetylene)

Ethylene is a planar molecule with a carbon-carbon double bond length (1.34 angstroms) that is significantly shorter than the corresponding single bond length (1.53 angstroms) in ethane. Acetylene has a linear H−C≡C−H geometry, and its carbon-carbon bond distance (1.20 angstroms) is even shorter than that of ethylene.

Bonding in Alkenes and Alkynes

The generally accepted bonding model for alkenes views the double bond as being composed of a σ (sigma) component and a π (pi) component. In the case of ethylene, each carbon is sp2 hybridized, and each is bonded to two hydrogens and the other carbon by σ bonds. Additionally, each carbon has a half-filled p orbital, the axis of which is perpendicular to the plane of the σ bonds. Side-by-side overlap of these two p orbitals generates a π bond. The pair of electrons in the π bond are equally likely to be found in the regions of space immediately above and below the plane defined by the atoms. Most of the important reactions of alkenes involve the electrons in the π component of the double bond because these are the electrons that are farthest from the positively charged nuclei and therefore the most weakly held.

σ component of a double bond π component of a double bond

The triple bond of an alkyne consists of one σ and two π components linking two sp hybridized carbons. In the case of acetylene, the molecule itself is linear with σ bonds between the two carbons and to each hydrogen. Each carbon has two p orbitals, the axes of which are perpendicular to each other. Overlap of two p orbitals, suitably aligned and on adjacent carbons, gives two π bonds.

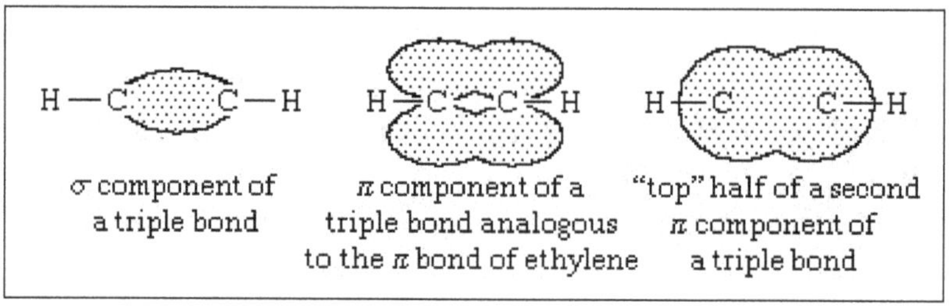

Nomenclature of Alkenes and Alkynes

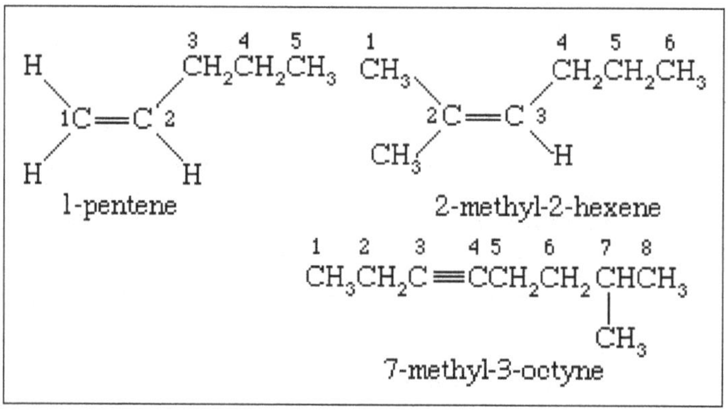

Ethylene and acetylene are synonyms in the IUPAC nomenclature system for ethene and ethyne, respectively. Higher alkenes and alkynes are named by counting the number of carbons in the longest continuous chain that includes the double or triple bond and appending an -ene (alkene) or -yne (alkyne) suffix to the stem name of the unbranched alkane having that number of carbons. The chain is numbered in the direction that gives the lowest number to the first multiply bonded carbon, and adding it as a prefix to the name. Once the chain is numbered with respect to the multiple bond, substituents attached to the parent chain are listed in alphabetical order and their positions identified by number.

Compounds that contain two double bonds are classified as dienes, those with three as trienes, and so forth. Dienes are named by replacing the -ane suffix of the corresponding alkane by -adiene and identifying the positions of the double bonds by numerical locants. Dienes are classified as cumulated, conjugated, or isolated according to whether the double bonds constitute a C=C=C unit, a C=C−C=C unit, or a C=C−$(CXY)_n$−C=C unit, respectively.

Double bonds can be incorporated into rings of all sizes, resulting in cycloalkenes. In naming substituted derivatives of cycloalkenes, numbering begins at and continues through the double bond.

$CH_3CH{=}C{=}CHCH_3$ 2,3-pentadiene (cumulated)

$CH_2{=}CHCH{=}CHCH_3$ 1,3-pentadiene (conjugated)

$CH_2{=}CHCH_2CH{=}CH_2$ 1,4-pentadiene (isolated)

Unlike rotation about carbon-carbon single bonds, which is exceedingly rapid, rotation about carbon-carbon double bonds does not occur under normal circumstances. Stereoisomerism is therefore possible in those alkenes in which neither carbon atom bears two identical substituents. In most cases, the names of stereoisomeric alkenes are distinguished by cis-trans notation. (An alternative method, based on the Cahn-Ingold-Prelog system and using E and Z prefixes, is also used.) Cycloalkenes in which the ring has eight or more carbons are capable of existing as cis or trans stereoisomers. trans-Cycloalkenes are too unstable to isolate when the ring has seven or fewer carbons.

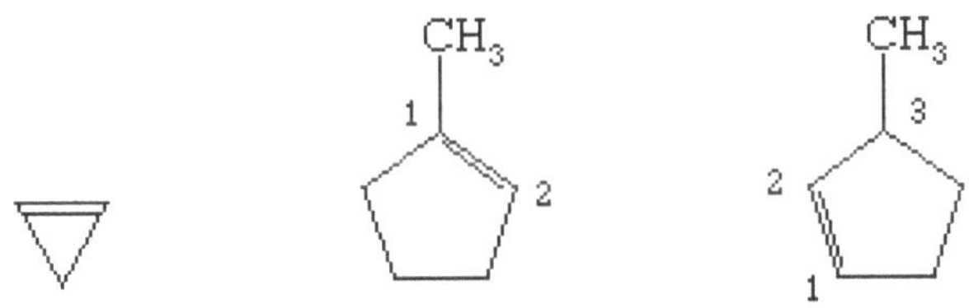

cyclopropene 1-methylcyclopentene 3-methylcyclopentene

Because the C–C≡C–C unit of an alkyne is linear, cycloalkynes are possible only when the number of carbon atoms in the ring is large enough to confer the flexibility necessary to accommodate this geometry. Cyclooctyne (C_8H_{12}) is the smallest cycloalkyne capable of being isolated and stored as a stable compound.

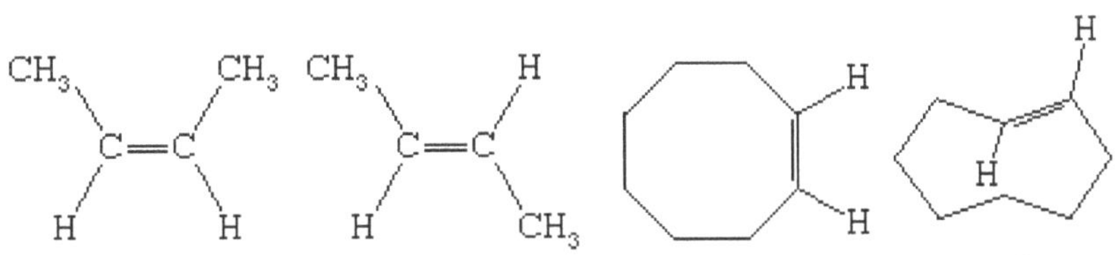

cis-2-butene *trans*-2-butene *cis*-cyclooctene *trans*-cyclooctene

Natural Occurrence

Ethylene is formed in small amounts as a plant hormone. The biosynthesis of ethylene involves an enzyme-catalyzed decomposition of a novel amino acid, and, once formed, ethylene stimulates the ripening of fruits.

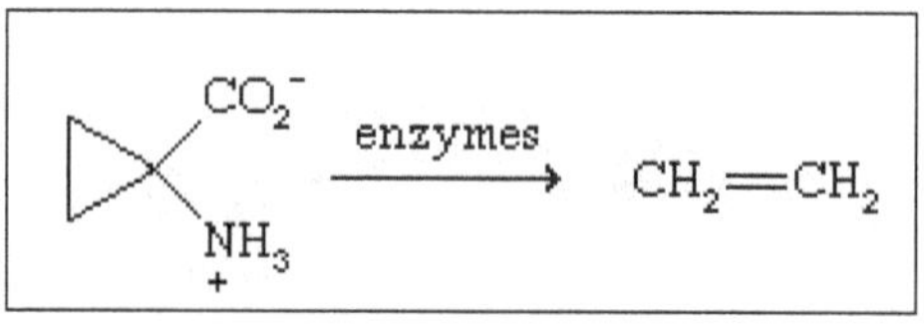

Alkenes are abundant in the essential oils of trees and other plants. (Essential oils are responsible for the characteristic odour, or "essence", of the plant from which they are obtained.) Myrcene and limonene, for example, are alkenes found in bayberry and lime oil, respectively. Oil of turpentine, obtained by distilling the exudate from pine trees, is a mixture of hydrocarbons rich in α-pinene. α-Pinene is used as a paint thinner as well as a starting material for the preparation of synthetic camphor, drugs, and other chemicals.

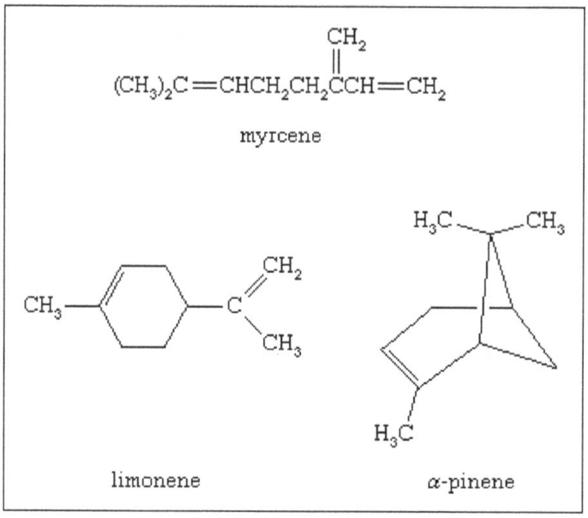

Other naturally occurring hydrocarbons with double bonds include plant pigments such as lycopene, which is responsible for the red colour of ripe tomatoes and water-melon. Lycopene is a polyene (meaning many double bonds) that belongs to a family of 40-carbon hydrocarbons known as carotenes.

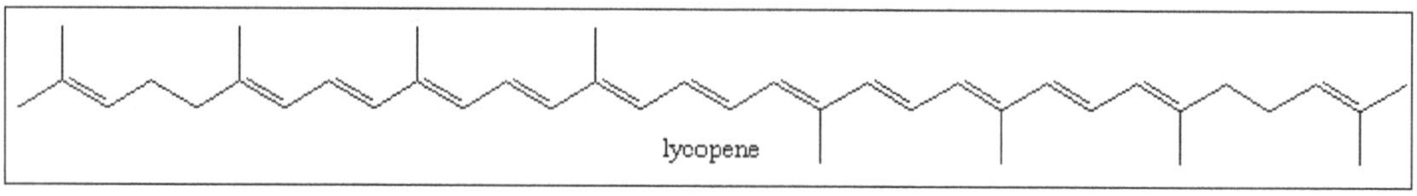

The sequence of alternating single and double bonds in lycopene is an example of a conjugated system. The degree of conjugation affects the light-absorption properties of unsaturated compounds. Simple alkenes absorb ultraviolet light and appear colourless. The wavelength of the light absorbed by unsaturated compounds becomes longer as the number of double bonds in conjugation with one another increases, with the result that polyenes containing regions of extended conjugation absorb visible light and appear yellow to red.

The hydrocarbon fraction of natural rubber (roughly 98 percent) is made up of a collection of polymer molecules, each of which contains approximately 20,000 C_5H_8 structural units joined together in a regular repeating pattern.

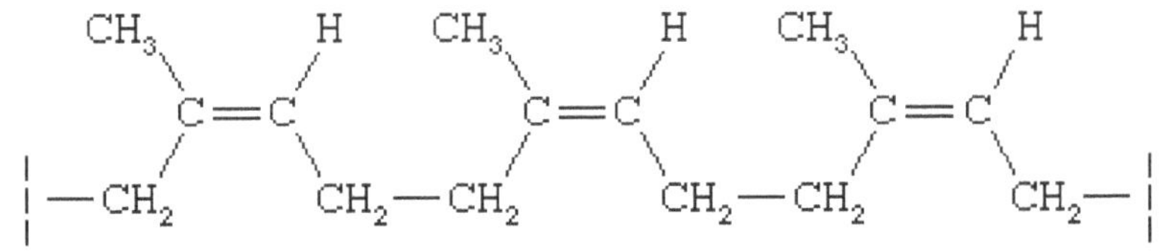

C_5H_8 units, three of which are shown, are linked together in natural rubber.

Natural products that contain carbon-carbon triple bonds, while numerous in plants and fungi, are far less abundant than those that contain double bonds and are much less frequently encountered.

Synthesis

The lower alkenes (through four-carbon alkenes) are produced commercially by cracking and dehydrogenation of the hydrocarbons present in natural gas and petroleum. The annual global production of ethylene averages around 75 million metric tons. Analogous processes yield approximately 2 million metric tons per year of 1,3-butadiene ($CH_2=CHCH=CH_2$). Approximately one-half of the ethylene is used to prepare polyethylene. Most of the remainder is utilized to make ethylene oxide (for the manufacture of ethylene glycol antifreeze and other products), vinyl chloride (for polymerization to polyvinyl chloride), and styrene (for polymerization to polystyrene). The principal application of propylene is in the preparation of polypropylene. 1,3-Butadiene is a starting material in the manufacture of synthetic rubber.

Higher alkenes and cycloalkenes are normally prepared by reactions in which a double bond is introduced into a saturated precursor by elimination (i.e., a reaction in which atoms or ions are lost from a molecule).

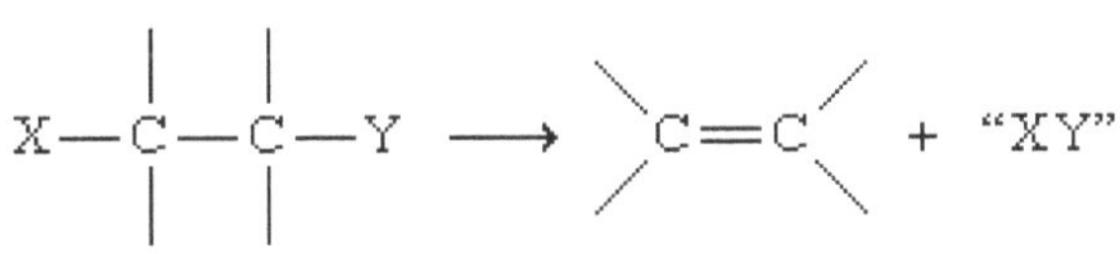

Examples include the dehydration of alcohols:

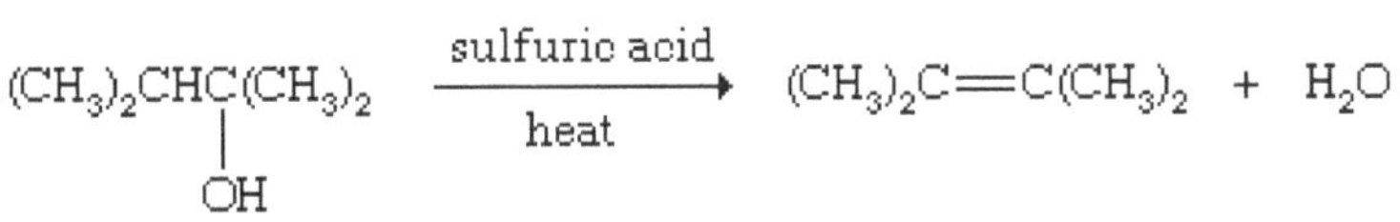

2-methyl-2-butanol 2-methyl-2-butene water

And the dehydrohalogenation (loss of a hydrogen atom and a halogen atom) of alkyl halides.

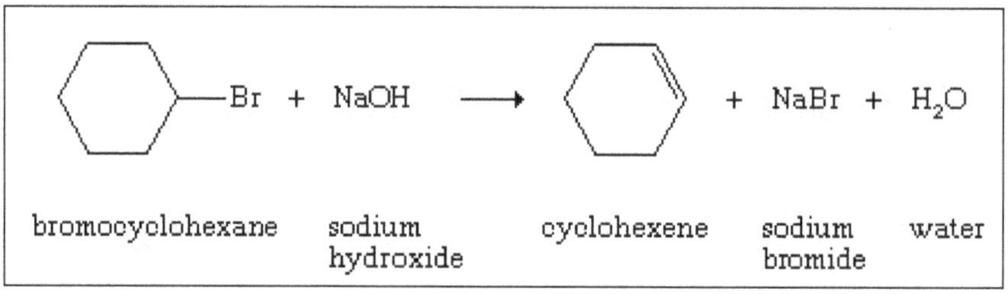

These usually are laboratory rather than commercial methods. Alkenes also can be prepared by partial hydrogenation of alkynes.

Acetylene is prepared industrially by cracking and dehydrogenation of hydrocarbons as described for ethylene. Temperatures of about 800 °C (1,500 °F) produce ethylene; temperatures of roughly 1,150 °C (2,100 °F) yield acetylene. Acetylene, relative to ethylene, is an unimportant industrial chemical. Most of the compounds capable of being derived from acetylene are prepared more economically from ethylene, which is a less expensive starting material. Higher alkynes can be made from acetylene or by double elimination of a dihaloalkane (i.e., removal of both halogen atoms from a disubstituted alkane).

$$CH_3CH_2CH(Br)CH_2Br + 2NaNH_2 \longrightarrow CH_3CH_2C{\equiv}CH + 2NaBr + 2NH_3$$

1,2-dibromobutane, sodium amide, 1-butyne, sodium bromide, ammonia

Physical Properties

The physical properties of alkenes and alkynes are generally similar to those of alkanes or cycloalkanes with equal numbers of carbon atoms. Alkynes have higher boiling points than alkanes or alkenes, because the electric field of an alkyne, with its increased number of weakly held π electrons, is more easily distorted, producing stronger attractive forces between molecules.

Table: Boiling points of alkenes and alkynes.

Name	Formula	Boiling point (°c)
Ethylene	$CH_2{=}CH_2$	−103.7
Acetylene	$HC{\equiv}CH$	−84.0
Propene	$CH_2{=}CHCH_3$	−47.6
Propyne	$HC{\equiv}CCH_3$	−23.2

1-butene	CH_2=$CHCH_2CH_3$	–6.1
Cis-2-butene	cis-CH_3CH=$CHCH_3$	+3.7
Trans-2-butene	trans-CH_3CH=$CHCH_3$	+0.9
2-methylpropene	CH_2=$C(CH_3)_2$	–6.6
1-butyne	HC≡CCH_2CH_3	+8.1
2-butyne	CH_3C≡CCH_3	+27.0
1-pentene	CH_2=$CHCH_2CH_2CH_3$	+30.2
1-pentyne	HC≡$CCH_2CH_2CH_3$	+40.2

Chemical Properties

Alkenes react with a much richer variety of compounds than alkanes. The characteristic reaction of alkanes is substitution; that of alkenes and alkynes is addition to the double or triple bond. Hydrogenation is the addition of molecular hydrogen (H_2) to a multiple bond, which converts alkenes to alkanes. The reaction occurs at a convenient rate only in the presence of certain finely divided metal catalysts, such as nickel (Ni), platinum (Pt), palladium (Pd), or rhodium (Rh).

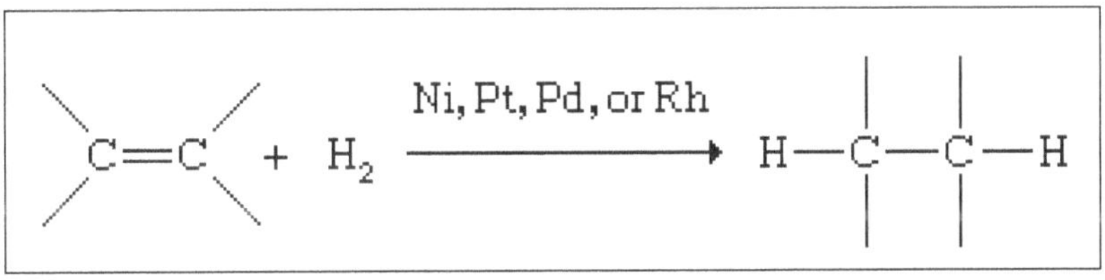

Hydrogenation is used to prepare alkanes and cycloalkanes and also to change the physical properties of highly unsaturated vegetable oils to increase their shelf life. In such processes the liquid oils are converted to fats of a more solid consistency. Butter substitutes such as margarine are prepared by partial hydrogenation of soybean oil.

Significant progress has been made in developing catalysts for enantioselective hydrogenation. An enantioselective hydrogenation is a hydrogenation in which one enantiomer of a chiral molecule (a molecule that can exist in two structural forms or enantiomers) is formed in greater amounts than the other. This normally involves converting one of the carbons of the double bond to a stereogenic centre.

Typical catalysts for enantioselective hydrogenation are based on enantiomerically homogeneous ligands bonded to rhodium. Enantioselectivities exceeding 90 percent of a single enantiomer are commonplace in enantioselective hydrogenations, a major application of which is in the synthesis of enantiomerically pure drugs.

The halogens bromine and chlorine add to alkenes to yield dihaloalkanes. Addition is rapid even at room temperature and requires no catalyst. The most important application of this reaction is the addition of chlorine to ethylene to give 1,2-dichloroethane, from which vinyl chloride is prepared.

$$\underset{\text{ethylene}}{CH_2{=}CH_2} + \underset{\text{chlorine}}{Cl_2} \longrightarrow \underset{\text{1,2-dichloroethane}}{ClCH_2CH_2Cl}$$

Compounds of the type HX, where X is a halogen or other electronegative group, also add to alkenes; the hydrogen atom of HX becomes bonded to one of the carbon atoms of the C=C unit, and the X atom becomes bonded to the other.

$$\underset{\text{2,3-dimethyl-2-butene}}{(CH_3)_2C{=}C(CH_3)_2} + \underset{\text{hydrogen chloride}}{HCl} \longrightarrow \underset{\text{2-chloro-2,3-dimethylbutane}}{(CH_3)_2CH\underset{\displaystyle Cl}{\underset{|}{C}}(CH_3)_2}$$

If HX is a strong acid, such as hydrochloric (HCl) or hydrobromic (HBr) acid, the reaction occurs rapidly; otherwise, an acid catalyst is required. One source of industrial ethanol, for example, is the reaction of ethylene with water in the presence of phosphoric acid.

When the two carbon atoms of a double bond are not equivalent, the H of the HX compound adds to the carbon that has the greater number of directly attached hydrogen atoms, and X adds to the one with the fewer. This generalization is called the Markovnikov rule, named after Russian chemist Vladimir Markovnikov, who proposed the rule in 1869. Thus, when sulfuric acid (H_2SO_4) adds to propylene, the product is isopropyl hydrogen sulfate, not n-propyl hydrogen sulfate ($CH_3CH_2CH_2OSO_3H$). This is the first step in the industrial preparation of isopropyl alcohol, which is formed when isopropyl hydrogen sulfate is heated with water.

The term regioselective describes the preference for a reaction that occurs in one direction rather than another, as in the addition of sulfuric acid to propylene. A regiospecific reaction is one that is 100 percent regioselective. The Markovnikov rule expresses the regioselectivity to be expected in the addition of unsymmetrical reagents (such as HX) to unsymmetrical alkenes (such as $H_2C{=}CHR$).

Boron hydrides, compounds of the type R_2BH, add to alkenes to give organoboranes (hydroboration), which can be oxidized to alcohols with hydrogen peroxide (H_2O_2) (oxidation). The net result is the same as if H and –OH add to the double bond with a regioselectivity opposite to the Markovnikov rule. The hydroboration-oxidation sequence is one of a large number of boron-based synthetic methods developed by American chemist Herbert C. Brown.

Vicinal diols, compounds with –OH groups on adjacent carbons, are formed when alkenes react with certain oxidizing agents, especially potassium permanganate ($KMnO_4$)

or osmium tetroxide (OsO_4). The most widely used methods employ catalytic amounts of OsO_4 in the presence of oxidizing agents such as tert-butyl hydroperoxide [$(CH_3)_3COOH$].

Alkenes are the customary starting materials from which epoxides, compounds containing a three-membered ring consisting of one oxygen atom and two carbon atoms, are made. The simplest epoxide, ethylene oxide (oxirane), is obtained by passing a mixture of ethylene and air (or oxygen) over a heated silver catalyst. Epoxides are useful intermediates for a number of transformations. Ethylene oxide, for example, is converted to ethylene glycol, which is used in the synthesis of polyester fibres and films and as the main component of automobile antifreeze. On a laboratory scale, epoxides are normally prepared by the reaction of an alkene and a peroxy acid.

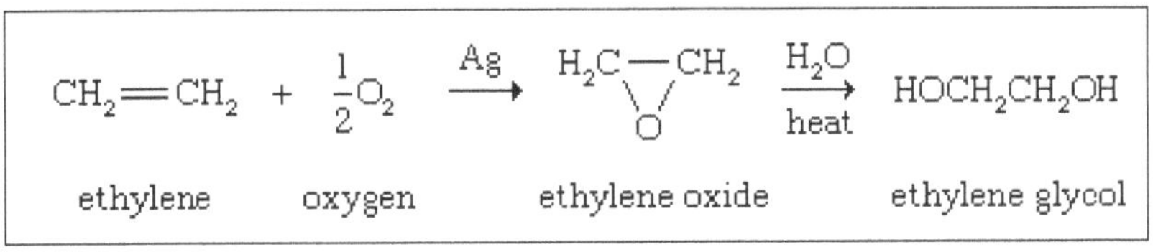

Conjugated dienes undergo a novel and useful reaction known as the Diels-Alder cycloaddition. In this reaction, a conjugated diene reacts with an alkene to form a compound that contains a cyclohexene ring. The unusual feature of the Diels-Alder cycloaddition is that two carbon-carbon bonds are formed in a single operation by a reaction that does not require catalysts of any kind. The German chemists Otto Diels and Kurt Alder received the Nobel Prize for Chemistry in 1950 for discovering and demonstrating the synthetic value of this reaction.

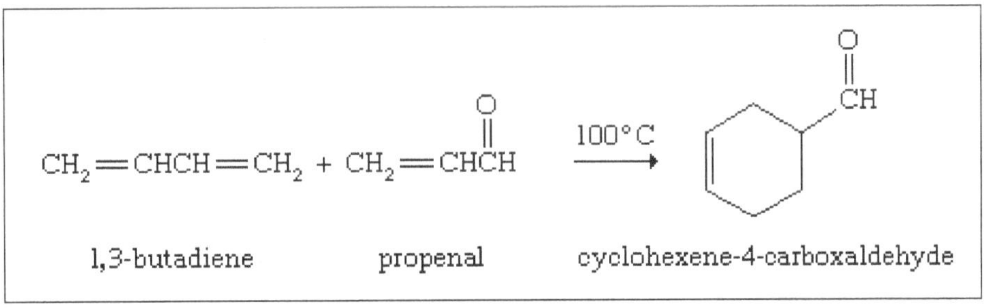

Alkynes undergo addition with many of the same substances that react with alkenes. Hydrogenation of alkynes can be controlled so as to yield either an alkene or an alkane. Two molecules of H_2 add to the triple bond to give an alkane under the usual conditions of catalytic hydrogenation.

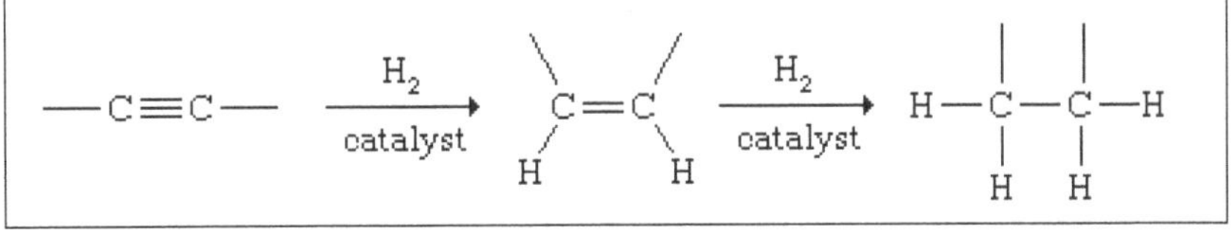

Special, less active (poisoned) catalysts have been developed that permit the reaction to be halted at the alkene stage, and the procedure is used as a method for the synthesis of alkenes. When stereoisomeric alkenes are possible reaction products, the cis isomer is formed almost exclusively.

Alkynes react with Br_2 or Cl_2 by first adding one molecule of the halogen to give a dihaloalkene and then a second to yield a tetrahaloalkane.

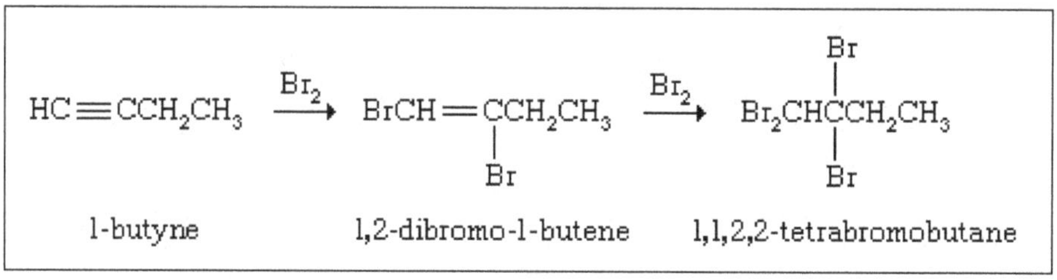

Compounds of the type HX, where X is an electronegative atom or group, also add to alkynes. When acetylene (HC≡CH) reacts with HCl, the product is vinyl chloride (CH_2 = CHCl), and, when HCN adds to acetylene, the product is acrylonitrile (CH_2 = CHCN). Both vinyl chloride and acrylonitrile are valuable starting materials for the production of useful polymers, but neither is prepared in significant quantities from acetylene, because each is available at lower cost from an alkene (vinyl chloride from ethylene and acrylonitrile from propylene).

Hydration of alkynes is unusual in that the initial product called an enol and characterized by an H−O−C=C− group, is unstable under the conditions of its formation and is converted to an isomer that contains a carbonyl group.

Although they are very weak acids, acetylene and terminal alkynes are much more acidic than alkenes and alkanes. Hydrogen attached to a triply bonded carbon can be removed by a very strong base such as sodium amide ($NaNH_2$) in liquid ammonia as the solvent.

$$\underset{\text{1-butyne}}{CH_3CH_2C{\equiv}CH} + \underset{\text{sodium amide}}{NaNH_2} \longrightarrow \underset{\text{1-butynylsodium}}{CH_3CH_2C{\equiv}CNa} + \underset{\text{ammonia}}{NH_3}$$

The sodium salt of the alkyne formed in this reaction is not normally isolated but is treated directly with an alkyl halide. The ensuing reaction proceeds with carbon-carbon bond formation and is used to prepare higher alkynes.

$$\underset{\text{1-butynylsodium}}{CH_3CH_2C{\equiv}CNa} + \underset{\text{bromomethane}}{CH_3Br} \longrightarrow \underset{\text{2-pentyne}}{CH_3CH_2C{\equiv}CCH_3} + \underset{\text{sodium bromide}}{NaBr}$$

Polymerization

A single alkene molecule, called a monomer, can add to the double bond of another to give a product, called a dimer, having twice the molecular weight. In the presence of an acid catalyst, the monomer 2-methylpropene (C_4H_8), for example, is converted to a mixture of C_8H_{16} alkenes (dimers) suitable for subsequent conversion to 2,2,4-trimethylpentane (isooctane).

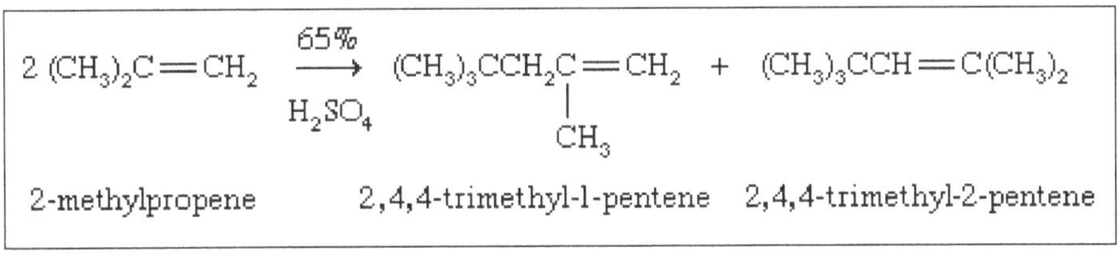

If the process is repeated, trimers, and eventually polymers—substances composed of a great many monomer units—are obtained.

Approximately one-half of the ethylene produced each year is used to prepare the polymer polyethylene. Polyethylene is a mixture of polymer chains of different lengths, where n, the number of monomer units, is on the order of 1,000–5,000.

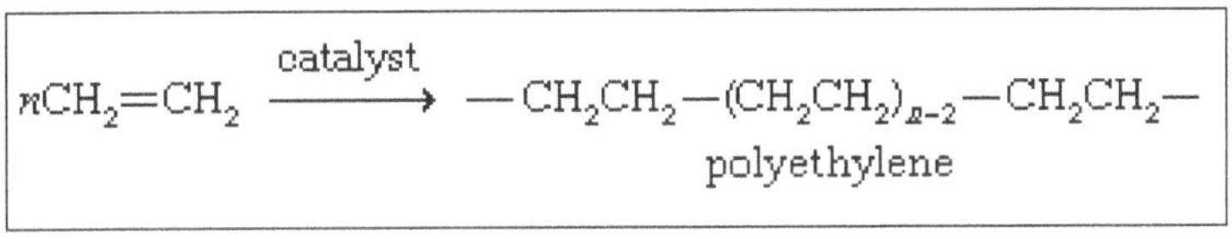

The distinguishing characteristic of polyethylene is its resistance to attack by most substances. Its resemblance to an alkane in this respect is not surprising, because the polymer chain is nearly void of functional groups. Its ends may have catalyst molecules attached or may terminate in a double bond by loss of a hydrogen atom at the next-to-last carbon. The properties of a particular sample of polyethylene depend mainly on the catalyst used and the conditions under which polymerization occur. A chain may be continuous, or it may sprout occasional branches of shorter chains. The more nearly continuous the chain, the greater is the density of the polymer.

Low-density polyethylene (LDPE) is obtained under conditions of free-radical polymerization, whereby polymerization is initiated by oxygen or peroxides under high pressure at roughly 200 °C (392 °F). Polyethylene, especially low-density polyethylene, is thermoplastic (softens and flows on heating) and can be extruded into sheets or films and molded into various shapes.

High-density polyethylene (HDPE) is obtained under conditions of coordination polymerization initiated by a mixture of titanium tetrachloride ($TiCl_4$) and triethylaluminum [$(CH_3CH_2)_3Al$]. Coordination polymerization was discovered by German chemist Karl Ziegler. Ziegler and Italian chemist Giulio Natta pioneered the development of Ziegler-Natta catalysts, for which they shared the 1963 Nobel Prize for Chemistry. The original Ziegler-Natta titanium tetrachloride-triethylaluminum catalyst has been joined by a variety of others. In addition to its application in the preparation of high-density polyethylene, coordination polymerization is the method by which ethylene oligomers, called linear α-olefins, and stereoregular polymers, especially polypropylene, are prepared.

Vinyl compounds, which are substituted derivatives of ethylene, can also be polymerized according to the following reaction:

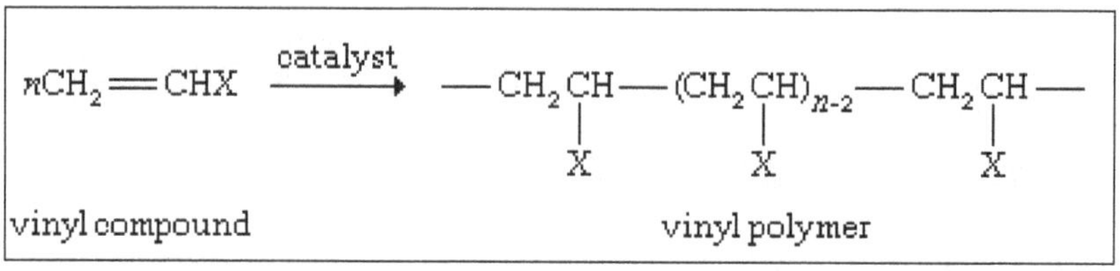

Polymerization of vinyl chloride (where X is Cl) gives polyvinyl chloride, or PVC, more than 27 million metric tons of which is used globally each year to produce pipes, floor tiles, siding for houses, gutters, and downspouts. Polymerization of styrene, X = C_6H_5 (a phenyl group derived from benzene), yields polystyrene, a durable polymer used to make luggage, refrigerator casings, and television cabinets and which can be foamed and used as a lightweight packaging and insulating material. If X = CH_3, the product is polypropylene, which is used to make films, molded articles, and fibres. Acrylonitrile, X = CN, gives polyacrylonitrile for use in carpet fibres and clothing.

Diene polymers have an important application as rubber substitutes. Natural rubber is a polymer of 2-methyl-1,3-butadiene (commonly called isoprene). Coordination polymerization conditions have been developed that convert isoprene to a polymer with properties identical to that of natural rubber.

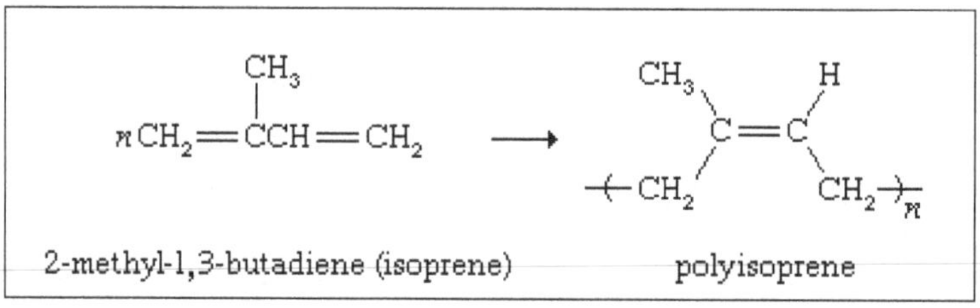

The largest portion of the synthetic rubber industry centres on styrene-butadiene rubber (SBR), which is a copolymer of styrene and 1,3-butadiene. Its major application is in automobile tires.

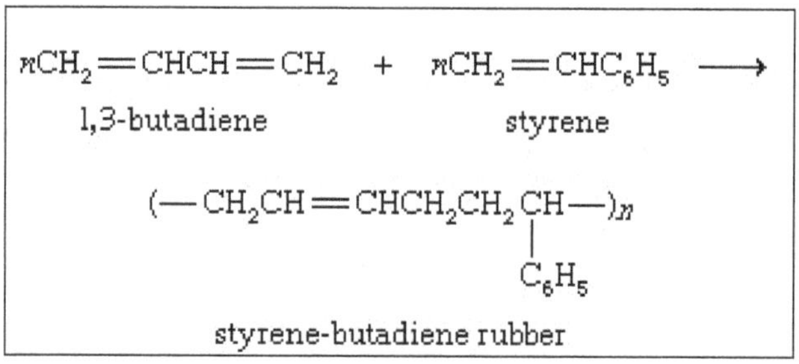

Alkyne polymerization is neither not nearly as developed nor as useful a procedure as alkene polymerization. The dimer of acetylene, vinylacetylene, is the starting material for the preparation of 2-chloro-1,3-butadiene, which in turn is polymerized to give the elastomer neoprene. Neoprene was the first commercially successful rubber substitute.

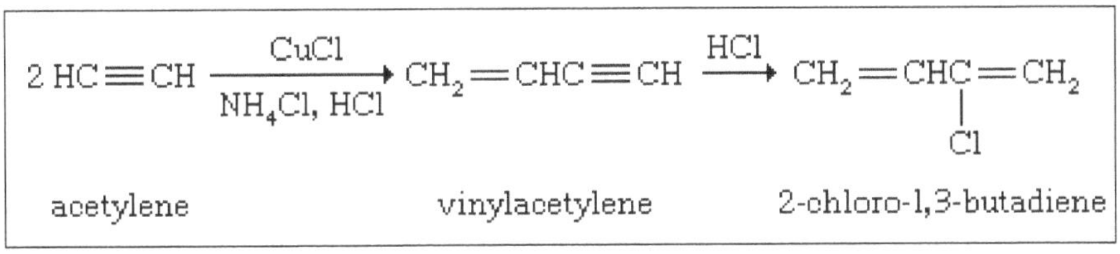

Aromatic Hydrocarbons

Benzene (C_6H_6), the simplest aromatic hydrocarbon, was first isolated in 1825 by English chemist Michael Faraday from the oily residues left from illuminating gas. In 1834 it was prepared from benzoic acid ($C_6H_5CO_2H$), a compound obtained by chemical degradation of gum benzoin, the fragrant balsam exuded by a tree that grows on the island of Java, Indonesia. Similarly, the hydrocarbon toluene ($C_6H_5CH_3$) received its name from tolu balsam, a substance isolated from a Central American tree and used in perfumery. Thus benzene, toluene, and related hydrocarbons, while not particularly pleasant-smelling themselves, were classified as aromatic because they were obtained from fragrant substances. Joseph Loschmidt, an Austrian chemist, recognized in 1861 that most aromatic substances have formulas that can be derived from benzene by replacing one or more hydrogens by other atoms or groups. The term aromatic thus came to mean any compound structurally derived from benzene. Use of the term expanded with time to include properties especially that of special stability, and eventually aromaticity came to be defined in terms of stability alone. The modern definition states that a compound is aromatic if it is significantly more stable than would be predicted on the basis of the most stable Lewis structural formula written for it. (This special stability is related to the number of electrons contained in a cyclic conjugated system.) All compounds that contain a benzene ring possess special stability and are classified as benzenoid aromatic compounds. Certain other compounds lack a benzene ring yet satisfy the criterion of special stability and are classified as nonbenzenoid aromatic compounds.

Arenes

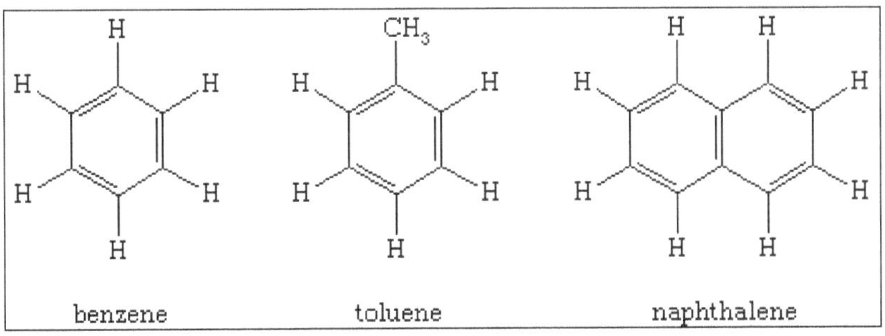

These compounds are hydrocarbons that contain a benzene ring as a structural unit. In addition to benzene, other examples include toluene and naphthalene.

(Hydrogen atoms connected to the benzene ring are shown for completeness in the above structural formulas. The more usual custom, which will be followed hereafter, omits them.)

Structure and Bonding

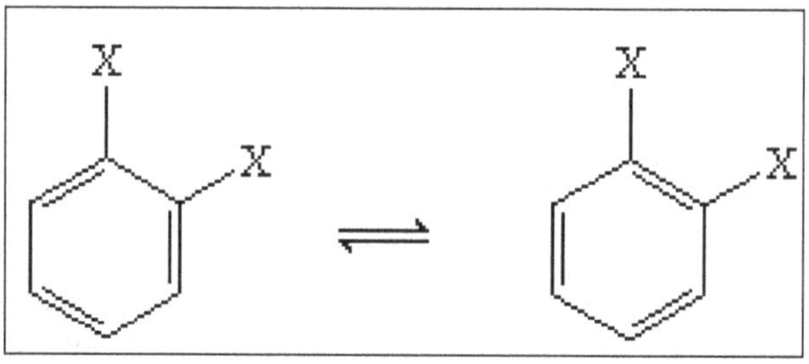

In 1865 the German chemist August Kekule von Stradonitz suggested the cyclic structure for benzene shown above. Kekule's structure, while consistent with the molecular formula and the fact that all of the hydrogen atoms of benzene are equivalent, needed to be modified to accommodate the observation that disubstitution of the ring at adjacent carbons did not produce isomers. Two isomeric products, as shown below, would be expected depending on the placement of the double bonds within the hexagon, but only one 1,2-disubstituted product was formed. In 1872 Kekule revised his proposal by assuming that two such isomers would interconvert so rapidly as to be inseparable from one another.

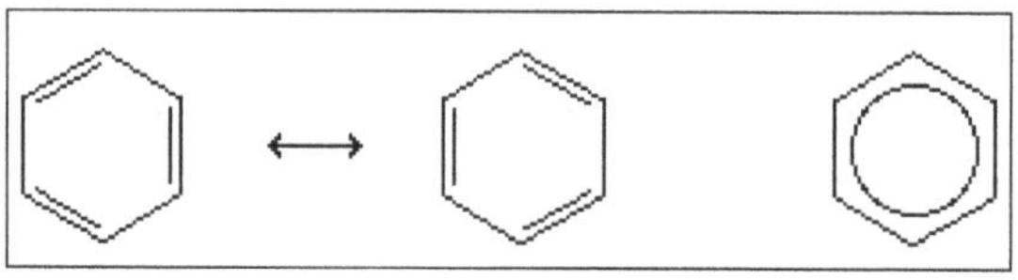

The next major advance in understanding was due largely to the American chemist Linus Pauling, who brought the concept of resonance—which had been introduced in the 1920s—to the question of structure and bonding in benzene. According to the resonance model, benzene does not exist as a pair of rapidly interconverting conjugated trienes but has a single structure that cannot be represented by formulations with localized electrons. The six π electrons (two for the π component of each double bond) are considered to be delocalized over the entire ring, meaning that each π electron is shared by all six carbon atoms rather than by two. Resonance between the two Kekule formulas is symbolized by an arrow of the type $\leftrightarrow$ to distinguish it from an interconversion process. The true structure of benzene is described as a hybrid of the two Kekule forms and is often simplified to a hexagon with an inscribed circle to represent the six delocalized π electrons. It is commonly said that a resonance hybrid is more stable than any of the contributing structures, which means, in the case of benzene, that each π electron, because it feels the attractive force of six carbons (delocalized), is more strongly held than if it were associated with only two of them (localized double bonds).

The orbital hybridization model of bonding in benzene is based on a σ bond framework of six sp^2 hybridized carbons. The six π electrons circulate above and below the plane of the ring in a region formed by the overlap of the p orbitals contributed by the six carbons.

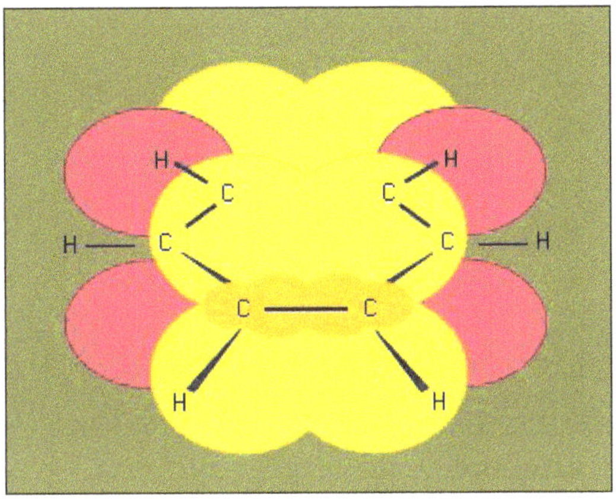

Benzene is the smallest of the organic aromatic hydrocarbons. It contains sigma bonds (represented by lines) and regions of high-pi electron density, formed by the overlapping of p orbitals (represented by the dark yellow shaded area) of adjacent carbon atoms, which give benzene its characteristic planar structure.

Benzene is a planar molecule with six C–C bond distances of equal length. The observed bond distance (1.40 angstroms) is midway between the sp^2-sp^2 single-bond distance (1.46 angstroms) and sp^2-sp^2 double-bond distance (1.34 angstroms) seen in conjugated dienes and is consistent with the bond order of 1.5 predicted by resonance theory. (Bond order is an index of bond strength. A bond order of 1 indicates that a single σ bond exists between two atoms, and a bond order of 2 indicates the presence of one σ and one π bond between two atoms. Fractional bond orders are possible for resonance structures, as in the case of benzene.) Benzene is a regular hexagon; all bond angles are 120°.

The special stability of benzene is evident in several ways. Benzene and its derivatives are much less reactive than expected. Arenes are unsaturated but resemble saturated hydrocarbons (i.e., alkanes) in their low reactivity more than they resemble unsaturated ones (alkenes and alkynes; see below Reactions). Thermodynamic estimates indicate that benzene is 30–36 kilocalories per mole more stable than expected for a localized conjugated triene structure.

Nomenclature

A number of monosubstituted derivatives of benzene have common names of long standing that have been absorbed into the IUPAC system. Examples include toluene ($C_6H_5CH_3$) and styrene ($C_6H_5CH = CH_2$). Disubstituted derivatives of benzene may have their substituents in a 1,2 (ortho, or o), 1,3 (meta, or m), or 1,4 (para, or p) relationship (where the numbers indicate the carbons to which the substituents are bonded) and may be named using either numerical locants or the ortho, meta, para notation.

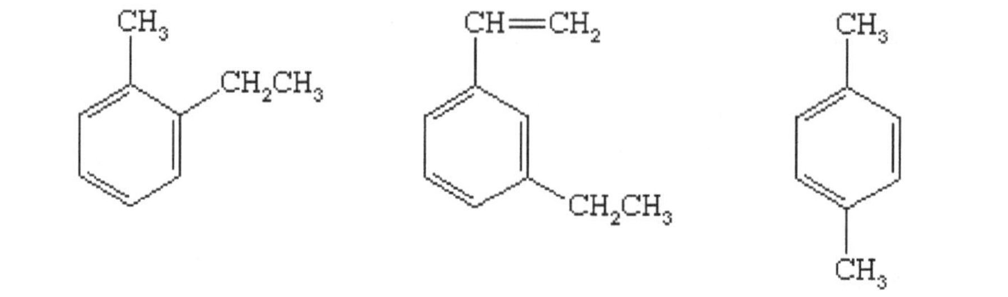

o-ethyltoluene 1-ethyl-2-methylbenzene | *m*-ethylstyrene 1-ethyl-3-vinylbenzene | *p*-xylene 1,4-dimethylbenzene

Two groups that contain benzene rings, C_6H_5–(phenyl) and $C_6H_5CH_2$–(benzyl), have special names, as in these examples:

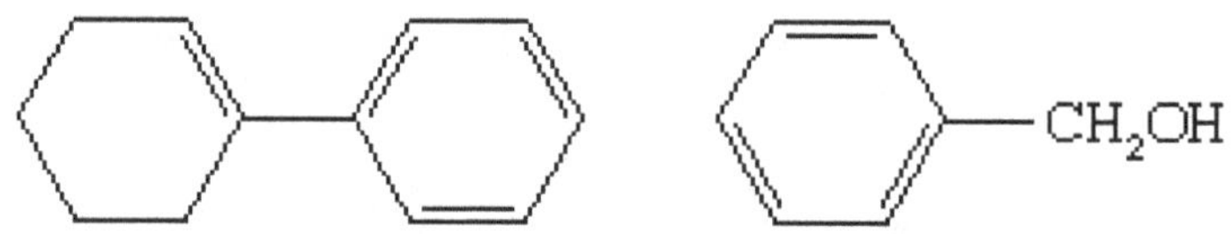

1-phenylcyclohexene | benzyl alcohol

Arenes in which two or more benzene rings share a common side are called polycyclic aromatic compounds. Each such assembly has a unique name, as the examples of naphthalene, anthracene, and phenanthrene illustrate.

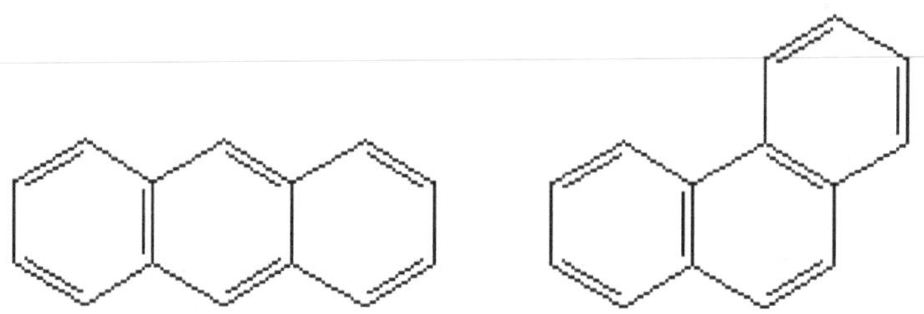

anthracene | phenanthrene

Certain polycyclic aromatic hydrocarbons are known to be carcinogenic and enter the environment when organic matter is burned. Benzo[a]pyrene, for example, is present in tobacco smoke and chimney soot and is formed when meat is cooked on barbecue grills.

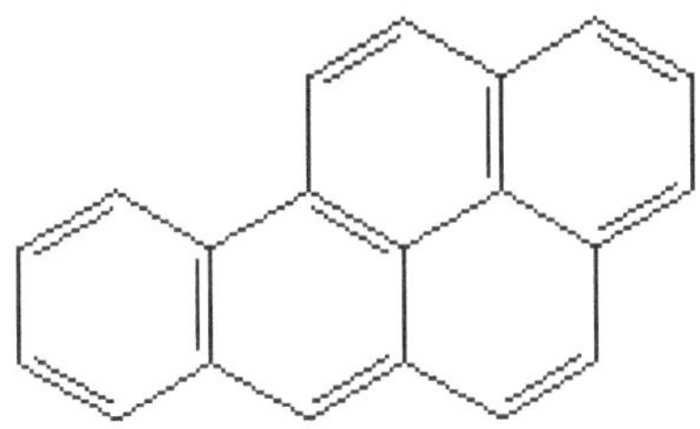

benzo[a]pyrene

Physical Properties

All arenes are either liquids or solids at room temperature; none are gases. Aromatic hydrocarbons are insoluble in water. Benzene was once widely used as a solvent, but evidence of its carcinogenic properties prompted its replacement by less hazardous solvents.

Table: Physical constants of benzene and selected arenes.

Name	Boiling point (°c)	Melting point (°c)
Benzene	80.1	+5.5
Toluene	110.6	−95
Ethylbenzene	136.2	−94
P-xylene	138.4	+13
Styrene	145	−30.6
Naphthalene	218	+80.3
Anthracene	342	+218
Phenanthrene	340	+100

Source and Synthesis

For a period of approximately 100 years encompassing the last half of the 19th century and the first half of the 20th century, coal was the main starting material for the large-scale production of aromatic compounds. When soft coal is heated in the absence of air, substances are formed that are volatile at the high temperatures employed (500–1,000 °C [930–1,800 °F], depending on the process), which when condensed give the material known as coal tar. Distillation of coal tar gives a number of fractions, the lowest boiling of which contains benzene, toluene, and other low-molecular-weight aromatic compounds. The higher-boiling fractions are sources of aromatic compounds of higher molecular weight. Beginning with the second half of the 20th century, petroleum replaced coal as the principal source of aromatic hydrocarbons. The stability of the benzene ring makes possible processes, known generally as catalytic reforming, in which alkanes are converted to arenes by a combination of isomerization and dehydrogenation events.

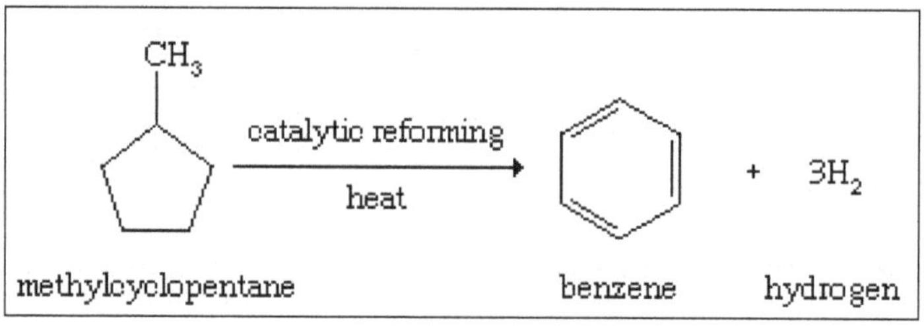

The arenes formed by catalytic reforming are used to boost the octane rating of gasoline and as starting materials for the synthesis of a variety of plastics, fibres, dyes, agricultural chemicals, and drugs.

Reactions

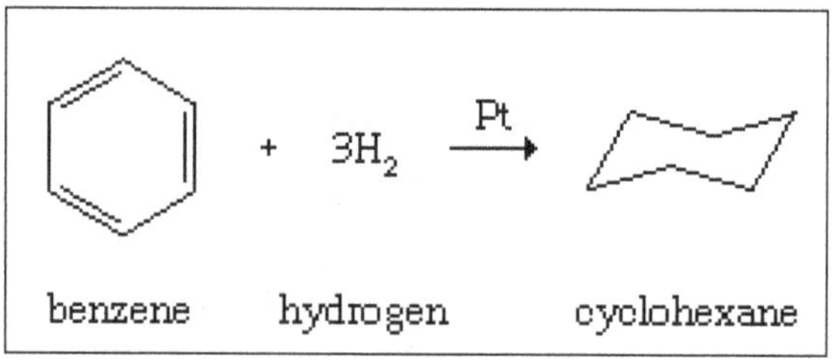

Like other hydrocarbons, arenes undergo combustion to form carbon dioxide and water, and like other unsaturated hydrocarbons, arenes undergo catalytic hydrogenation.

nitration:	C_6H_6 benzene	+	$HONO_2$ nitric acid	$\xrightarrow{H_2SO_4}$	$C_6H_5NO_2$ nitrobenzene	+	H_2O water
halogenation: (X = Cl or Br)	C_6H_6 benzene	+	X_2 halogen	$\xrightarrow{FeX_3}$	C_6H_5X halobenzene	+	HX hydrogen halide
sulfonation:	C_6H_6 benzene	+	SO_3 sulfur trioxide	$\longrightarrow$	$C_6H_5SO_3H$ benzenesulfonic acid		
alkylation: (R = alkyl group)	C_6H_6 benzene	+	RCl alkyl halide	$\xrightarrow{AlCl_3}$	C_6H_5R alkylbenzene	+	HCl hydrogen chloride
acylation:	C_6H_6 benzene	+	$R\overset{\overset{O}{\|}}{C}Cl$ acyl chloride	$\xrightarrow{AlCl_3}$	$C_6H_5\overset{\overset{O}{\|}}{C}R$ acylbenzene	+	HCl hydrogen chloride

However, many species that react with alkenes by addition react with arenes by replacing one of the hydrogens on the ring (substitution). This behaviour is most pronounced with species known as electrophiles (electron seekers), and the characteristic reaction of an arene is electrophilic aromatic substitution. Representative electrophilic aromatic substitutions, shown with benzene as the arene, include nitration, halogenation, sulfonation, alkylation, and acylation.

Alkylation and acylation reactions of aromatic compounds that are catalyzed by aluminum chloride ($AlCl_3$) are referred to as Friedel-Crafts reactions after French chemist and mineralogist Charles Friedel and American chemist James M. Crafts, who discovered this reaction at the Sorbonne in 1877. Further substitution is possible, and under certain circumstances all six hydrogen atoms of benzene are capable of being replaced. The products of electrophilic aromatic substitution in benzene and its derivatives are employed in subsequent transformations to give a variety of useful products.

The benzene ring is relatively resistant toward oxidation with the exception of its combustion. Arenes that bear alkyl side chains, when treated with strong oxidizing agents, undergo oxidation of the side chain while the ring remains intact.

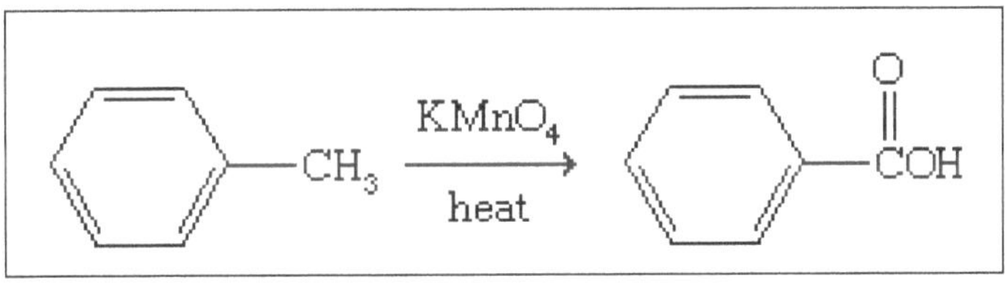

Under conditions of biological oxidation by the cytochrome P-450 enzyme system in the liver, benzene and polycyclic aromatic hydrocarbons undergo epoxidation of their ring. The epoxides that form react with deoxyribonucleic acid (DNA), and it is believed that this process is responsible for the carcinogenic properties of polycyclic aromatic hydrocarbons.

Nonbenzenoid Aromatic Compounds

Once it became clear that the special stability of benzene and related compounds was associated with the cyclic nature of its conjugated system of double bonds, organic chemists attempted to synthesize both larger and smaller analogs. The earliest targets were cyclobutadiene (C_4H_4) and cyclooctatetraene (C_8H_8).

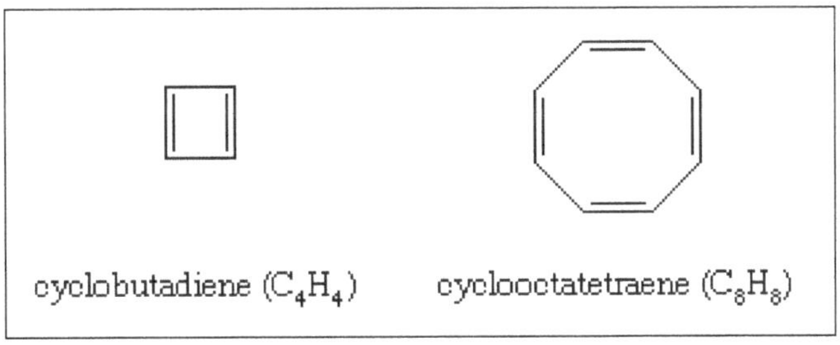

Of the two, cyclooctatetraene proved more accessible. It was first prepared in 1911 by chemical degradation of the alkaloid pseudopelletierine by the German chemist Richard Willstätter. A more direct synthesis from acetylene was developed in the 1940s. While cyclooctatetraene is a stable substance, thermochemical measurements show that it does not possess the special stability required to be classified as an aromatic hydrocarbon. Structural studies reveal that, unlike benzene in which all of the ring bonds are of equal length (1.40 angstroms), cyclooctatetraene has four short (1.33 angstroms) and four long (1.46 angstroms) carbon-carbon distances consistent with a pattern of alternating single and double bonds. Cyclooctatetraene, moreover, has a nonplanar tub-shaped structure, which, because it is not planar, does not permit the eight π electrons to be delocalized over all the carbon atoms. Classifying cyclooctatetraene as an aliphatic hydrocarbon is also consistent with numerous observations concerning its chemical reactivity, which is characterized by a tendency to undergo alkenelike addition rather than arenelike substitution.

Cyclobutadiene resisted attempts at chemical synthesis until the 1950s when evidence for its intermediacy in certain reactions was obtained. The high reactivity of cyclobutadiene was interpreted as evidence against its aromaticity. Subsequent low-temperature spectroscopic studies revealed that cyclobutadiene has a rectangular structure with alternating single and double bonds unlike the square shape required for an electron-delocalized aromatic molecule.

Annulenes and the Hückel Rule

Insights into the requirements for aromaticity were provided by German physicist Erich Hückel in 1931. Limiting his analysis to planar, monocyclic, completely conjugated polyenes, Hückel calculated that compounds of this type are aromatic if they contain $4n + 2$ π electrons, where n is a whole number. According to the Hückel rule, 2, 6, 10, 14... π electrons ($n = 0, 1, 2, 3,...,$ etc.) confer aromaticity to this class of compounds, but 4, 8, 12, 16... π electrons do not. Benzene, which has 6 π electrons, is aromatic, but cyclobutadiene, which has 4, and cyclooctatraene, which has 8, are nonaromatic.

Monocyclic, completely conjugated polyenes (the hydrocarbons treated by the Hückel rule) are referred to as annulenes, and individual annulenes are differentiated by a numerical prefix equal to the number of π electrons. Beyond benzene, [10]-annulene is the first hydrocarbon to satisfy the Hückel rule. A structure in which all of the double bonds are cis, however, would be a regular 10-sided polygon requiring bond angles of 144° (instead of the 120° angles required for sp^2 hybridized carbon) and would suffer considerable angle strain. The destabilization owing to angle strain apparently exceeds the stabilization associated with aromaticity and makes all-cis-cyclodecapentaene a highly reactive substance. An isomer in which two of the double bonds are trans should, in principle, be free of angle strain. It is destabilized, however, by a repulsive force between two hydrogen atoms that are forced together in the interior of the ring, and for this reason it is relatively reactive.

five *cis* double bonds three *cis* and two *trans* double bonds
cyclodecapentaene stereoisomers

[18]-Annulene is predicted to be aromatic by the Hückel rule ($4n + 2 = 18$ when $n = 4$). The structure shown has a shape that makes it free of angle strain and is large enough so that repulsive forces between hydrogen atoms in the interior are minimal. Thermochemical measurements indicate resonance energy of roughly 100 kilocalories per mole, and structural studies reveal that the molecule is planar with all its bond distances falling in the range 1.37–1.43 angstroms. In terms of its chemical reactivity, however, [18]-annulene resembles an alkene more than it resembles benzene.

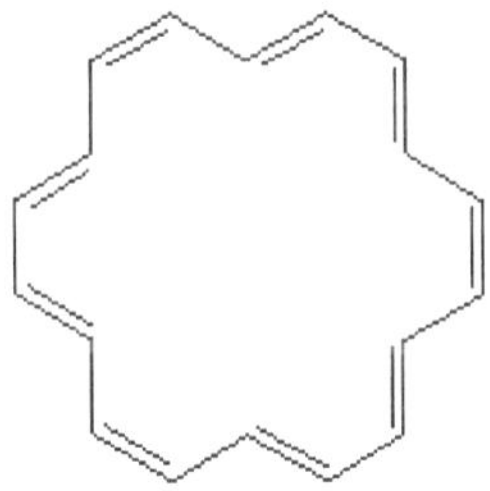

[18]-annulene

Polycyclic Nonaromatic Compounds

The Hückel rule is not designed to apply to polycyclic compounds. Nevertheless, a similar dependence on the number of π electrons is apparent. The bicyclic hydrocarbon azulene has the same number of π electrons (10) as naphthalene and, like naphthalene, is aromatic. Pentalene and heptalene, analogs with 8 and 12 π electrons, respectively, are not aromatic. Both are relatively unstable, highly reactive substances.

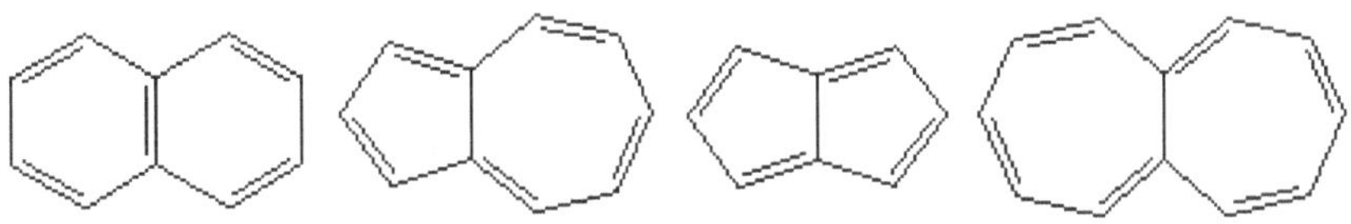

naphthalene azulene pentalene heptalene
aromatic not aromatic

ISOLOBAL PRINCIPLE

The isolobal principle (more formally known as the isolobal analogy) is a strategy used in organometallic chemistry to relate the structure of organic and inorganic molecular fragments in order to predict bonding properties of organometallic compounds. Roald Hoffmann described molecular fragments as isolobal "if the number, symmetry properties, approximate energy and shape of the frontier orbitals and the number of electrons in them are similar – not identical, but similar". One can predict the bonding and reactivity of a lesser-known species from that of a better-known species if the two molecular fragments have similar frontier orbitals, the highest occupied molecular orbital (HOMO) and the lowest unoccupied molecular orbital (LUMO). Isolobal compounds are analogues to isoelectronic compounds that share the same number of valence electrons and structure. A graphic representation of isolobal structures, with the isolobal pairs connected through a double-headed arrow with half an orbital below, is found in figure.

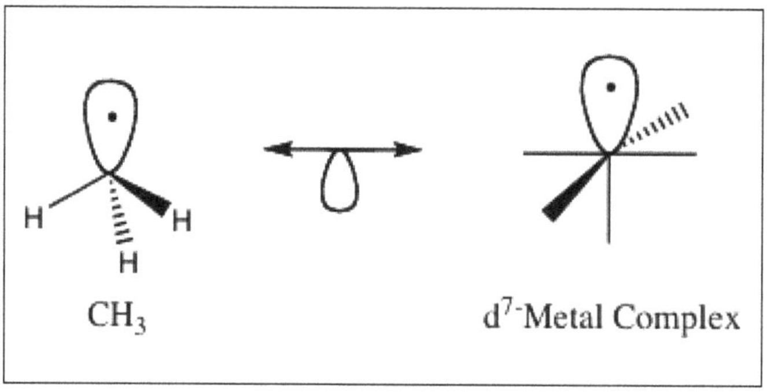

Basic example of the isolobal analogy.

For his work on the isolobal analogy, Hoffmann was awarded the Nobel Prize in Chemistry in 1981, which he shared with Kenichi Fukui. In his Nobel Prize lecture, Hoffmann stressed that the isolobal analogy is a useful, yet simple, model and thus is bound to fail in certain instances.

Construction of Isolobal Fragments

To begin to generate an isolobal fragment, the molecule needs to follow certain criteria. Molecules based around main group elements should satisfy the octet rule when all bonding and nonbonding molecular orbitals (MOs) are filled and all antibonding MOs are empty. For example, methane is a simple molecule from which to form a main group fragment. The removal of a hydrogen atom from methane generates a methyl radical. The molecule retains its molecular geometry as the frontier orbital points in the direction of the missing hydrogen atom. Further removal of hydrogen results in the formation of a second frontier orbital. This process can be repeated until only one bond remains to the molecule's central atom. Figure demonstrates this example of step-by-step generation of isolobal fragments.

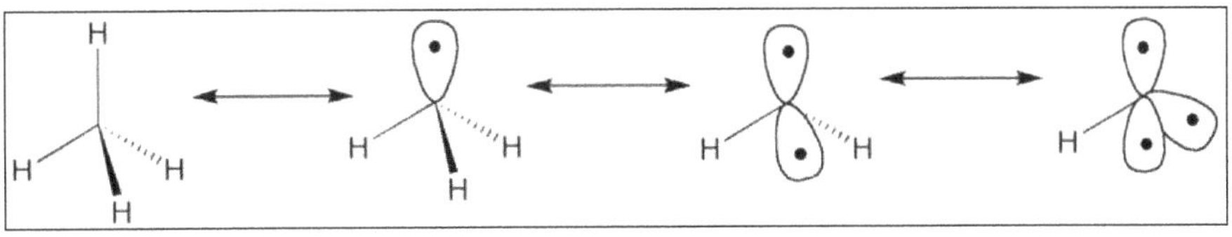

Construction of frontier or bitals from methane.

The isolobal fragments of octahedral complexes, such as ML_6, can be created in a similar fashion. Transition metal complexes should initially satisfy the eighteen electron rule, have no net charge, and their ligands should be two electron donors (Lewis bases). Consequently, the metal center for the ML_6 starting point must be d^6. Removal of a ligand is analogous to the removal of hydrogen of methane in the previous example resulting in a frontier orbital, which points toward the removed ligand. Cleaving the bond between the metal center and one ligand results in a ML^-_5 radical complex. In order to satisfy the zero-charge criteria the metal center must be changed. For example, a MoL_6 complex is d^6 and neutral. However, removing a ligand to form the first frontier orbital would result in a MoL^-_5 complex because Mo has obtained an additional electron making it d^7. To remedy this, Mo can be exchanged for Mn, which would form a neutral d^7 complex in this case, as shown in figure. This trend can continue until only one ligand is left coordinated to the metal center.

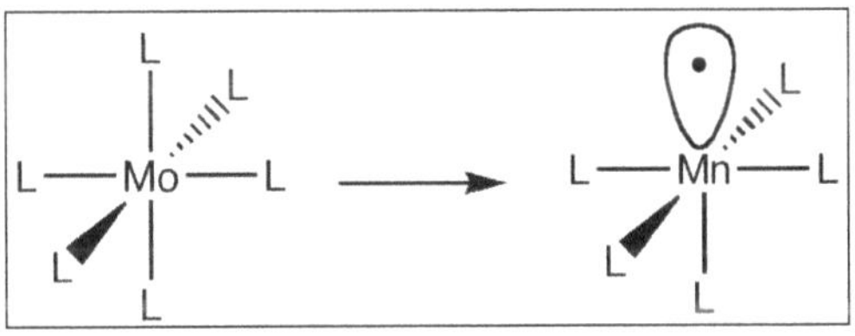

Production of a frontier orbital in an octahedral complex. Since the process is not charge producing, the metal center must change from d^6 Mo to d^7 Mn to retain the neutral charge.

Relationship between Tetrahedral and Octahedral Fragments

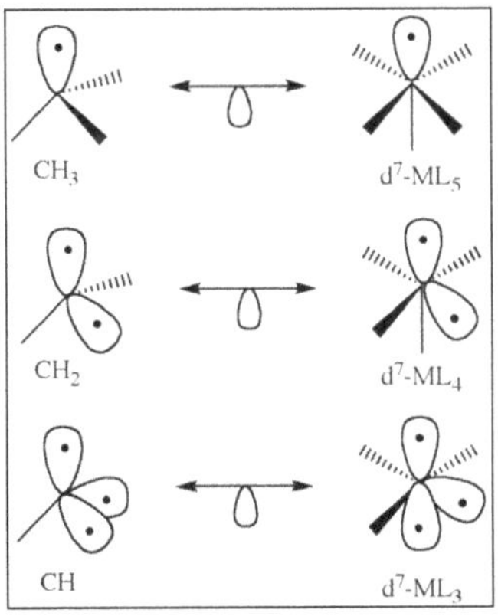

Isolobal fragments of tetrahedral and octahedral geometries.

Isolobal fragments of tetrahedral and octahedral molecules can be related. Structures with the same number of frontier orbitals are isolobal to one another. For example, the methane with two hydrogen atoms removed, CH_2 is isolobal to a d^7 ML_4 complex formed from an octahedral starting complex.

MO Theory Dependence

Any sort of saturated molecule can be the starting point for generating isolobal fragments. The molecules' bonding and nonbonding molecular orbitals (MOs) should be filled and the antibonding MOs empty. With each consecutive generation of an isolobal fragment, electrons are removed from the bonding orbitals and a frontier orbital is created. The frontier orbitals are at a higher energy level than the bonding and nonbonding MOs. Each frontier orbital contains one electron. For example, consider figure, which shows the production of frontier orbitals in tetrahedral and octahedral molecules.

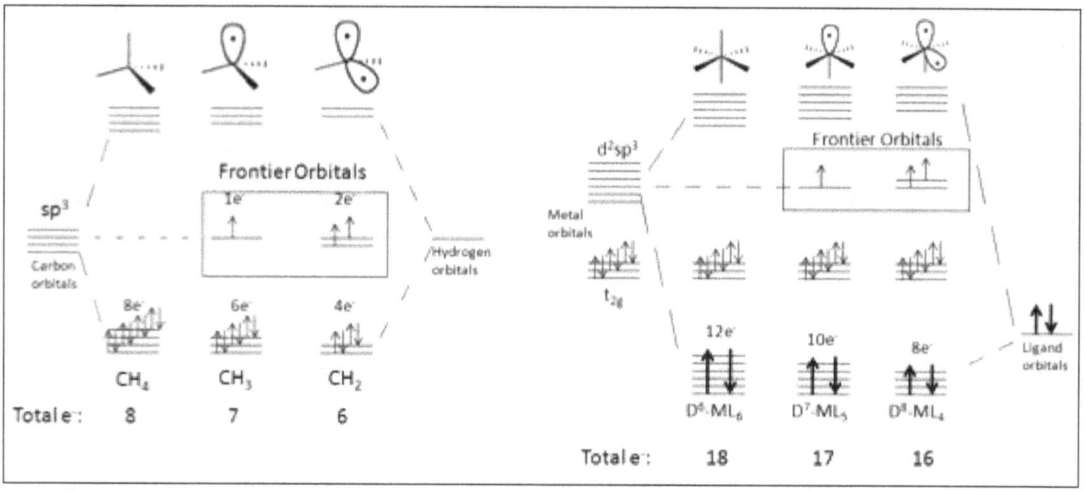

Molecular orbital diagram depiction of frontier orbitals in methane and a basic ML_6 metal complex.

When a fragment is formed from CH_4, one of the sp^3 hybrid orbitals involved in bonding becomes a nonbonding singly occupied frontier orbital. The frontier orbital's increased energy level is also shown in the figure. Similarly when starting with a metal complex such as d^6-ML_6, the d^2sp^3 hybrid orbitals are affected. Furthermore, the t_{2g} nonbonding metal orbitals are unaltered.

Extensions of the Analogy

The isolobal analogy has applications beyond simple octahedral complexes. It can be used with a variety of ligands, charged species and non-octahedral complexes.

Ligands

Typical ligands used in the isolobal analogy are two-electron donors such as phosphines, halogens or carbonyls. However, other types of ligands can be employed. If ligands donate multiple pairs of electrons, they will occupy multiple coordination sites. For example, the cyclopentadienyl anion is a six-electron donor, so it occupies three coordination sites. Polydentate ligands can also be used in the analogy, such as ethylenediamine, a bidentate ligand, or triethylenetetramine, a tetradentate ligand.

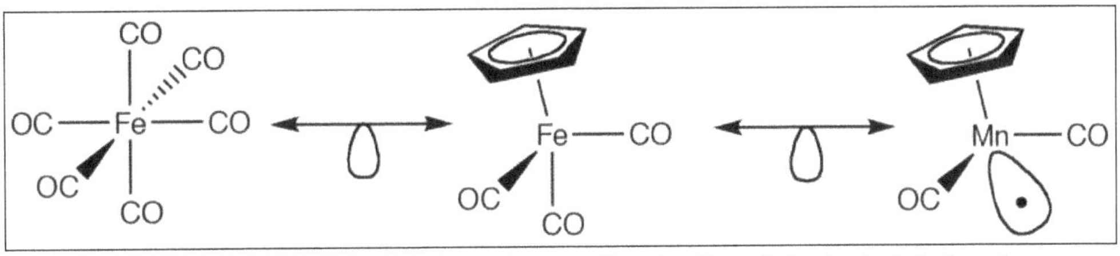

Example of cyclopentadiene, a multiply-coordinating ligand, in the isolobal analogy.

Isoelectronic Fragments

The isolobal analogy can also be used with isoelectronic fragments having the same coordination number, which allows charged species to be considered. For example, $Re(CO)_5$ is isolobal with CH_3 and therefore, $[Ru(CO)_5]^+$ and $[Mo(CO)_5]^-$ are also isolobal with CH_3. Any 17-electron metal complex would be isolobal in this example.

In a similar sense, the addition or removal of electrons from two isolobal fragments results in two new isolobal fragments. Since $Re(CO)_5$ is isolobal with CH_3, $[Re(CO)_5]^+$ is isolobal with CH^+_3.

Non-octahedral Complexes

Table: Isolobal relationship between octahedral and square planar complexes.

Octahedral ML_n	Square-planar ML_{n-2}
d^6 : $Mo(CO)_5$	d^8 : $[PdCl_3]^-$
d^8 : $Os(CO)_4$	d^{10} : $Ni(PR_3)_2$

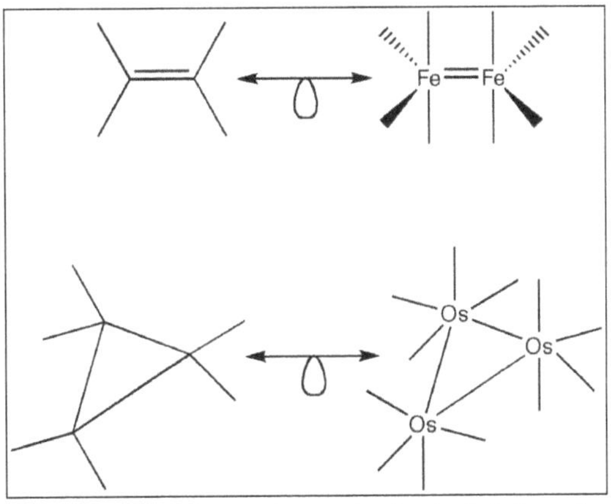

Examples of non-basic shapes in the isolobal analogy.

The analogy applies to other shapes besides tetrahedral and octahedral geometries. The derivations used in octahedral geometry are valid for most other geometries.

The exception is square-planar because square-planar complexes typically abide by the 16-electron rule. Assuming ligands act as two-electron donors the metal center in square-planar molecules is d^8. To relate an octahedral fragment, ML_n, where M has a d^x electron configuration to a square planar analogous fragment, the formula ML_{n-2} where M has a d^{x+2} electron configuration should be followed.

Applications and Examples

Uses of the isolobal analogy include providing a shortcut to understanding electronic structure, predicting reactivity and reaction mechanisms, and a method of classifying molecules. Applications are typically utilized to make connections between well-known systems and less familiar systems. For example, the possibility of unsynthesized compounds can be imagined from those of known molecular conformations. The isolobal analogy does not guarantee these products are capable of being produced, but only proposes a possibility. Consider the molecule $Fe(CO)_3$ complexed with cyclobutadiene. $Fe(CO)_3$ is isolobal with CH^+. Therefore, one can predict that CH^+ will coordinate with cyclobutadiene in a similar fashion that $Fe(CO)_3$ will. Thus the molecule $C_5H^+_5$ can be envisioned regardless of its actual existence.

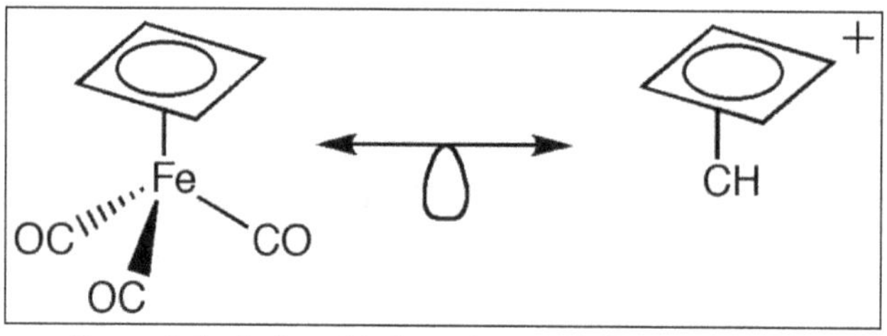

Potential relationship based on isolobal principles.

Predicting the reactivity of complexes can also be accomplished using the isolobal analogy. From the simple expectation of two CH_3 radicals reacting to form ethane one can use the analogy to predict M–C or M–M bonding such as $(CH_3)M(CO)_5$ and $M_2(CO)_{10}$, where M is d^7.

Another application of the isolobal analogy is assisting in predicting reaction mechanisms. As in the other applications the mechanisms of well-known reactions can be used to help predict mechanistic pathways of lesser-known reactions. There is no limit on the potential comparisons between organic and inorganic complexes. The analogy can flow in either direction (organic to inorganic) or within each division (organic to organic).

Arteaga Müller utilize the isolobal analogy to relate imido half-sandwich complexes with isoelectronic dicyclopentadienyl complexes. The isolobal relationship of the imido and the cyclopentadienyl ligands is the key to this comparison. The study found the reactivity of these two types of complexes to be similar although their catalytic abilities differed in some respects. This study shows that the isolobal analogy does not make perfect predictions between two isolobal fragments, as Hoffman warned in his Nobel Lecture.

Wu apply the isolobal analogy to explore relationships involving structures, energies and magnetic properties between polyhedral boron carbonyls and their hydrocarbon relatives. As determined in this study, although isolobal, these two sets of molecules have significant differences in their strain energy.

Goldman and Tyler used the isolobal analogy to determine the most likely mechanism for a deletion reaction. One of the products of the irradiation of $CpW(CO)_3Me$ in the presence of PPh_3 is $CpW(CO)^-_3$. The mechanism of said reaction was studied and theorized to be isolobal to the disproportionation of metal-metal bonded dimers involving 19-valence electron intermediates. The reactions are composed of isolobal fragments and the key intermediates of both reactions are isolobal. Thus, the reaction pathways are mechanistically isolobal.

18-ELECTRON RULE

The 18-electron rule is a rule used primarily for predicting and rationalizing formulas for stable metal complexes, especially organometallic compounds. The rule is based on the fact that the valence shells of transition metals consist of nine valence orbitals (one's orbital, three p orbitals and five d orbitals), which collectively can accommodate 18 electrons as either bonding or nonbonding electron pairs. This means that the combination of these nine atomic orbitals with ligand orbitals creates nine molecular orbitals that are either metal-ligand bonding or non-bonding. When a metal complex has 18 valence electrons, it is said to have achieved the same electron configuration as the noble gas in the period. The rule and its exceptions are similar to the application of the octet rule to main group elements. The rule is not helpful for complexes of metals that are not transition metals, and interesting or useful transition metal complexes will violate the rule because of the consequences deviating from the rule bears on reactivity. The rule was first proposed by American chemist Irving Langmuir in 1921.

Applicability

The rule usefully predicts the formulas for low-spin complexes of the Cr, Mn, Fe, and Co triads. Well-known examples include ferrocene, iron pentacarbonyl, chromium carbonyl, and nickel carbonyl.

Ligands in a complex determine the applicability of the 18-electron rule. In general, complexes that obey the rule are composed at least partly of π-acceptor ligands (also known as π-acids). This kind of ligand exerts a very strong ligand field, which lowers the energies of the resultant molecular orbitals so that they are favorably occupied.

Typical ligands include olefins, phosphines, and CO. Complexes of π-acids typically feature metal in a low-oxidation state. The relationship between oxidation state and the nature of the ligands is rationalized within the framework of π backbonding.

Consequences for Reactivity

Compounds that obey the 18-electron rule are typically "exchange inert". Examples include $[Co(NH_3)_6]Cl_3$, $Mo(CO)_6$, and $[Fe(CN)_6]^{4-}$. In such cases, in general ligand exchange occurs via dissociative substitution mechanisms, wherein the rate of reaction is determined by the rate of dissociation of a ligand. On the other hand, 18-electron compounds can be highly reactive toward electrophiles such as protons, and such reactions are associative in mechanism, being acid-base reactions.

Complexes with fewer than 18 valence electrons tend to show enhanced reactivity. Thus, the 18-electron rule is often a recipe for non-reactivity in either a stoichiometric or a catalytic sense.

Duodectet Rule

Current computational findings suggest valence p-orbitals on the metal participate in metal-ligand bonding, albeit weakly. Some new theoretical treatments do not count the metal p-orbitals in metal-ligand bonding, although these orbitals are still included as polarization functions. This results in a duodectet (12-electron) rule for one s-orbital and five d-orbitals only, but the modified rule has yet to be adopted by the general chemistry community.

Exceptions

π-donor or σ-donor ligands with small interactions with the metal orbitals lead to a weak ligand field which increases the energies of t_{2g} orbitals. These molecular orbitals become non-bonding or weakly anti-bonding orbitals (small Δ_{oct}). Therefore, addition or removal of electron has little effect on complex stability. In this case, there is no restriction on the number of d-electrons and complexes with 12–22 electrons are possible. Small Δ_{oct} makes filling $e_g{}^*$ possible (>18 e^-) and π-donor ligands can make t_{2g} antibonding (<18 e^-). These types of ligand are located in the low-to-medium part of the spectrochemical series. For example: $[TiF_6]^{2-}$ (Ti(IV), d^0, 12 e^-), $[Co(NH_3)_6]^{3+}$ (Co(III), d^6, 18 e^-), $[Cu(OH_2)_6]^{2+}$ (Cu(II), d^9, 21 e^-).

In terms of metal ions, Δ_{oct} increases down a group as well as with increasing oxidation number. Strong ligand fields lead to low-spin complexes which cause some exceptions to the 18-electron rule.

16-Electron Complexes

An important class of complexes that violate the 18e rule are the 16-electron complexes with metal d^8 configurations. All high-spin d^8 metal ions are octahedral (or tetrahedral), but the low-spin d^8 metal ions are all square planar. Important examples of square-planar low-spin d^8 metal Ions are Rh(I), Ir(I), Ni(II), Pd(II), and Pt(II). At picture below is shown the splitting of the d subshell in low-spin square-planar complexes. Examples are especially prevalent for derivatives of the cobalt and nickel triads. Such compounds are typically square-planar. The most famous example is Vaska's complex (IrCl(CO)(PPh_3)$_2$), $[PtCl_4]^{2-}$, and Zeise's salt $[PtCl_3(\eta^2\text{-}C_2H_4)]^-$. In such complexes, the d_{z^2} orbital is doubly occupied and nonbonding.

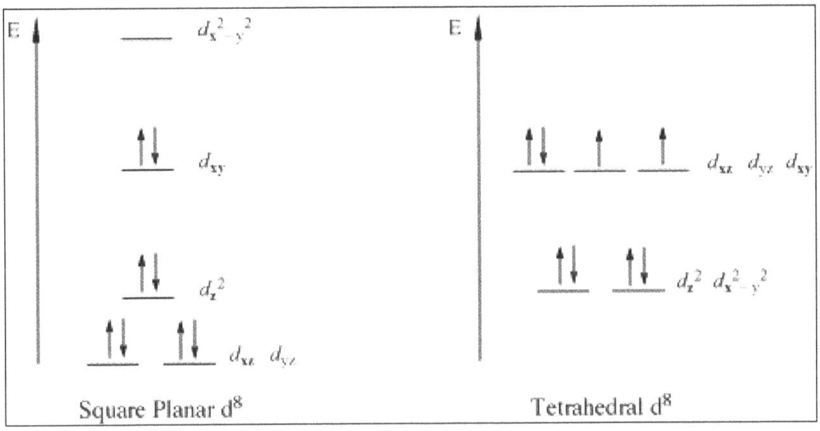

Many catalytic cycles operate via complexes that alternate between 18-electron and square-planar 16-electron configurations. Examples include Monsanto acetic acid synthesis, hydrogenations, hydroformylations, olefin isomerizations, and some alkene polymerizations.

Other violations can be classified according to the kinds of ligands on the metal center.

Bulky Ligands

Bulky ligands can preclude the approach of the full complement of ligands that would allow the metal to achieve the 18 electron configuration. Examples:

- Ti(neopentyl)$_4$ (8 e^-)
- $Cp^*_2Ti(C_2H_4)$ (16 e^-)
- $V(CO)_6$ (17 e^-)
- $Cp^*Cr(CO)_3$ (17 e^-)
- Pt(P^{t}Bu3)$_2$ (14 e^-)
- Co(norbornyl)$_4$ (13 e^-)
- $[FeCp_2]^+$ (17 e^-)

Sometimes such complexes engage in agostic interactions with the hydrocarbon framework of the bulky ligand. For example:

- $W(CO)_3[P(C_6H_{11})_3]_2$ has 16 e^- but has a short bonding contact between one C–H bond and the W center.
- $Cp(PMe_3)V(CHCMe_3)$ (14 e^-, diamagnetic) has a short V–H bond with the 'alkylidene-H', so the description of the compound is somewhere between $Cp(PMe_3)V(CHCMe_3)$ and $Cp(PMe_3)V(H)(CCMe_3)$.

High-spin Complexes

High-spin metal complexes have singly occupied orbitals and may not have any empty orbitals into which ligands could donate electron density. In general, there are few or no π-acidic ligands in the complex. These singly occupied orbitals can combine with the singly occupied orbitals of radical ligands (e.g., oxygen), or addition of a strong field ligand can cause electron-pairing, thus creating a vacant orbital that it can donate into. Examples:

- $CrCl_3(THF)_3$ (15 e^-)
- $[Mn(H_2O)_6]^{2+}$ (17 e^-)
- $[Cu(H_2O)_6]^{2+}$ (21 e^-)

Complexes containing strongly π-donating ligands often violate the 18-electron rule. These ligands include fluoride (F^-), oxide (O^{2-}), nitride (N^{3-}), alkoxides (RO^-), and imides (RN^{2-}). Examples:

- $[CrO_4]^{2-}$ (16 e^-)
- $Mo(=NR)_2Cl_2$ (12 e^-)

In the latter case, there is substantial donation of the nitrogen lone pairs to the Mo (so the compound could also be described as a 16 e^- compound). This can be seen from the short Mo–N bond length, and from the angle Mo–N–C(R), which is nearly 180°. Counter-examples:

- trans-$WO_2(Me_2PCH_2CH_2PMe_2)_2$ (18 e^-)
- Cp^*ReO_3 (18 e^-)

In these cases, the M=O bonds are "pure" double bonds (i.e., no donation of the lone pairs of the oxygen to the metal), as reflected in the relatively long bond distances.

π-Donating Ligands

Ligands where the coordinating atom bears nonbonding lone pairs often stabilize unsaturated complexes. Metal amides and alkoxides often violate the 18e rule.

Combinations of Effects

The above factors can sometimes combine. Examples include:

- Cp^*VOCl_2 (14 e^-)
- $TiCl_4$ (8 e^-)

Higher Electron Counts

Some complexes have more than 18 electrons. Examples:

- Cobaltocene (19 e^-)
- Nickelocene (20 e^-)
- The hexaaquacopper(II) ion $[Cu(H_2O)_6]^{2+}$ (21 e^-)
- $TM(CO)_8^-$ (TM = Sc, Y, La) (20 e^-)

Often, cases where complexes have more than 18 valence electrons are attributed to electrostatic forces – the metal attracts ligands to itself to try to counterbalance its positive charge, and the number of electrons it ends up with is unimportant. In the case of the metallocenes, the chelating nature of the cyclopentadienyl ligand stabilizes its bonding to the metal. Somewhat satisfying are the two following observations: cobaltocene is a strong electron donor, readily forming the 18-electron cobaltocenium cation; and nickelocene tends to react with substrates to give 18-electron complexes, e.g. $CpNiCl(PR_3)$ and free CpH.

In the case of nickelocene, the extra two electrons are in orbitals which are weakly metal-carbon antibonding; this is why it often participates in reactions where the M–C bonds are broken and the electron count of the metal changes to 18.

The 20-electron systems $TM(CO)_8^-$ (TM = Sc, Y, La) have a cubic (O_h) equilibrium geometry and a singlet ($^1A_{1g}$) electronic ground state. There is one occupied valence MO with a_{2u} symmetry, which is formed only by ligand orbitals without a contribution from the metal AOs. But the adducts $TM(CO)_8^-$ (TM=Sc, Y, La) fulfill the 18-electron rule when one considers only those valence electrons, which occupy metal–ligand bonding orbitals.

PI BACKBONDING

π backbonding, also called π backdonation, is a concept from chemistry in which electrons move from an atomic orbital on one atom to an appropriate symmetry antibonding orbital on a π-acceptor ligand. It is especially common in the organometallic chemistry of transition metals with multi-atomic ligands such as carbon monoxide, ethylene or the nitrosonium cation. Electrons from the metal are used to bond to the ligand, in the process relieving the metal of excess negative charge. Compounds where π backbonding occurs include $Ni(CO)_4$ and Zeise's salt. IUPAC offers the following definition for backbonding:

> "A description of the bonding of π-conjugated ligands to a transition metal which involves a synergic process with donation of electrons from the filled π-orbital or lone electron pair orbital of the ligand into an empty orbital of the metal (donor–acceptor bond), together with release (back donation) of electrons from an nd orbital of the metal (which is of π-symmetry with respect to the metal–ligand axis) into the empty π*-antibonding orbital of the ligand".

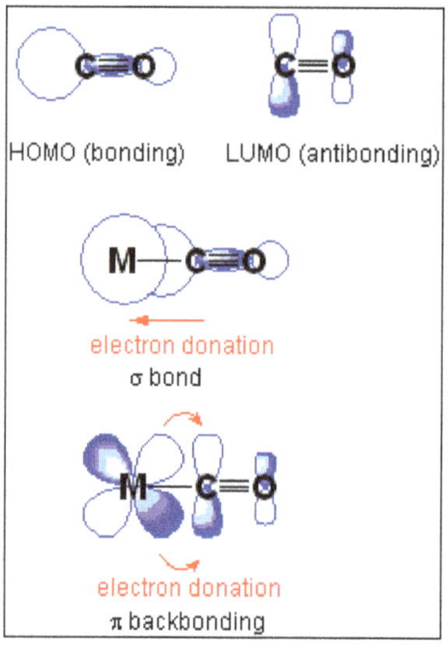

(Top) the HOMO and LUMO of CO. (Middle) an example of a sigma bonding orbital in which CO donates electrons to a metal's center from its HOMO. (Bottom) an example where the metal center donates electrons through a d orbital to CO's LUMO.

Metal Carbonyls, Nitrosyls and Isocyanides

The electrons are partially transferred from a d-orbital of the metal to anti-bonding molecular orbitals of CO (and its analogues). This electron-transfer (i) strengthens the metal–C bond and (ii) weakens the C–O bond. The strengthening of the M–CO bond is reflected in increases of the vibrational frequencies for the M–C bond (often outside of the range for the usual IR spectrophotometers). Furthermore, the M–CO bond length is shortened. The weakening of the C–O bond is indicated by a decrease in the wavenumber of the ν_{CO} band(s) from that for free CO (2143 cm^{-1}), for example to 2060 cm^{-1} in $Ni(CO)_4$ and 1981 cm^{-1} in $Cr(CO)_6$, and 1790 cm^{-1} in the anion $[Fe(CO)_4]^{2-}$. For this reason, IR spectroscopy is an important diagnostic technique in metal–carbonyl chemistry.

Many ligands other than CO are strong "backbonders". Nitric oxide is an even stronger π-acceptor than is CO and ν_{NO} is a diagnostic tool in metal–nitrosyl chemistry. Isocyanides, RNC, are another class of ligands that are capable of π-backbonding. In contrast with CO, the σ-donor lone pair on the C atom of isocyanides is antibonding in nature and upon complexation the CN bond is strengthened and the ν_{CN} increased. At the same time, π-backbonding lowers the ν_{CN}. Depending on the balance of σ-bonding versus π-backbonding, the ν_{CN} can either be raised (for example, upon complexation with weak π-donor metals, such as Pt(II)) or lowered (for example, upon complexation with strong π-donor metals, such as Ni(0)). For the isocyanides, an additional parameter is the MC=N–C angle, which deviates from 180° in highly electron-rich systems. Other ligands have a weak π-backbonding ability, which creates a labilization effect of CO, which is described by the cis effect.

Metal–alkene and Metal–alkyne Complexes

As in metal–carbonyls, electrons are partially transferred from a d-orbital of the metal to antibonding molecular orbitals of the alkenes and alkynes. This electron transfer (i) strengthens the metal–ligand bond and (ii) weakens the C–C bonds within the ligand. In the case of metal-alkenes and alkynes, the strengthening of the $M–C_2R_4$ and $M–C_2R_2$ bond is reflected in bending of the C–C–R angles which assume greater sp^3 and sp^2 character, respectively. Thus strong π backbonding causes a metal-alkene complex to assume the character of a metallacyclopropane. Electronegative substituents exhibit greater π backbonding. Thus, strong π backbonding ligands are tetrafluoroethylene, tetracyanoethylene, and hexafluoro-2-butyne.

Metal-phosphine Complexes

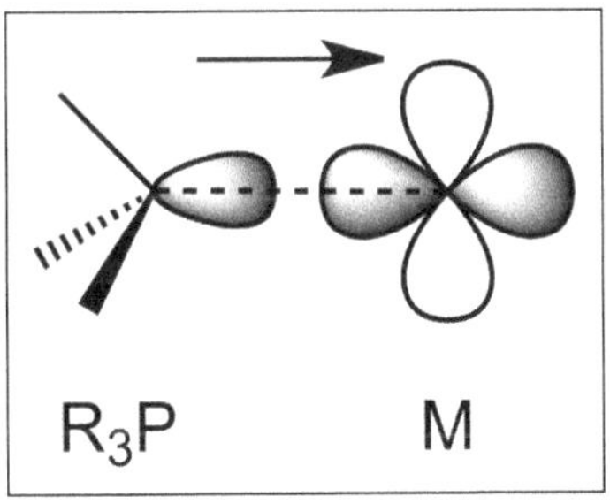

R3P–M σ bonding.

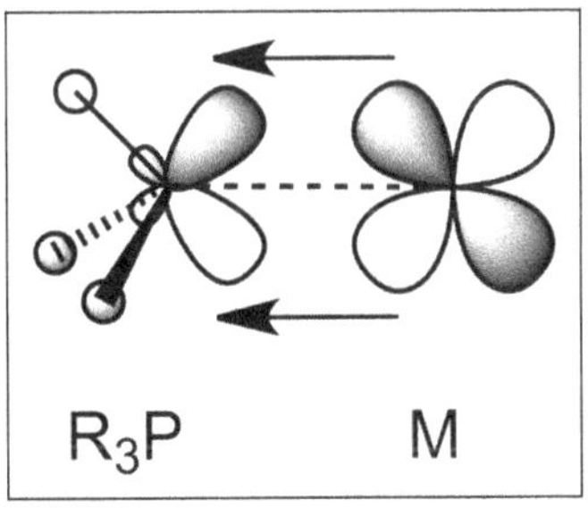

R_3P–M π backbonding.

Phosphines accept electron density from metal p or d orbitals into combinations of P–C σ* antibonding orbitals that have π symmetry. When phosphines bond to electron-rich metal atoms, backbonding would be expected to lengthen P–C bonds as P–C σ* orbitals become populated by electrons. The expected lengthening of the P–C distance is often hidden by an opposing effect: as the phosphorus lone pair is donated to the metal, P(lone pair)–R(bonding pair) repulsions decrease, which acts to shorten the P–C bond. The two effects have been deconvoluted by comparing the structures of pairs of metal-phosphine complexes that differ only by one electron. Oxidation of

R_3P–M complexes results in longer M–P bonds and shorter P–C bonds, consistent with π-backbonding. In early work, phosphine ligands were thought to utilize 3d orbitals to form M–P pi-bonding, but it is now accepted that d-orbitals on phosphorus are not involved in bonding as they are too high in energy.

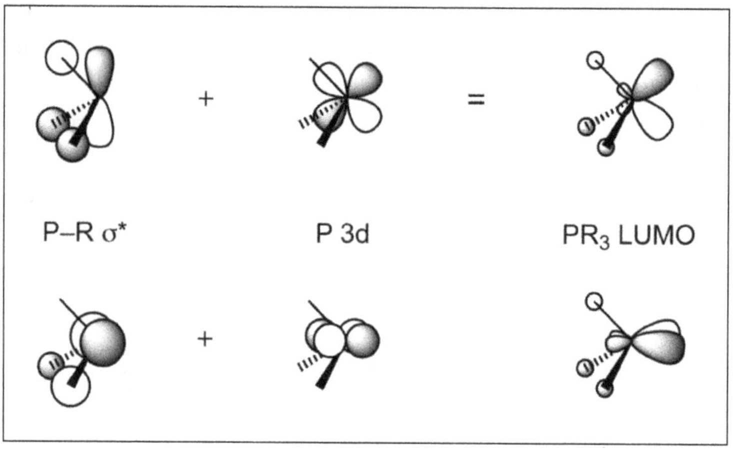

POLYHEDRAL SKELETAL ELECTRON PAIR THEORY

In chemistry the polyhedral skeletal electron pair theory (PSEPT) provides electron counting rules useful for predicting the structures of clusters such as borane and carborane clusters. The electron counting rules were originally formulated by Kenneth Wade and were further developed by Michael Mingos and others; they are sometimes known as Wade's rules or the Wade–Mingos rules. The rules are based on a molecular orbital treatment of the bonding. These rules have been extended and unified in the form of the Jemmis mno rules.

Predicting Structures of Cluster Compounds

Different rules (4n, 5n, or 6n) are invoked depending on the number of electrons per vertex.

The 4n rules are reasonably accurate in predicting the structures of clusters having about 4 electrons per vertex, as is the case for many boranes and carboranes. For such clusters, the structures are based on deltahedra, which are polyhedra in which every face is triangular. The 4n clusters are classified as closo-, nido-, arachno- or hypho-, based on whether they represent a complete (closo-) deltahedron, or a deltahedron that is missing one (nido-), two (arachno-) or three (hypho-) vertices.

However, hypho clusters are relatively uncommon due to the fact that the electron count is high enough to start to fill antibonding orbitals and destabilize the 4n structure. If the electron count is close to 5 electrons per vertex, the structure often changes to one governed by the 5n rules, which are based on 3-connected polyhedra.

As the electron count increases further, the structures of clusters with 5n electron counts become unstable, so the 6n rules can be implemented. The 6n clusters have structures that are based on rings.

A molecular orbital treatment can be used to rationalize the bonding of cluster compounds of the 4n, 5n, and 6n types.

4n Rules

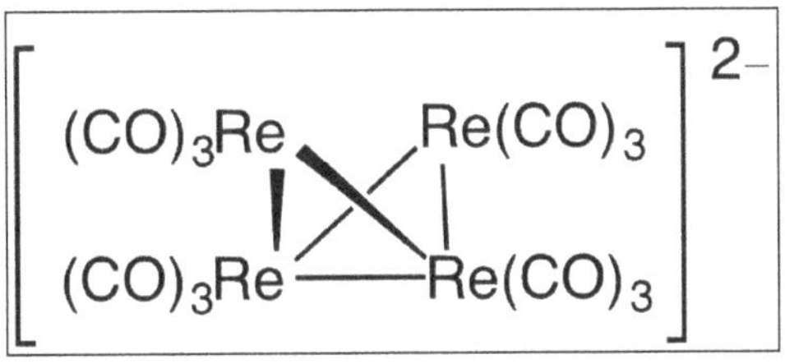

The structure of the butterfly cluster $[Re_4(CO)_{12}]^{2-}$ conforms to the predictions of PSEPT.

The following polyhedra are closo polyhedra, and are the basis for the 4n rules; each of these have triangular faces. The number of vertices in the cluster determines what polyhedron the structure is based on.

Number of vertices	Polyhedron
4	Tetrahedron
5	Trigonal bipyramid
6	Octahedron
7	Pentagonal bipyramid
8	D_{2d} (trigonal) dodecahedron (snub disphenoid)
9	Tricapped trigonal prism
10	Bicapped square antiprism
11	Edge-contracted icosahedron (octadecahedron)
12	Icosahedron (bicapped pentagonal antiprism)

Using the electron count, the predicted structure can be found. n is the number of vertices in the cluster. The 4n rules are enumerated in the following table.

Electron count	Name	Predicted structure
4n – 2	Bicapped closo	n – 2 vertex closo polyhedron with 2 capped (augmented) faces
4n	Capped closo	n – 1 vertex closo polyhedron with 1 face capped

4n + 2	closo	closo polyhedron with n vertices
4n + 4	nido	n + 1 vertex closo polyhedron with 1 missing vertex
4n + 6	arachno	n + 2 vertex closo polyhedron with 2 missing vertices
4n + 8	hypho	n + 3 vertex closo polyhedron with 3 missing vertices
4n + 10	klado	n + 4 vertex closo polyhedron with 4 missing vertices

When counting electrons for each cluster, the number of valence electrons is enumerated. For each transition metal present, 10 electrons are subtracted from the total electron count. For example, in $Rh_6(CO)_{16}$ the total number of electrons would be $6 \times 9 + 16 \times 2 - 6 \times 10 = 86 - 6 \times 10 = 26$. Therefore, the cluster is a closo polyhedron because n = 6, with 4n + 2 = 26.

Other rules may be considered when predicting the structure of clusters:

- For clusters consisting mostly of transition metals, any main group elements present are often best counted as ligands or interstitial atoms, rather than vertices.
- Larger and more electropositive atoms tend to occupy vertices of high connectivity and smaller more electronegative atoms tend to occupy vertices of low connectivity.
- In the special case of boron hydride clusters, each boron connected to 3 or more vertices has one terminal hydride, while boron connected to 2 other vertices has 2 terminal hydrogens. If more hydrogens are present, they are placed in open face positions to even out the coordination number of the vertices.
- For the special case of transition metal clusters, ligands are added to the metal centers to give the metals reasonable coordination numbers, and if any hydrogen atoms are present they are placed in bridging positions to even out the coordination numbers of the vertices.

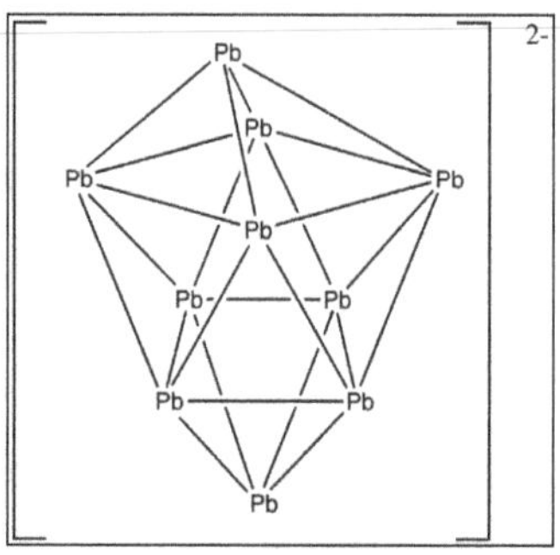

In general, closo structures with n vertices are n-vertex polyhedra.

- To predict the structure of a nido cluster, the closo cluster with n + 1 vertices is used as a starting point; if the cluster is composed of small atoms a high connectivity vertex is removed, while if the cluster is composed of large atoms a low connectivity vertex is removed.

- To predict the structure of an arachno cluster, the closo polyhedron with n + 2 vertices is used as the starting point, and the n + 1 vertex nido complex is generated by following the rule above; a second vertex adjacent to the first is removed if the cluster is composed of mostly small atoms, a second vertex not adjacent to the first is removed if the cluster is composed mostly of large atoms.

$Pb^{2-}{}_{10}$

Example of $Pb^{2-}{}_{10}$:

- Electron count: 10 × Pb + 2 (for the negative charge) = 10 × 4 + 2 = 42 electrons.
- Since n = 10, 4n + 2 = 42, so the cluster is a closo bicapped square antiprism.

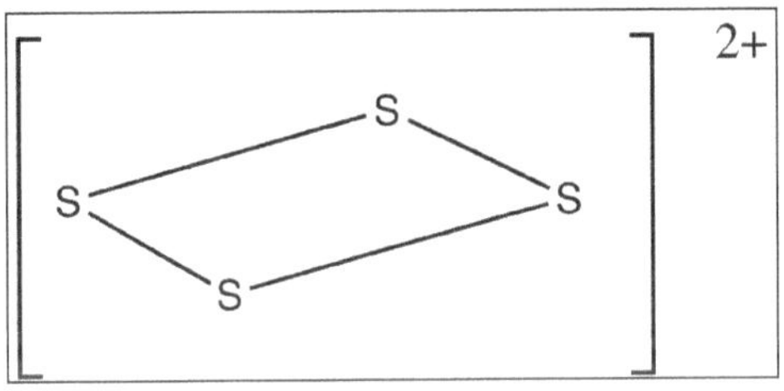

$S^{2+}{}_4$

Example of $S^{2+}{}_4$:

- Electron count: 4 × S – 2 (for the positive charge) = 4 × 6 – 2 = 22 electrons.
- Since n = 4, 4n + 6 = 22, so the cluster is arachno.
- Starting from an octahedron, a vertex of high connectivity is removed, and then a non-adjacent vertex is removed.

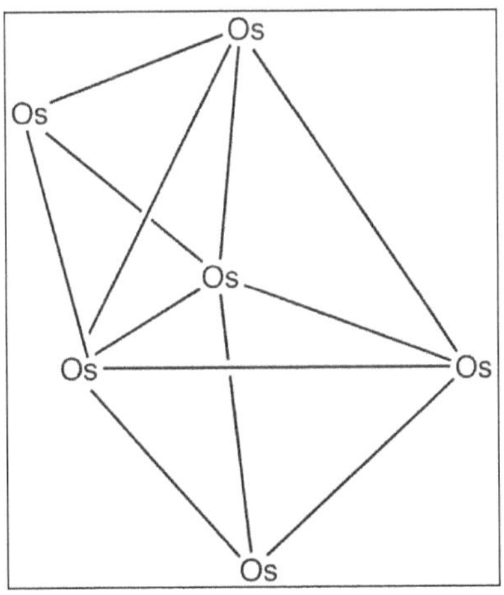

$Os_6(CO)_{18}$, carbonyls omitted:

Example of $Os_6(CO)_{18}$:

- Electron count: 6 × Os + 18 × CO – 60 (for 6 osmium atoms) = 6 × 8 + 18 × 2 – 60 = 24.
- Since n = 6, 4n = 24, so the cluster is capped closo.
- Starting from a trigonal bipyramid, a face is capped. The carbonyls have been omitted for clarity.

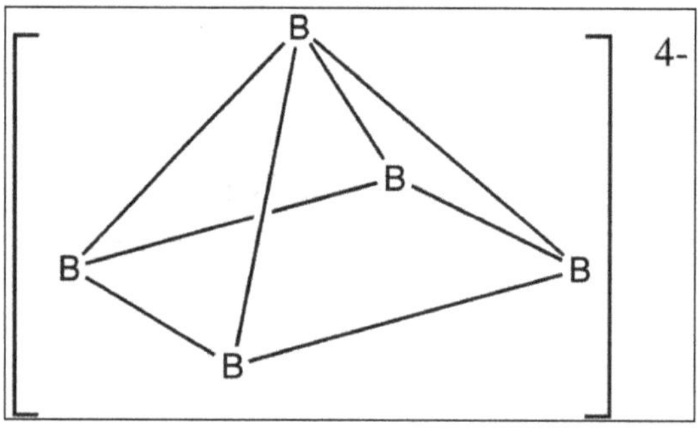

$B_5H_5^{4-}$, hydrogens omitted:

Example of $B_5H_5^{4-}$:

- Electron count: 5 × B + 5 × H + 4 (for the negative charge) = 5 × 3 + 5 × 1 + 4 = 24.
- Since n = 5, 4n + 4 = 24, so the cluster is nido.
- Starting from an octahedron, one of the vertices is removed.

The rules are useful in also predicting the structure of carboranes.

Example of $C_2B_7H_{13}$:

- Electron count = 2 × C + 7 × B + 13 × H = 2 × 4 + 3 × 7 + 13 × 1 = 42.
- Since n in this case is 9, 4n + 6 = 42, the cluster is arachno.

The bookkeeping for deltahedral clusters is sometimes carried out by counting skeletal electrons instead of the total number of electrons. The skeletal orbital (electron pair) and skeletal electron counts for the four types of deltahedral clusters are:

- n-vertex closo: n + 1 skeletal orbitals, 2n + 2 skeletal electrons.
- n-vertex nido: n + 2 skeletal orbitals, 2n + 4 skeletal electrons.
- n-vertex arachno: n + 3 skeletal orbitals, 2n + 6 skeletal electrons.
- n-vertex hypho: n + 4 skeletal orbitals, 2n + 8 skeletal electrons.

The skeletal electron counts are determined by summing the total of the following number of electrons:

- 2 from each BH unit.
- 3 from each CH unit.
- 1 from each additional hydrogen atom (over and above the ones on the BH and CH units).
- The anionic charge electrons.

5n Rules

The 4n rule mainly deals with clusters with electron counts of 4n + k, in which approximately 4 electrons are on each vertex. As more electrons are added per vertex, the number of the electrons per vertex approaches 5. Rather than adopting structures based on deltahedra, the 5n-type clusters have structures based on a different series of polyhedra known as the 3-connected polyhedra, in which each vertex is connected to 3 other vertices. The 3-connected polyhedra are the duals of the deltahedra. The common types of 3-connected polyhedra are listed below.

Number of vertices	Type of 3-connected polyhedron
4	Tetrahedron
6	Trigonal prism
8	Cube
10	Pentagonal prism
12	D_{2d} pseudo-octahedron (dual of snub disphenoid)
14	Dual of triaugmented triangular prism (K_5 associahedron)
16	Square truncated trapezohedron
18	Dual of edge-contracted icosahedron
20	Dodecahedron

The 5n rules are as follows:

Total electron count	Predicted structure
5n	n-vertex 3-connected polyhedron
5n + 1	n – 1 vertex 3-connected polyhedron with one vertex inserted into an edge
5n + 2	n – 2 vertex 3-connected polyhedron with two vertices inserted into edges
5n + k	n – k vertex 3-connected polyhedron with k vertices inserted into edges

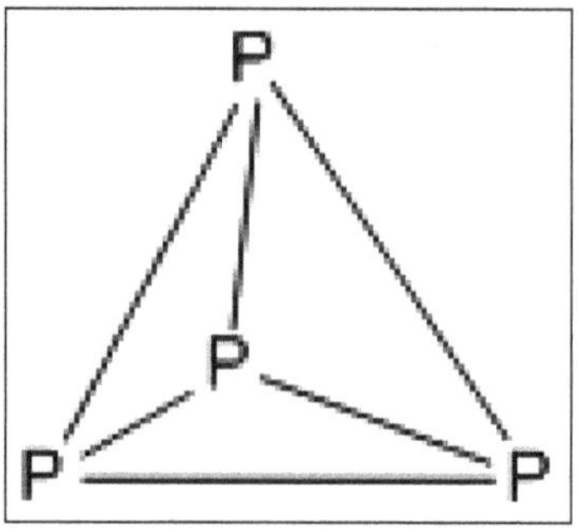

5n cluster: P_4 :

Example of P_4:

- Electron count: 4 × P = 4 × 5 = 20.
- It is a 5n structure with n = 4, so it is tetrahedral.

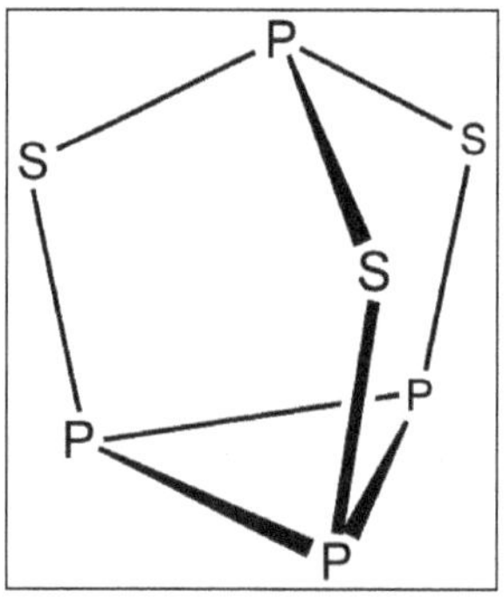

5n + 3 cluster: P_4S_3:

Example of P_4S_3:

- Electron count 4 × P + 3 × S = 4 × 5 + 3 × 6 = 38.
- It is a 5n + 3 structure with n = 7. Three vertices are inserted into edges.

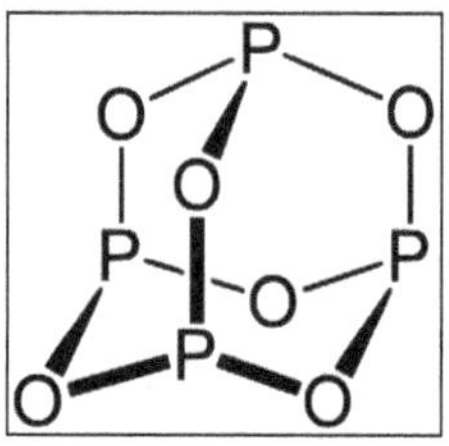

Example of P_4O_6:

- Electron count 4 × P + 6 × O = 4 × 5 + 6 × 6 = 56.
- It is a 5n + 6 structure with n = 10. Six vertices are inserted into edges.

6n Rules

As more electrons are added to a 5n cluster, the number of electrons per vertex approaches 6. Instead of adopting structures based on 4n or 5n rules, the clusters tend to have structures governed by the 6n rules, which are based on rings. The rules for the 6n structures are as follows.

Total electron count	Predicted structure
6n – k	n-membered ring with $k/2$ transannular bonds
6n – 4	n-membered ring with 2 transannular bonds
6n – 2	n-membered ring with 1 transannular bond
6n	n-membered ring
6n + 2	n-membered chain (n-membered ring with 1 broken bond)

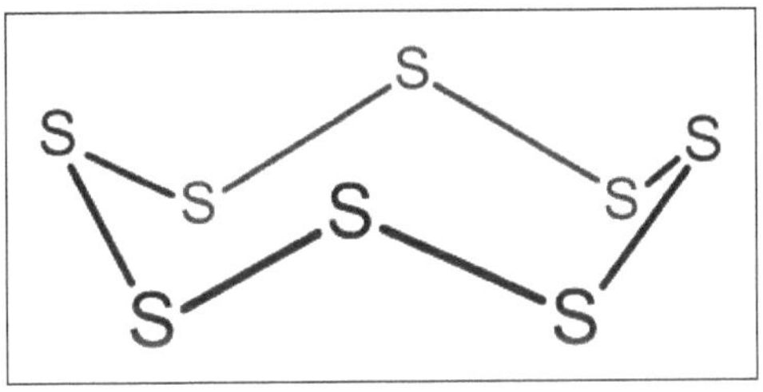

S_8 crown

Example of S_8:

- Electron count = 8 × S = 8 × 6 = 48 electrons.
- Since n = 8, 6n = 48, so the cluster is an 8-membered ring.

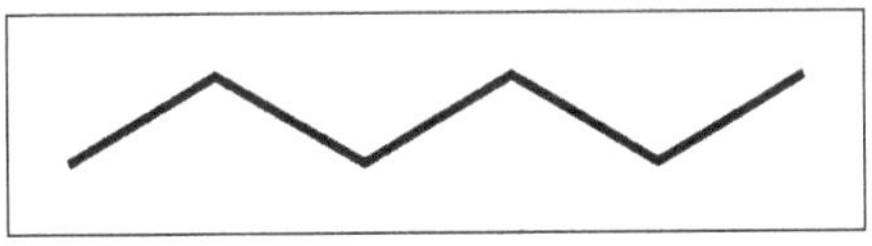

6n + 2 cluster: hexane.

Example of Hexane (C_6H_{14}):

- Electron count = 6 × C + 14 × H = 6 × 4 + 14 × 1 = 38.
- Since n = 6, 6n = 36 and 6n + 2 = 38, so the cluster is a 6-membered chain.

Isolobal Vertex Units

Provided a vertex unit is isolobal with BH then it can, in principle at least, be substituted for a BH unit, even though BH and CH are not isoelectronic. The CH^+ unit is isolobal, hence the rules are applicable to carboranes. This can be explained due to a frontier orbital treatment. Additionally there are isolobal transition-metal units. For example, $Fe(CO)_3$ provides 2 electrons. The derivation of this is briefly as follows:

- Fe has 8 valence electrons.
- Each carbonyl group is a net 2 electron donor after the internal σ- and π-bonding are taken into account making 14 electrons.
- 3 pairs are considered to be involved in Fe–CO σ-bonding and 3 pairs are involved in π-backbonding from Fe to CO reducing the 14 to 2.

Bonding in Cluster Compounds

Polyhedron

B_2H_6: The bonding in diborane is best described by treating each B as sp^3-hybridized. Two sp^3-hybrid orbitals on each boron form the bonds to the terminal hydrogens. The remaining sp3-orbitals create the bonds with the bridging hydrogens. Because the angles in the diborane structure are not tetrahedral the orbitals also likely contain some sp^2 character.

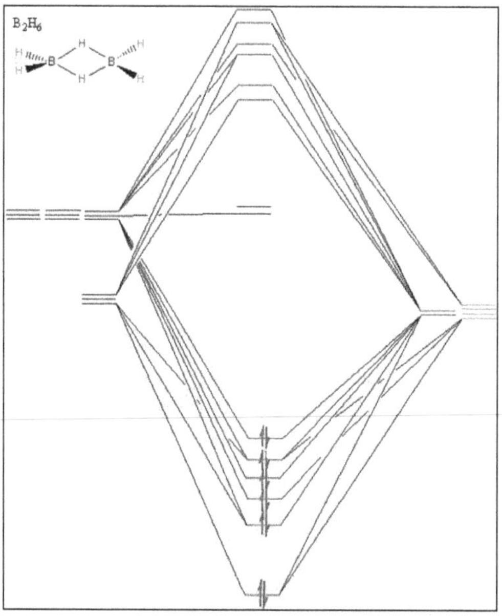

Molecular-orbital (MO) diagram of B_2H_6. Atoms and their corresponding orbitals are colored the same. Green MOs symbolize bonding, while red symbolize anti-bonding.

Closo-$B_6H^{2-}{}_6$: The boron atoms lie on each vertex of the octahedron and are sp hybridized. One sp-hybrid radiates away from the structure forming the bond with the hydrogen atom. The other sp-hybrid radiates into the center of the structure forming a large bonding molecular orbital at the center of the cluster. The remaining two unhybridized orbitals lie along the tangent of the sphere like structure creating more bonding and antibonding orbitals between the boron vertices. The orbital diagram breaks down as follows:

- The 18 framework molecular orbitals, (MOs), derived from the 18 boron atomic orbitals are:
  - 1 bonding MO at the center of the cluster and 5 antibonding MOs from the 6 sp-radial hybrid orbitals.
  - 6 bonding MOs and 6 antibonding MOs from the 12 tangential p-orbitals.

The total skeletal bonding orbitals is therefore 7, i.e. n + 1.

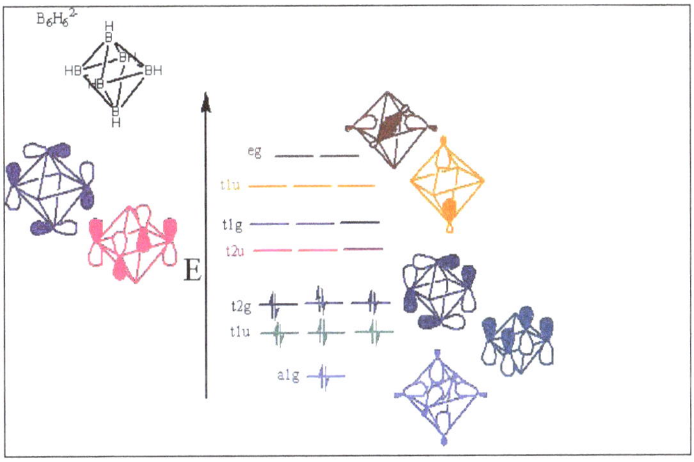

MO diagram of $B_6H^{2-}_6$ showing the orbitals responsible for forming the cluster. Pictorial representations of the orbitals are shown; the MO sets of T and E symmetry will each have two or one additional pictorial representation, respectively, that are not shown here.

Main Group Atom Clusters

The bonding in other main group cluster compounds follows similar rules as those described for the boron cluster bonding. The atoms at the vertex hybridize in a way which allows the lowest energy structure to form.

The 18 framework molecular orbitals, (MOs), derived from the 18 boron atomic orbitals are:

- 1 bonding MO at the center of the cluster and 5 antibonding MOs from the 6 sp-radial hybrid orbitals.
- 6 bonding MOs and 6 antibonding MOs from the 12 tangential p-orbitals.

The total skeletal bonding orbitals is therefore 7, i.e. n + 1.

Transition Metal Clusters

Transition metal clusters use the d orbitals for bonding so have up to nine bonding orbitals, instead of only the four present in boron and main group clusters. There is also more bonding flexibility in transition metal clusters depending on whether vertex metal electron pairs are involved in cluster bonding or appear as lone pairs. The cluster chlorides and carbonyls of transition metals will be briefly discussed here as they represent opposite ends of the spectrochemical series and show important features of the differences between transition metal clusters with different ligands. In chloride clusters the energy splitting of the valence d orbitals increases upon formation of the cluster. The number and symmetry of these orbitals are dependent upon the type and structure of each individual cluster complex. Conversely in the carbonyl clusters the energy splitting of the valence d orbitals is greater before the formation of the cluster.

MO Diagram Clusters Metal Chlorides and Metal Carbonyls

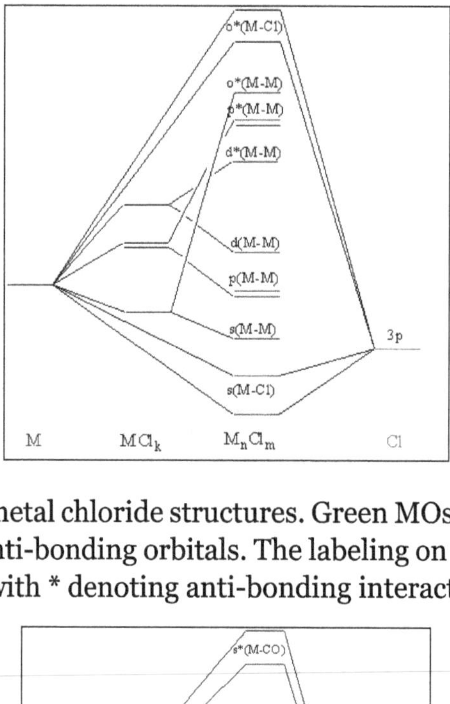

General MO diagram of metal chloride structures. Green MOs represent bonding orbitals while red represent anti-bonding orbitals. The labeling on the MOs is as follows: s = σ, p = π, d = δ bonding, with * denoting anti-bonding interactions.

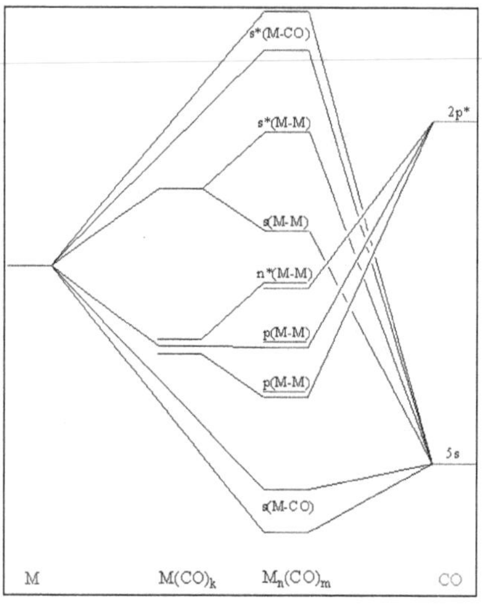

General MO diagram for metal carbonyl clusters. Green MOs represent bonding orbitals while red represent anti-bonding orbitals. The labeling on the MOs is as follows: s = σ, p = π, d = δ bonding with * denoting anti-bonding interactions

LIGAND FIELD THEORY

The ligand field theory (LFT) fleshes out the ideas of crystal field theory with molecular orbital theory concepts. It provides a method for understanding M–L bonding and antibonding orbitals; however, it has been strongly disputed by computational studies in favor of valence bond models that incorporate hypervalency. Still, LFT provides a more complete picture of complex bonding than crystal field theory. Furthermore, the portions of LFT under dispute have nothing to do with CFT, so "no harm no foul".

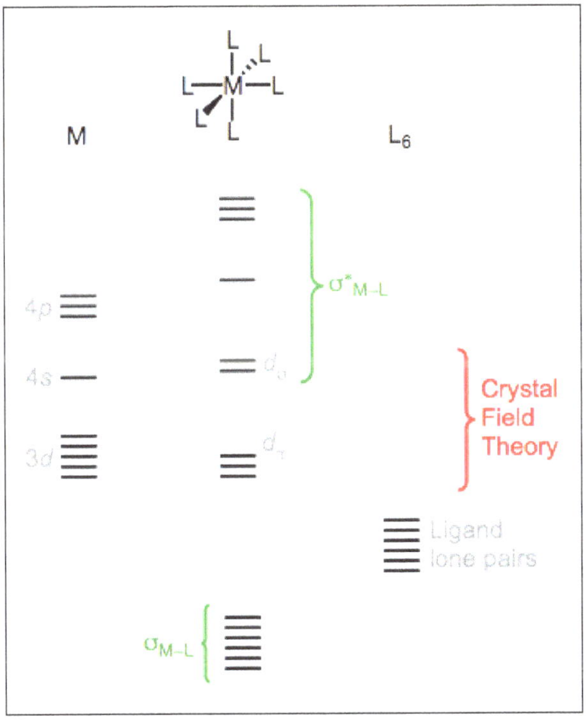

The M–L bonding molecular orbitals mostly have ligand character, while the antibonding orbitals mostly reside on the metal.

Notice that the perturbations of the metallic d orbitals are consistent with the ideas of crystal field theory for the octahedral geometry. The labels dσ and dπ will be useful for us later, and indicate how the d orbitals overlap with the incoming ligand orbitals—the dσ orbitals overlap in a head-on, sigma-type manner, and the dπ orbitals overlap in a side-on, pi-type manner. The dσ orbitals are destabilized through electrostatic repulsion, while the dπ orbitals essentially remain non-bonding. What LFT adds to these CFT ideas is a description of the fate of the remaining unfilled metal valence orbitals and the filled ligand orbitals. According to the LFT picture, orbital overlap between the 4s, 4p and symmetry-matched 4d (i.e., the 4dσ) AOs and the six ligand HOMOs results in six bonding and six antibonding MOs. Notice that the math works out:

1 4s + 3 4p + 2 3dσ + 6 ligand HOMOs = 6 bonding MOs + 6 antibonding MOs

As necessary, the number of atomic orbitals in (12) equals the number of molecular orbitals out (12). But more important than the number of MOs is the character of each orbital—where is the electron density primarily located in these MOs? Ligand field theory provides a logical answer: the bonding MOs primarily possess ligand character, while the antibonding MOs are primarily localized on the metal. The guiding principle here is that MOs are composed primarily of those AOs to which they are closest in energy. In the end, we arrive at a critically important, albeit expected insight: the metal is electrophilic (owner of many unfilled MOs) while the ligands are nucleophilic (owners of filled MOs).

This may seem like a relatively simple picture at first glance. All we did, it seems, was tack the 4s, 4p and ligand orbitals on to crystal field theory. Yet this groundwork is necessary to build a truly powerful theory of the electronic structure of organometallic complexes.

FRONTIER MOLECULAR ORBITAL THEORY

The frontier molecular orbitals of a compound are at the "frontier" of electron occupation—the highest-energy occupied and lowest-energy unoccupied molecular orbitals (the HOMO and LUMO). The HOMO is logically viewed as nucleophilic or electron donating, while the LUMO is electrophilic and electron accepting. Furthermore, both chemical reactions and resonance can be explained by interactions (overlap) between a filled HOMO and an empty LUMO on one or more molecules. FMO theory uses these foundational ideas to explain the structure and reactivity of molecules, and at least in organic chemistry, the theory has been insanely successful.

When it comes to organometallic complexes, we can gain considerable insight by noticing that orbitals on the frontier are usually the metallic d orbitals. Assuming that the metal is d6, the frontier falls between the dπ and dσ orbitals. Without a need for any quantum chemical calculations (only CFT and the number of d electrons are required), we can thus begin to anticipate the shapes of the frontier orbitals of organometallic complexes. Combined with a solid understanding of organic FMO theory and the frontier orbitals of organic ligands, we can recognize important orbital interactions in organometallic complexes that can explain a wide variety of observations.

Before digging in to an example, let's introduce some terminology commonly used by organometallic chemists related to FMO theory. The terms acidity and basicity are used in the context of FMO theory very broadly. Acidity refers to the desire of a ligand, metal center, or specific orbital to accept electron density (from any viable electron source, not just a Brønsted base). Basicity refers to the desire of a ligand, metal center, or specific orbital to donate electrons (to any viable electron sink, not just a Brønsted acid). FMO theory also distinguishes between σ-acids, σ-bases, π-acids, and π-bases. The first two, σ-X, want to accept or donate electrons in a σ-type fashion—that is, aligned head on with another orbital. The latter two, π-X, want to accept or donate electrons in a π-type fashion—that is, aligned side by side with another orbital. Most commonly, these terms are applied to ligands to indicate their expected behavior in a fancy-schmancy but quite descriptive way, and the terms can also be applied to the metal center.

π-acid (π-acceptor)	Low-lying π*	Electrophilic
π-base (π-donor)	High-lying π	Nucleophilic
σ-acid (σ-acceptor)	Low-lying σ*	Electrophilic
σ-base (σ-donor)	High-lying σ	Nucleophilic

Sigma and pi acids and bases, with their most reactive frontier MOs and typical electronic behavior.

Tungsten(0) hexacarbonyl and the other metal(0) carbonyls are interesting species. Let's deconstruct W(CO)6.

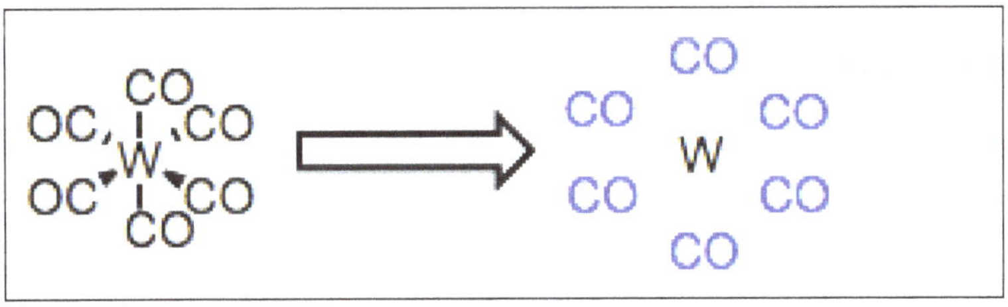

Tungsten(0) hexacarbonyl.

We can make a very good argument that this compound ought to fall apart as soon as it's made. The complex is neutral, but the deconstructed ligands and metal are all neutral too, and entropy loves the right-hand side of the figure. Yet, the complex is reasonably stable.

FMO theory provides a good answer: there is more to the W–CO bond than meets the eye. Like almost all metal–carbon bonds, we expect W and C to be bound primarily through an interaction between a filled non-bonding orbital on carbon (HOMO) and an empty dσ orbital on the metal (LUMO). That's depicted below, and it's typical organometallic FMO theory. Interactions like this help explain why, for instance, extremely electron-poor ligands like PF3 make horrible ligands and bind to metals very weakly.

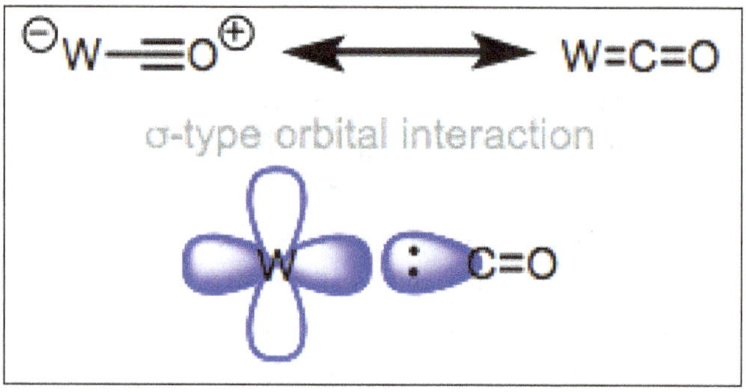

A typical orbital interaction between a metal and an organic ligand.

The right-hand resonance structure, which depicts two bonds between tungsten and carbon, suggests that there is another bonding interaction between W and C, perhaps one where the LUMO of the ligand and the HOMO of the metal get involved? Can we draw a viable HOMO–LUMO interaction by reversing the typical roles of metal and ligand? The LUMO of the ligand is a π* orbital, and the HOMOs of the metal are the d orbitals not aligned along the octahedral axes (the dπ orbitals). The symmetry of these orbitals is just right. See the lower orbital interaction in the figure.

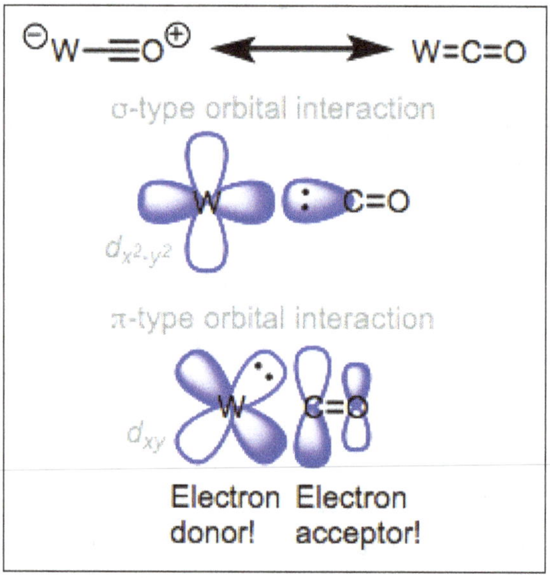

There's a new interaction in town and the metal is an electron donor in the pi-type interaction.

Please note that the tungsten atom is not rotated at all; two different d orbitals are depicted in the two interactions. The second role-reversing interaction is important for π-acidic ligands like carbon monoxide, and is called backbonding. The interesting thing about backbonding is that when considering it, we must imagine the metal as electron donor and the ligand as electron acceptor. There is an important lesson here: consider the HOMO and LUMO of both the metal and ligands and their possible interactions. When it comes to tungsten(0) carbonyl, donation from the electron-rich (or π-basic) metal center to the π-acidic CO ligands is what holds everything together.

References

- Hydrocarbon, science: britannica.com, Retrieved 27 April, 2020
- Arteaga Müller, R.; Sánchez Nieves, J.; Ramos, J.; Royo, P.; Mosquera, M. E. G. (2008). "Isolobal Zwitterionic Niobium and Tantalum Imido and Zirconium Monocyclopentadienyl Complexes: Theoretical and Methyl Methacrylate Polymerization Studies". Organometallics. 27 (7): 1417–1426. doi:10.1021/om701068h
- Ligand-Field-Theory-and-Frontier-Molecular-Orbital-Theory, Inorganic-Chemistry-Supplemental-Modules-Inorganic-Chemistry-Organometallic-Chemistry-Structural-Fundamentals: chem.libretexts.org, Retrieved 19 June, 2020
- Jensen, William B. (2005). "The Origin of the 18-Electron Rule". J. Chem. Educ. 82 (1): 28. Bibcode:2005JChEd..82...28J. doi:10.1021/ed082p28
- Fey, N.; Orpen, A. G.; Harvey, J. N. (2009). "Building ligand knowledge bases for organometallic chemistry: Computational description of phosphorus(III)-donor ligands and the metal–phosphorus bonds". Coord. Chem. Rev. 253 (5–6): 704–722. doi:10.1016/j.ccr.2008.04.017

3

Organometallic Compounds of Main Group Elements

Organometallic compounds of main group elements belong to the s-block and p-block elements. It includes organomagnesium compounds, organoaluminium compounds, organoboron compounds, organolithium reagent, Ziese's salt, etc. The topics elaborated in this chapter will help in gaining a better perspective of organometallic compounds of main group elements.

ORGANOMAGNESIUM COMPOUNDS

Organomagnesium Compounds are compounds that contain a carbon-magnesium bond. Two types of organomagnesium compounds are known: (1) complete compounds, such as magnesium dialkyls or magnesium diaryls, R_2Mg; and (2) mixed compounds, such as alkyl magnesium or aryl magnesium halides, RMgX (X = Cl, Br, I). Complete organomagnesium compounds are crystalline substances and are highly sensitive to the effects of oxygen, moisture, and carbon dioxide (they are self-igniting). Such compounds were first produced in the mid-19th century using the reaction of organomercuric compounds, R_2Hg, with magnesium; no use was found for them in organic synthesis.

In 1900 the French chemist F. Grignard developed a simple method of preparing mixed organomagnesium compounds and demonstrated a wide range of possible uses for them in organic synthesis. He established that metallic magnesium in absolute (anhydrous) ether reacts with alkyl or aryl halides, RX, to form compounds that pass into the ether solution. These compounds, known as Grignard reagents, are extremely unstable in free form. Therefore, they are not isolated but rather are used in the form of solutions, which are stable in the absence of moisture and atmospheric oxygen.

Subsequently, methods were developed for the preparation of Grignard reagents in hydrocarbon media (for example, benzene, xylene, and toluene) and in the absence of a solvent, which resulted in the possibility of using organomagnesium compounds in industry. However, the most widespread method has proved to be RMgX synthesis using ether solvents. An increase in the solvating properties of the solvent simplifies the formation of Grignard reagents. For example, vinyl halides, $CH_2 = CHX$, do not react with magnesium in ether but rather form organomagnesium compounds in tetrahydrofuran. Ethynyl magnesium halides may be produced by the interaction of alkyl magnesium halides and acetylene derivatives:

$$C_6H_5C \equiv CH + C_2H_5MgBr - C_6H_5C \equiv CMgBr + C_2H_6$$

Organomagnesium compounds are widely used in the preparation of various types of organic compounds. The industrial synthesis of certain organosilicon compounds, aromatic substances, and Pharmaceuticals is carried out with the aid of organomagnesium compounds.

Organomagnesium compounds are less reactive than the corresponding organolithium ones because, due to the higher electronegativity of magnesium compared to lithium, the carbon–magnesium bond is less polarized than the carbon–lithium one. Nevertheless, magnesium derivatives in hydrocarbons can initiate the polymerization of monomers which are more reactive towards carbanions than styrene and dienes such as acrylic and methacrylic monomers and also vinylpyridines. The marked withdrawing effect of the substituents favours the attack of the double bond. Grignard reagents, such as phenylmagnesium bromide or n-butylmagnesium bromide, initiate acrylic and related monomers and produce stereoregular polymers. Symmetrical organomagnesium derivatives are also efficient in the polymerization of acrylics in toluene, but they are less stereospecific than Grignard reagents. Natta and co-workers have performed the stereospecific polymerization of 2-vinylpyridine, initiating the reaction with various ether-free Grignard reagents dispersed in toluene. Elimination of ether is achieved by heating the compounds under vacuum. Soum et al. found that initiation by dibenzylmagnesium, prepared from the corresponding mercury derivatives, produces isotactic living poly(2-vinylpyridine), but also produces side products resulting from nucleophilic attack on the pyridine ring by the benzylic carbanion. They also showed that initiation occurs from one of the two benzylic groups, the second one being inactivated by the reaction of the first one. Polymerization proceeds with the benzylmagnesium cation as countercation. When benzylpicolylmagnesium is used as initiator, side reactions are absent and well-defined living isotactic poly(2-vinylpyridine) is obtained in toluene.

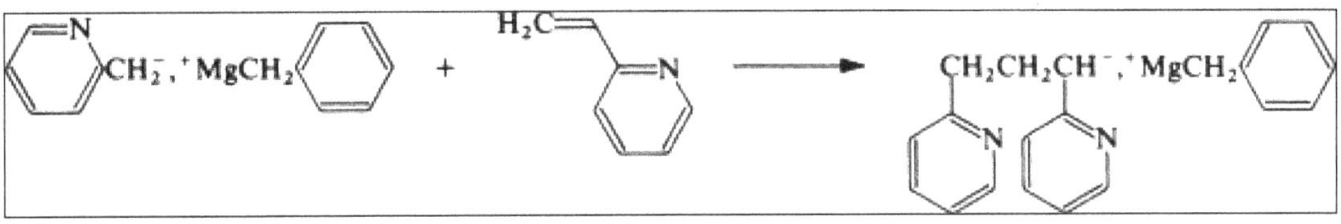

It is worthy of note that these compounds, like other magnesium derivatives, are strongly aggregated in hydrocarbon solutions, but that the corresponding living chain is non-aggregated due to the complexing power of the pyridine rings.

Benzylpicolylmagnesium has also been used to polymerize methyl methacrylate; the resulting polymer is atactic, whereas initiation by dibenzylmagnesium is isospecific, thus showing the influence of the pyridine ring on the stereoregulating mechanism.

Grignard Reagent

A Grignard reagent or Grignard compound is a chemical compound with the generic formula R–Mg–X, where X is a halogen and R is an organic group, normally an alkyl or aryl. Two typical examples are methylmagnesium chloride H_3C–Mg–Cl and phenylmagnesium bromide (C_6H_5)–Mg–Br. They are a subclass of the organomagnesium compounds.

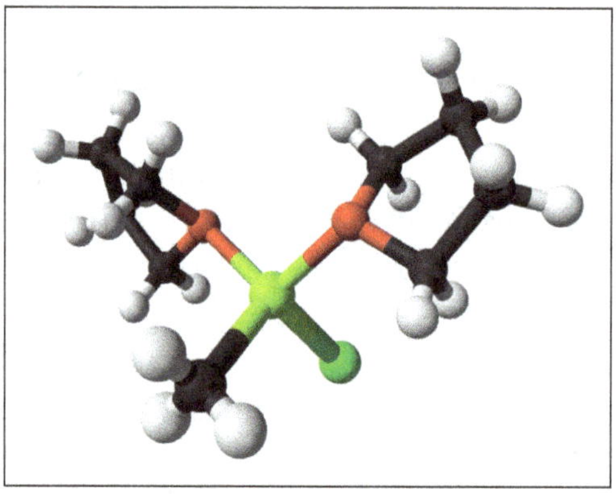

Usually Grignard reagents are written as RMgX, but in fact the magnesium (II) centre is tetrahedral when dissolved in Lewis basic solvents, as shown here for the bis-adduct of methylmagnesium chloride and THF.

Grignard compounds are popular reagents in organic synthesis for creating new carbon-carbon bonds. For example, when reacted with another halogenated compound R'–X' in the presence of a suitable catalyst, they typically yield R–R' and the magnesium halide MgXX' as a byproduct; and the latter is insoluble in the solvents normally used. In this aspect, they are similar to organolithium reagents.

Pure Grignard reagents are extremely reactive solids. They are normally handled as solutions in solvents such as diethyl ether or tetrahydrofuran; which are relatively stable as long as water is excluded. In such a medium, a Grignard reagent is invariably present as a complex with the magnesium atom connected to the two ether oxygens by coordination bonds.

Synthesis

Grignard reagents are prepared by treating an organic halide (normally organobromine) with magnesium metal. Cyclic or acyclic ethers are required to stabilize the organomagnesium compound. Water and air, which rapidly destroy the reagent by protonolysis or oxidation, are excluded using air-free techniques. Although the reagents still need to be dry, ultrasound can allow Grignard reagents to form in wet solvents by activating the magnesium such that it consumes the water.

As is common for reactions involving solids and solution, the formation of Grignard reagents is often subject to an induction period. During this stage, the passivating oxide on the magnesium is removed. After this induction period, the reactions can be highly exothermic. This exothermicity must be considered when a reaction is scaled-up from laboratory to production plant. Most organohalides will work, but carbon-fluorine bonds are generally unreactive, except with specially activated magnesium (through Rieke metals).

Magnesium

Typically the reaction to form Grignard reagents involves the use of magnesium ribbon. All magnesium is coated with a passivating layer of magnesium oxide, which inhibits reactions with the organic halide. Many methods have been developed to weaken this passivating layer, thereby exposing highly reactive magnesium to the organic halide. Mechanical methods include crushing of the Mg pieces in situ, rapid stirring, and sonication. Iodine, methyl iodide, and 1,2-dibromoethane are common activating agents. The use of 1,2-dibromoethane is advantageous as its action can be monitored by the observation of bubbles of ethylene. Furthermore, the side-products are innocuous:

$$Mg + BrC_2H_4Br \rightarrow C_2H_4 + MgBr_2$$

The amount of Mg consumed by these activating agents is usually insignificant. A small amount of mercuric chloride will amalgamate the surface of the metal, enhancing its reactivity. Addition of preformed Grignard reagent is often used as the initiator.

Specially activated magnesium, such as Rieke magnesium, circumvents this problem. The oxide layer can also be broken up using ultrasound, using a stirring rod to scratch the oxidized layer off, or by adding a few drops of iodine or 1,2-Diiodoethane. Another option is to use sublimed magnesium or magnesium anthracene.

Mechanism

In terms of mechanism, the reaction proceeds through single electron transfer:

$$R - X + Mg \rightarrow R - X^{\bullet -} + Mg^{\bullet +}$$

$$R - X^{\bullet -} \rightarrow R^{\bullet} + X^{-}$$

$$R^{\bullet} + Mg^{\bullet +} \rightarrow RMg^{+}$$

$$RMg^{+} + X^{-} \rightarrow RMgX$$

Testing Grignard Reagents

Because Grignard reagents are so sensitive to moisture and oxygen, many methods have been developed to test the quality of a batch. Typical tests involve titrations with weighable, anhydrous protic reagents, e.g. menthol in the presence of a color-indicator.

The interaction of the Grignard reagent with phenanthroline or 2,2'-bipyridine causes a color change.

Mg Transfer Reaction (Halogen–Mg Exchange)

An alternative preparation of Grignard reagents involves transfer of Mg from a preformed Grignard reagent to an organic halide. This method offers the advantage that the Mg transfers tolerates many functional groups. A typical reaction involves isopropylmagnesium chloride and aryl bromide or iodides:

$$\mathrm{i-PrMgCl + ArCl \rightarrow i-PrCl + ArMgCl}$$

Reactions of Grignard Reagents

Carbonyl Compounds

Grignard reagents react with a variety of carbonyl derivatives:

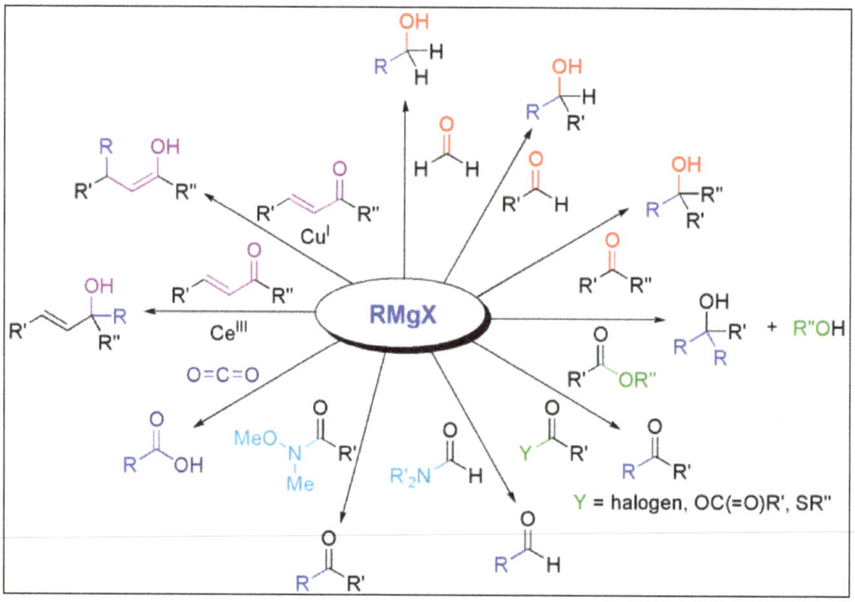

The most common application of Grignard reagents is the alkylation of aldehydes and ketones, i.e. the Grignard reaction:

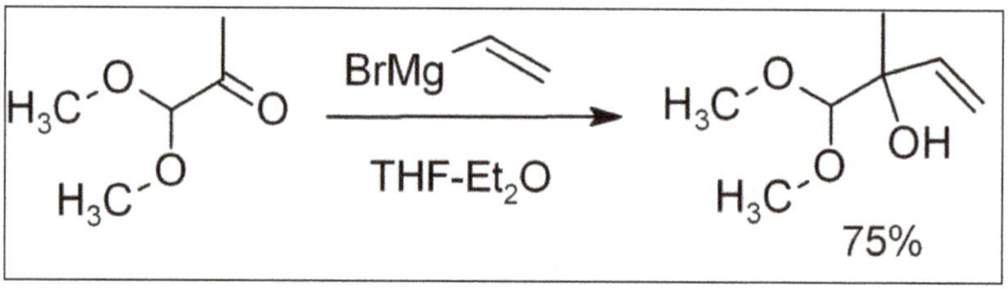

Note that the acetal function (a protected carbonyl) does not react.

Such reactions usually involve an aqueous acidic workup, though this step is rarely shown in reaction schemes. In cases where the Grignard reagent is adding to an aldehyde or a prochiral ketone, the Felkin-Anh model or Cram's Rule can usually predict which stereoisomer will be formed. With easily deprotonated 1,3-diketones and related acidic substrates, the Grignard reagent RMgX functions merely as a base, giving the enolate anion and liberating the alkane RH.

Grignard reagents are nucleophiles in nucleophilic aliphatic substitutions for instance with alkyl halides in a key step in industrial Naproxen production:

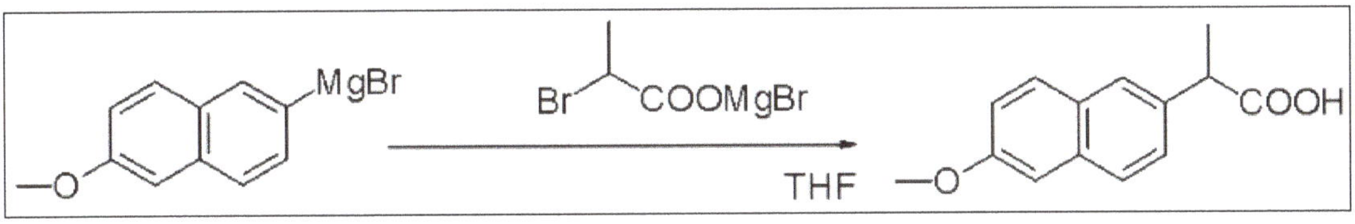

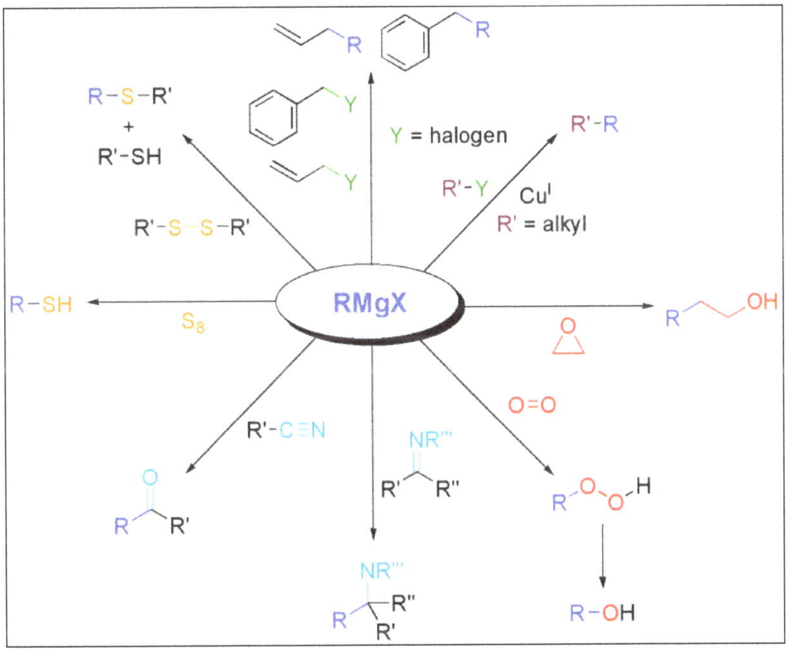

Reactions as a Base

Grignard reagents serve as a base for protic substrates (this scheme does not show workup conditions, which typically includes water). Grignard reagents are basic and react with alcohols, phenols, etc. to give alkoxides (ROMgBr). The phenoxide derivative is susceptible to formylation paraformaldehyde to give salicylaldehyde.

Alkylation of Metals and Metalloids

Like organolithium compounds, Grignard reagents are useful for forming carbon–heteroatom bonds.

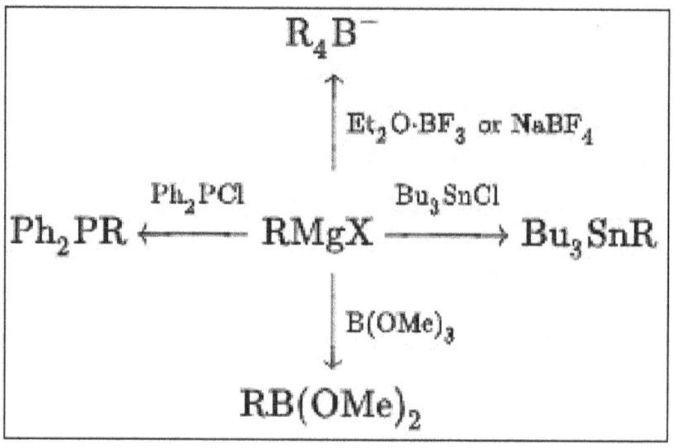

Grignard reagents react with many metal-based electrophiles. For example, they undergo transmetallation with cadmium chloride ($CdCl_2$) to give dialkylcadmium:

$$2\ RMgX + CdCl_2 \rightarrow R_2Cd + 2\ Mg(X)Cl$$

Schlenk Equilibrium

Most Grignard reactions are conducted in ethereal solvents, especially diethyl ether and THF. With the chelating diether dioxane, some Grignard reagents undergo a redistribution reaction to give diorganomagnesium compounds (R = organic group, X = halide):

$$2\ RMgX + \text{dioxane} \rightleftharpoons R_2Mg + MgX_2\ (\text{dioxane})$$

This reaction is known as the Schlenk equilibrium.

Coupling with Organic Halides

Grignard reagents do not typically react with organic halides, in contrast with their high reactivity with other main group halides. In the presence of metal catalysts, however, Grignard reagents participate in C-C coupling reactions. For example, nonylmagnesium bromide reacts with methyl p-chlorobenzoate to give p-nonylbenzoic acid, in the presence of Tris(acetylacetonato)iron(III) ($Fe(acac)_3$), after workup with NaOH to hydrolyze the ester, shown as follows. Without the $Fe(acac)_3$, the Grignard reagent would attack the ester group over the aryl halide.

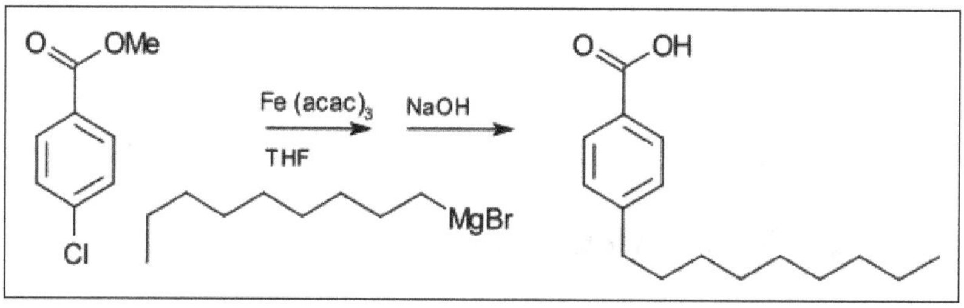

For the coupling of aryl halides with aryl Grignard reagents, nickel chloride in tetrahydrofuran (THF) is also a good catalyst. Additionally, an effective catalyst for the couplings of alkyl halides is dilithium tetrachlorocuprate (Li_2CuCl_4), prepared by mixing lithium chloride (LiCl) and copper(II) chloride ($CuCl_2$) in THF. The Kumada-Corriu coupling gives access to [substituted] styrenes.

Oxidation

Treatment of a Grignard reagent with oxygen gives the magnesium organoperoxide. Hydrolysis of this material yields hydroperoxides or alcohol. These reactions involve radical intermediates.

$$R-MgX+O_2 \rightarrow R\cdot+O_2^{-}+MgX^{+} \rightarrow R-O-O-MgX+H3O^{+} \rightarrow R-O-O-H+HO-MgX+H^{+}$$

$$\downarrow R-MgX$$

$$R-O-MgX+H3O^{+} \rightarrow R-O-H+HO-MgX+H^{+}$$

The simple oxidation of Grignard reagents to give alcohols is of little practical import as yields are generally poor. In contrast, two-step sequence via a borane (vide supra) that is subsequently oxidized to the alcohol with hydrogen peroxide is of synthetic utility.

The synthetic utility of Grignard oxidations can be increased by a reaction of Grignard reagents with oxygen in presence of an alkene to an ethylene extended alcohol. This modification requires aryl or vinyl Grignards. Adding just the Grignard and the alkene does not result in a reaction demonstrating that the presence of oxygen is essential. The only drawback is the requirement of at least two equivalents of Grignard although this can partly be circumvented by the use of a dual Grignard system with a cheap reducing Grignard such as n-butylmagnesium bromide.

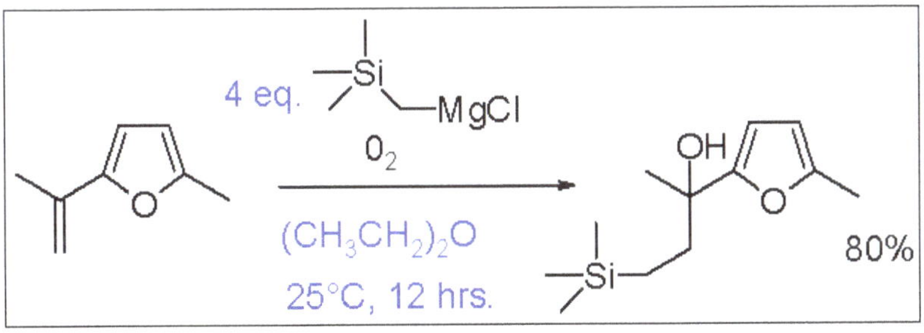

Elimination

In the Boord olefin synthesis, the addition of magnesium to certain β-haloethers results in an elimination reaction to the alkene. This reaction can limit the utility of Grignard reactions.

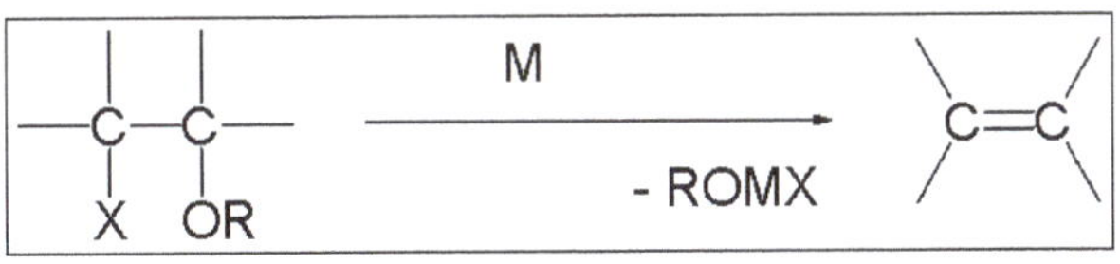

Industrial Use

An example of the Grignard reaction is a key step in the (non-stereoselective) industrial production of Tamoxifen (currently used for the treatment of estrogen receptor positive breast cancer in women):

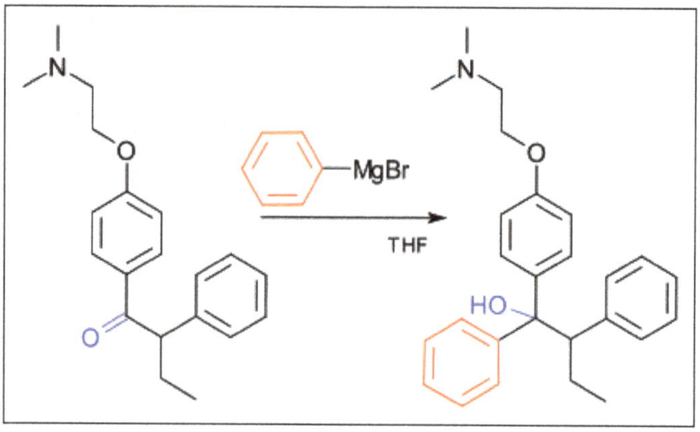

ORGANOALUMINIUM COMPOUNDS

Organoaluminium chemistry is the study of compounds containing bonds between carbon and aluminium bond. It is one of the major themes within organometallic chemistry. Illustrative organoaluminium compounds are the dimer trimethylaluminium, the monomer triisobutylaluminium, and the titanium-aluminium compound called Tebbe's reagent. The behavior of organoaluminium compounds can be understood in terms of the polarity of the C–Al bond and the high Lewis acidity of the three-coordinated species. Industrially, these compounds are mainly used for the production of polyolefins.

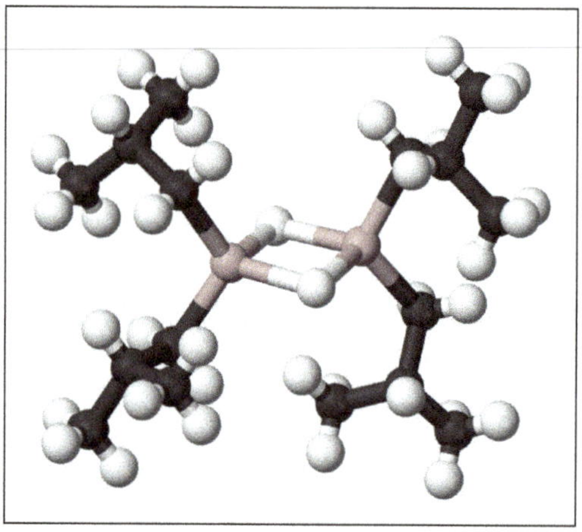

The ball-and-stick model of diisobutylaluminium hydride, showing aluminium as pink, carbon as black, and hydrogen as white.

Structure and Bonding

Aluminium (III) Compounds

Organoaluminium compounds generally feature three- and four-coordinate Al centers, although higher coordination numbers are observed with inorganic ligands such as fluoride. In accord with the usual trends, four-coordinate Al prefers to be tetrahedral. In contrast to boron, aluminium is a larger atom and easily accommodates four carbon ligands. The triorganoaluminium compounds are thus usually dimeric with a pair of bridging alkyl ligands, e.g., $Al_2(C_2H_5)_4(\mu\text{-}C_2H_5)_2$. Thus, despite its common name of triethylaluminium, this compound contains two aluminium centres, and six ethyl groups. When the organoaluminium compound contains hydride or halide, these smaller ligands tend to occupy the bridging sites. Three coordination occurs when the R groups is bulky, e.g. $Al(Mes)_3$ (Mes = 2,4,6-$Me_3C_6H_2$ or mesityl) or isobutyl.

Ligand Exchange in Trialkylaluminium Compounds

The trialkylaluminium dimers often participate in dynamic equilibria, resulting in the interchange of bridging and terminal ligands as well as ligand exchange between dimers. Even in noncoordinating solvents, Al-Me exchange is fast, as confirmed by proton NMR spectroscopy. For example, at −25 °C the ^1H NMR spectrum of Me_6Al_2 comprises two signals in 1:2 ratios, as expected from the solid state structure. At 20 °C, only one signal is observed because exchange of terminal and bridging methyl groups is too fast to be resolved by NMR. The high Lewis acidity of the monomeric species is related to the size of the Al(III) center and its tendency to achieve an octet configuration.

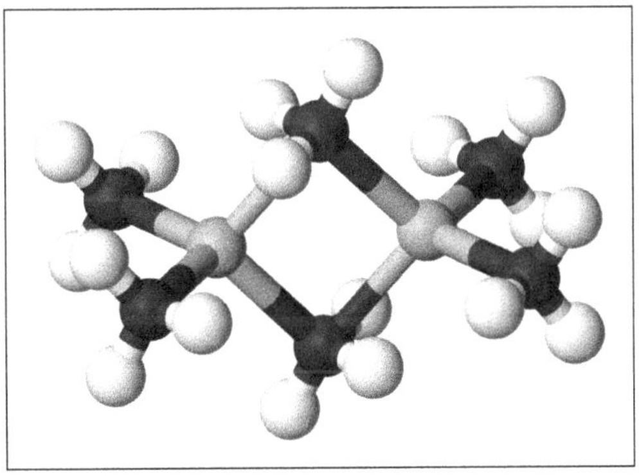

Structure of trimethylaluminium, a compound that features five-coordinate carbon.

Low Valent Organoaluminium Compounds

The first organoaluminium compound with an Al-Al bond was reported in 1988 as $(((Me_3Si)_2CH)_2Al)_2$ (a dialane). They are typically prepared reduction of the dialkylaluminium chlorides by metallic potassium:

$$(R_2AlCl)_2 + 2K \rightarrow R_2Al - AlR_2 + 2KCl$$

Another notable group of alanes are tetraalanes containing four Al(I) centres. These compounds adopt a tetrahedrane core, as illustrated by $(Cp^*Al)_4$ and $((Me_3Si_3C)Al)_4$. The cluster $[Al_{12}(i\text{-}Bu)_{12}]^{2-}$ was obtained from related investigations on the reduction of organoaluminium compounds. This dianion adopts an icosahedral structure reminiscent of dodecaborate ($[B_{12}H_{12}]^{2-}$). Its formal oxidation state is less than one.

Preparation

Alkyl Halides and Aluminium

Industrially, simple aluminium alkyls of the type Al_2R_6 (R = Me, Et) are prepared in a two-step process beginning with the alkylation of aluminium powder:

$$2\,Al + 3CH_3CH_2Cl \rightarrow (CH_3CH_2)_3Al_2Cl_3$$

The reaction resembles the synthesis Grignard reagents. The product, $(CH_3CH_2)_3Al_2Cl_3$, is called ethylaluminium sesquichloride. The term sesquichloride refers to the fact that, on average, the Cl:Al ratio is 1.5. These sesquichlorides can be converted to the triorganoaluminium derivatives by reduction:

$$2(CH_3CH_2)_3Al_2Cl_3 + 6Na \rightarrow (CH_3CH_2)_6Al_2 + 2\,Al + 6NaCl$$

This method is used for production of trimethylaluminium and triethylaluminium.

Hydroalumination

Aluminium powder reacts directly with certain terminal alkenes in the presence of hydrogen. The process entails two steps, the first producing dialkylaluminium hydrides. Such reactions are typically conducted at elevated temperatures and require activation by trialkylaluminium reagents:

$$3\,Al + 3/2\;H_2 + 6CH_2 = CHR \rightarrow [HAl(CH_2CHR)_2]_3$$

For nonbulky R groups, the organoaluminium hydrides are typically trimeric. In a subsequent step, these hydrides are treated with more alkene to effect hydroalumiunation:

$$2[HAl(CH_2CHR)_2]_3 + 3CH_2 = CHR \rightarrow 3[Al_2(CH_2CHR)_3$$

Diisobutylaluminium hydride, which is dimeric, is prepared by hydride elimination from triisobutylaluminium:

$$2i - Bu_3Al \rightarrow (i - Bu_2AlH)_2 + 2\left(CH_3\right)_2 C = CH_2$$

Carboalumination

Organoaluminum compounds can react with alkenes and alkynes, resulting in the net addition of one organyl group and the metal fragment across the multiple bonds (carboalumination). This process can proceed in a purely thermal manner or in the presence of a transition metal catalyst. For the uncatalyzed process, monoaddition is only possible when the alkene is substituted. For ethylene, carboalumination leads to a Poisson distribution of higher alkylaluminum species. The reaction is regioselective for 1-alkenes. The so-called ZACA reaction first reported by E.-i. Negishi is an example of an asymmetric carboalumination of alkenes catalyzed by a chiral zirconocene catalyst.

The methylalumination of alkynes in the presence of Cp_2ZrCl_2 is employed for the synthesis of stereodefined trisubstituted olefin fragments, a common substructure in terpene and polyketide natural products. The synthesis of (E)-4-iodo-3-methylbut-3-en-1-ol shown below is a typical application of this reaction:

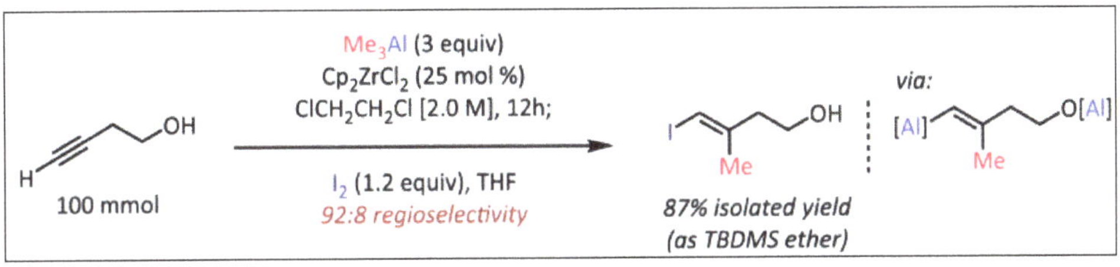

For terminal alkynes, the reaction generally proceeds with good regioselectivity (>90:10 rr) and complete syn selectivity, even in the presence of propargylic or homopropargylic heteroatom substituents. Unfortunately, extension of the zirconocene-catalyzed methylalumination to alkylalumination with higher alkyls results in lower yields and poor regioselectivities.

Laboratory Preparations

Although the simple members are commercially available at low cost, many methods have been developed for their synthesis in the laboratory, including metathesis or transmetalation. Metathesis of aluminium trichloride with RLi or RMgX gives the trialkyl:

$$AlCl_3 + 3BuLi \rightarrow Bu_3Al + 3LiCl$$

Transmetalation:

$$2Al + 3HgPh_2 \rightarrow 2AlPh_3 + 3Hg$$

Reactions

The high reactivity of organoaluminium compounds toward electrophiles is attributed to the charge separation between aluminium and carbon atom.

Lewis Acidity

Organoaluminium compounds are hard acids and readily form adducts with bases such as pyridine, THF and tertiary amines. These adducts are tetrahedral at Al.

Electrophiles

The Al–C bond is polarized such that the carbon is highly basic. Acids react to give alkanes. For example, alcohols give alkoxides:

$$AlR'_3 + ROH \rightarrow 1/n\ (R'_2Al - OR)_n + R'H$$

A wide variety of acids can be employed beyond the simple mineral acids. Amines give amido derivatives. With carbon dioxide, trialkylaluminium compounds give the dialkylaluminium carboxylate, and subsequently alkyl aluminium dicarboxylates:

$$AlR_3 + CO_2 \rightarrow R_2AlO_2CR$$

$$R_2AlO_2CR + CO_2 \rightarrow RAl(O_2CR)_2$$

The conversion is reminiscent of the carbonation of Grignard reagents.

Similarly, the reaction between trialkylaluminum compounds and carbon dioxide has been used to synthesise alcohols, olefins, or ketones.

With oxygen one obtains the corresponding alkoxides, which can be hydrolysed to the alcohols:

$$AlR_3 + 3/2\ O_2 \rightarrow Al(OR)_3$$

A structurally characterized organoaluminum peroxide is [{HC[C(Me)N-C_6H_5]$_2$}Al(R)-O-O-CMe_3] [R=CH($SiMe_3$)$_2$].

The reaction between pure trialalkylaluminum compounds and water, alcohols, phenols, amines, carbon dioxide, sulfur oxides, nitrogen oxides, halogens, and halogenated hydrocarbons can be violent.

Alkene Polymerization

Industrially, organoaluminium compounds are used as catalysts for alkene polymerization to polyolefins, for example the catalyst methylaluminoxane.

Trimethylaluminium

Trimethylaluminium is one of the simplest examples of an organoaluminium compound. Despite its name it has the formula $Al_2(CH_3)_6$ (abbreviated as Al_2Me_6 or TMA), as it exists as a dimer. This colorless liquid is an industrially important compound but must be handled with care due to its pyrophoricity; it evolves white smoke (aluminium oxides) when the vapor is released into the air.

Structure and Bonding

Al_2Me_6 exists mostly as a dimer at room temperature and pressure, analogous in structure and bonding to diborane. As with diborane, the molecules are connected by 2

3-center-2-electron bonds: the shared methyl groups bridge between the two aluminium atoms. The Al-C(terminal) and Al-C(bridging) distances are 1.97 and 2.14 Å, respectively. The carbon atoms of the bridging methyl groups are each surrounded by five neighbors: three hydrogen atoms and two aluminium atoms. The methyl groups interchange readily intramolecularly and intermolecularly.

3-Centered-2-electron bonds are an example of "electron-deficient" molecules and tend to undergo reactions with Lewis bases that would give products consisting of 2-centered-2-electron bonds. For example, upon treatment with amines gives adducts R_3N-$AlMe_3$. Another reaction that gives products that follow the octet rule is the reaction of Al_2Me_6 with aluminium trichloride to give $(AlMe_2Cl)_2$.

The monomer species $AlMe_3$, which has an aluminium atom bonded to three methyl groups, occurs at high temperature and low pressure. VSEPR Theory predicts and electron diffraction confirms that it has a trigonal planar (threefold) symmetry, as observed in BMe_3.

Synthesis and Applications

TMA is prepared via a two-step process that can be summarized as follows:

$$2\,Al + 6\,CH_3Cl + 6\,Na \rightarrow Al_2(CH_3)_6 + 6\,NaCl$$

TMA is mainly used for the production of methylaluminoxane, an activator for Ziegler–Natta catalysts for olefin polymerisation. TMA is also employed as a methylation agent. Tebbe's reagent, which is used for the methylenation of esters and ketones, is prepared from TMA.

TMA is also used in semiconductor fabrication to deposit thin film, high-k dielectrics such as Al_2O_3 via the processes of chemical vapor deposition or atomic layer deposition.

TMA forms a complex with the tertiary amine DABCO, which is safer to handle than TMA itself.

In combination with 20 to 100 mol % Cp_2ZrCl_2 (zirconocene dichloride), the $(CH_3)_2Al$-CH_3 adds "across" alkynes to give vinyl aluminum species that are useful in organic synthesis in a reaction known as carboalumination.

The NASA ATREX mission (Anomalous Transport Rocket Experiment) employed the white smoke that TMA forms on air contact to study the high altitude jet stream.

Semiconductor Grade TMA

TMA is the preferred metalorganic source for metalorganic vapour phase epitaxy (MOVPE) of aluminium-containing compound semiconductors, such as AlAs, AlN, AlP, AlSb, AlGaAs, AlInGaAs, AlInGaP, AlGaN, AlInGaN, AlInGaNP, etc. Criteria for TMA quality focus on (a) elemental impurities, (b) oxygenated and organic impurities.

Ethylaluminium Sesquichloride

Ethylaluminium sesquichloride, also called EASC, is an industrially important organoaluminium compound used primarily as a precursor to triethylaluminium and as a catalyst component in Ziegler–Natta type systems for olefin and diene polymerizations. Other applications include use in alkylation reactions and as a catalyst component in linear oligomerization and cyclization of unsaturated hydrocarbons. EASC is a colourless liquid, spontaneously combustible in air and reacts violently when in contact with water and many other compounds.

Production

Methyl, ethyl, and other alkyl or aralkyl halides that are not dehydrohalogenated readily can react with aluminium metal in an exothermic process to form organoaluminium sesquihalides in high yields. An important example is the reaction of ethyl chloride with aluminium to form ethylaluminium sesquichloride.

$$3C_2H_5Cl + 2Al \rightarrow (C_2H_5)_3Al_2Cl_3$$

The reaction is carried out with aluminium in the form of turnings, shavings, granules, or powder. Oxygen and moisture must be rigorously excluded. The reaction can be initiated with a small amount of mercury or iodine. It also can be started by treating the aluminium with an alkylaluminium halide.

The products are equilibrium mixtures of the codimer ($R_2AlX \bullet RAlX_2$) and homodimers [$(R_2AlX)_2$ and $(RAlX_2)_2$], in which the two aluminium atoms of each component are halogen-bridged.

When byproduct reactions take place to a significant extent, the excess Al – Cl content in the $R_3Al_2Cl_3$ product can be decreased by addition of the calculated amount of triethylaluminium. Overall, however, it is critical to control reaction conditions as slight excursions can result in catastrophic events.

Reactions

Conversion to Trialkylaluminium Compounds

The alkylaluminium sesquihalides convert to dialkylaluminium halide or trialkylaluminium materials upon treatment with active metals, such as sodium or magnesium. For example, diethylaluminium chloride or triethylaluminium can be produced from ethylaluminium sesquichloride by sodium reduction:

$$2(C_2H_5)_3Al_2Cl_3 + 3\,Na \rightarrow 3\,(C_2H_5)_2AlCl + Al + 3NaCl$$

$$3(C_2H_5)_2AlCl + 3Na \rightarrow 2(C_2H_5)_3Al + Al + 3NaCl$$

A magnesium-aluminium alloy can provide the reduction function simultaneously with reaction between aluminium and the alkyl halide:

$$4C_2H_5Cl + Al_2Mg \rightarrow 2(C_2H_5)_2AlCl + Mg_2Cl$$

The sesquichloride reduction process is currently the most economical route available for production of trimethylaluminium. Trialkylaluminium products made by this process generally contain trace levels of residual chloride but do not contain aluminium hydrides other low-level components found in the aluminium-hydrogen-olefin processes.

Reaction of Acids with Trialkylaluminiums to produce Alkylaluminium Chlorides

Diethylaluminium chloride (DEAC), ethylaluminium sesquichloride (EASC), and ethylaluminium dichloride (EADC) can all be prepared by the reaction of triethylaluminium with HCl (where n = 1,1.5 or 2):

$$(C_2H_5)_3Al_3 + n\ HCl \rightarrow (C_2H_5)_{3-n}AlCl_n + n\ C_2H_6$$

The products and their derivatives are used as components of catalysts for the production of polyolefins and some elastomers.

ORGANOBORON COMPOUNDS

Organoborane or organoboron compounds are chemical compounds of boron and carbon that are organic derivatives of BH3, for example trialkyl boranes. Organoboron chemistry or organoborane chemistry is the chemistry of these compounds. Organoboron compounds are important reagents in organic chemistry enabling many chemical transformations, the most important one called hydroboration.

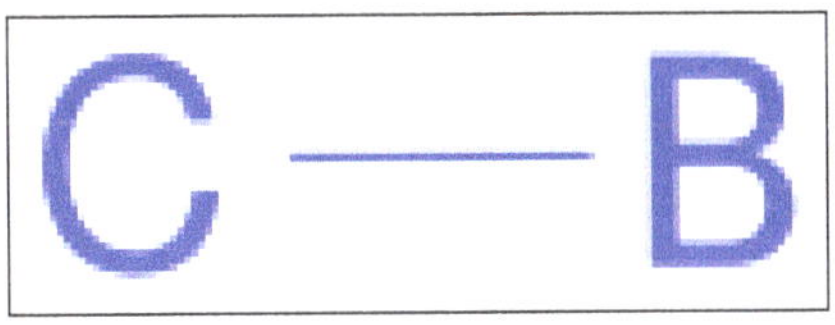

Organoboron.

Properties of the B-C Bond

The C-B bond has low polarity (the difference in electronegativity 2.55 for carbon and 2.04 for boron), and therefore alkyl boron compounds are in general stable though easily oxidized.

In part because its lower electronegativity, boron often forms electron-deficient compounds, such as the triorganoboranes. Vinyl groups and aryl groups donate electrons and make boron less electrophilic and the C-B bond gains some double bond character. Like the parent borane, diborane, organoboranes are classified in organic chemistry as strong electrophiles because boron is unable to gain a full octet of electrons. Unlike diborane however, most organoboranes do not form dimers.

Synthesis

Grignard Reagents

Simple organoboranes such as triethylborane or tris(pentafluorophenyl)boron can be prepared from trifluoroborane (as the ether complex) and the ethyl or pentafluorophenyl Grignard reagent. The borates (R_4B^- are generated via addition of R^--equivalents (RMgX, RLi, etc.) to R_3B).

Alkenes

Alkenes insert into B-H bonds of boranes in a process called hydroboration. The process involves anti-Makovnikov addition. Hydroboration of alkenes or alkynes with borane (BH_3) or borane equivalents leads to the conversion of only 33% of the starting olefin to product after oxidation or protonolysis—the remaining olefin is incorporated into boron-containing byproducts. One organoboron reagent that is often employed in synthesis is 9-BBN. Hydroborations take place stereospecifically in a syn mode that is on the same face of the alkene. In this concerted reaction the transition state is represented as a square with the corners occupied by carbon, carbon, hydrogen and boron with maximum overlap between the two olefin p-orbitals and the empty boron orbital.

By Borylation

Metal-catalyzed C-H Borylation reactions are transition metal catalyzed organic reactions that produce an organoboron compound through functionalization of aliphatic and aromatic C-H bonds. A common reagent in this type of reaction is bis(pinacolato)diboron.

Reactions

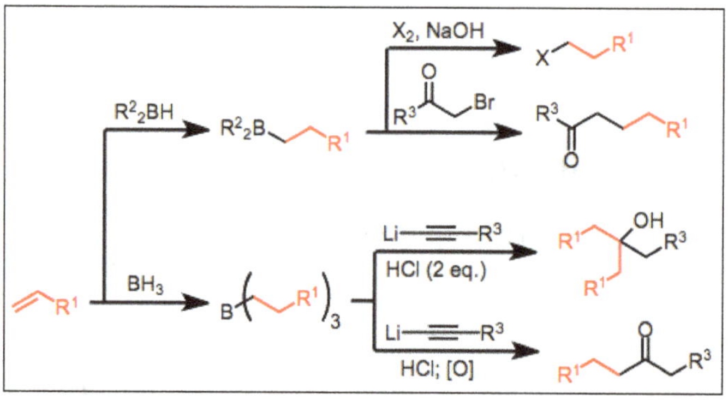

Boron–carbon bonds are polarized toward carbon. Thus, the carbon attached to boron is nucleophilic. This property is harnessed to transfer one of the R groups to an electrophilic center either inter- or (more often) intramolecularly. In the latter case, the nucleophilic R group is able to undergo 1,2-migration towards an electrophilic carbon attached to boron. The resulting reorganized borane can then be oxidized or subjected to protonolysis to afford organic products.

Hydroboration-oxidation

In organic synthesis the hydroboration reaction is taken further to generate other functional groups in the place of the boron group. The hydroboration-oxidation reaction offers a route to alcohols by oxidation of the borane with hydrogen peroxide or to the carbonyl group with the stronger oxidizing agent chromium oxide.

Rearrangements

Carbon monoxide is found to react with trialkylboranes. What follows is a 1,2-rearrangement whereby an alkyl substituent migrates from boron to the carbon of the carbonyl group. Homologated primary alcohols result from the treatment of organoboranes with carbon monoxide and a hydride.

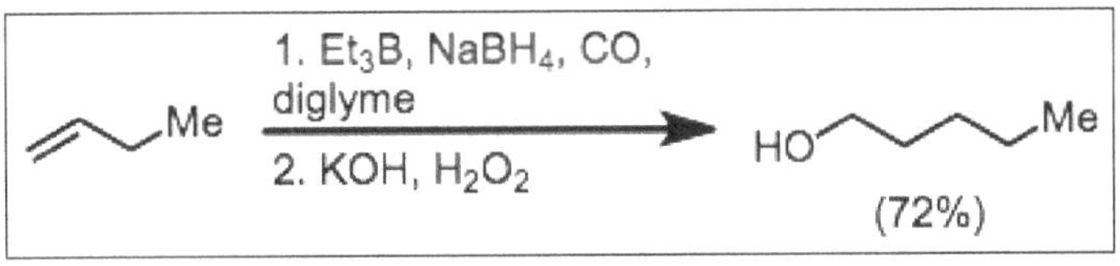

Allylboration

Asymmetric allylboration demonstrates another useful application of organoboranes in carbon–carbon bond formation. In this example from Nicolaou's synthesis of the epothilones, asymmetric allylboration (using an allylborane derived from chiral alpha-pinene) is used in conjunction with TBS protection and ozonolysis. Overall, this provides a two-carbon homologation sequence that delivers the required acetogenin sequence.

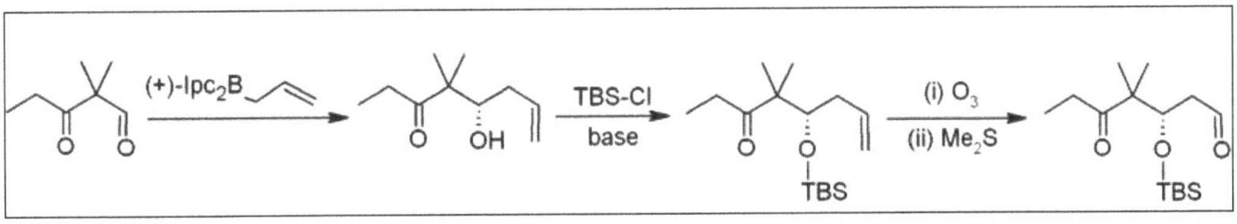

As Reducing Agent

Borane hydrides such as 9-BBN and L-selectride (lithium tri-sec-butylborohydride) are reducing agents. An example of an asymmetric catalyst for carbonyl reductions is the

CBS catalyst. This catalyst is also based on boron, the purpose of which is coordination to the carbonyl oxygen atom.

Borates

Trialkylboranes, BR_3, can be oxidized to the corresponding borates, $B(OR)_3$. One method for the determination of the amount of C-B bonds in a compound is by oxidation of R_3B with trimethylamine oxide (Me_3NO) to $B(OR)_3$. The trimethylamine (Me_3N) formed can then be titrated.

Boronic acids $RB(OH)_2$ react with potassium bifluoride $K[HF_2]$ to form trifluoroborate salts $K[RBF_3]$ which are precursors to nucleophilic alkyl and aryl boron difluorides, $ArBF_2$. The salts are more stable than the boronic acids themselves and used for instance in alkylation of certain aldehydes:

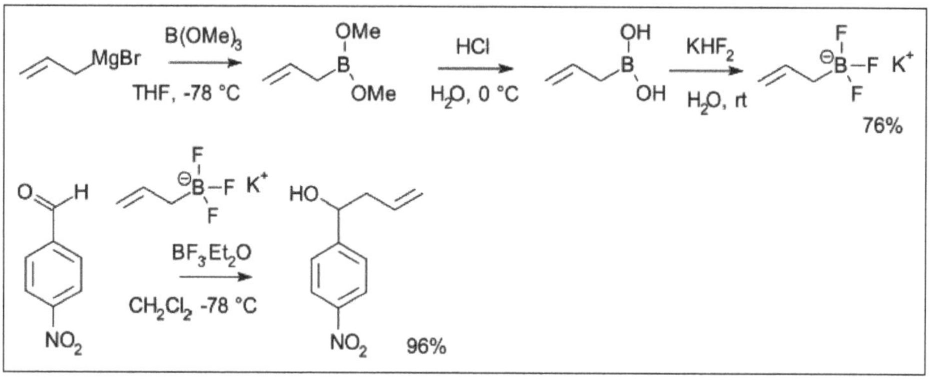

Suzuki Reaction and Related Reactions

Organoboron compounds also lend themselves to transmetalation reactions, especially with organopalladium compounds. This reaction type is exemplified in the Suzuki reaction, which involves coupling of aryl- or vinyl-boronic acid with an aryl- or vinyl-halide catalyzed by a palladium(o) complex:

$$R_1 - BY_2 + R_2 - X \xrightarrow[\text{Base}]{\text{Pd catalyst}} R_1 - R_2$$

This reaction is an important method for making carbon-carbon bonds.

Classes of Organoboron Compounds

Organoboranes and Hydrides

Among the most studied classes of organoboron compounds have the formula BR_nH_{3-n}. These compounds are used as catalysts, reagents, and synthetic intermediates. The trialkyl and triaryl derivatives feature trigonal planar boron center that is typically only weakly Lewis acidic. Except for a few very bulky derivatives, the hydrides ($BRnH_{3-n}$ for n = 1 or 2) exist as dimers, reminiscent of the structure of diborane itself. Trisubstituted derivatives, e.g. triethylboron are monomers.

Borinic and Boronic Acids and Esters ($BR_n(OR)_{3-n}$)

Compounds of the type $BR_n(OR)_{3-n}$ are called borinic esters (n = 2), boronic esters (n = 1), and borates (n = 0). Boronic acids are used in Suzuki reaction. Trimethyl borate, which is debatably not an organoboron compound, is an intermediate in the production of sodium borohydride.

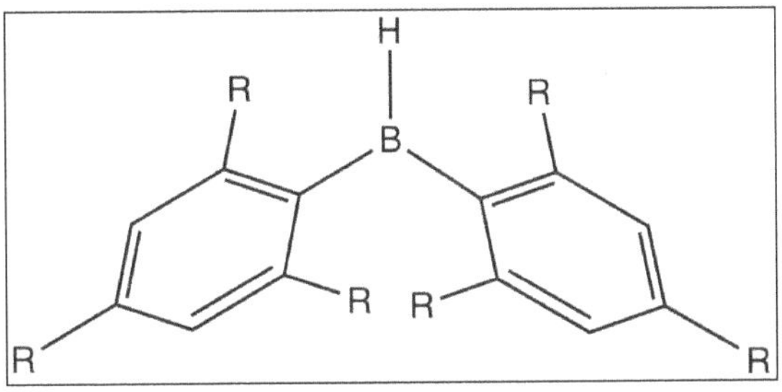

Structure of a rare monomeric boron hydride, R = i-Pr.

Boron Clusters

Boron is renowned for forming cluster compounds, e.g. dodecaborate $[B_{12}H_{12}]^{2-}$. Many organic derivatives are known for such clusters. One example is $[B_{12}(CH_3)_{12}]^{2-}$ and its radical derivative $[B_{12}(CH_3)_{12}]^{-}$. Related cluster compounds with carbon vertices are called carboranes are. The best known is orthocarborane, with the formula $C_2B_{10}H_{12}$. Although they have few commercial applications, carboranes have attracted much attention because they are so structurally unusual. Anionic derivatives, dicarbollides, e.g., $[C_2B_9H_{11}]^{2-}$ are ligands that behave like cyclopentadienide.

Bora-substituted Aromatic Compounds

In borabenzene, one CH center in benzene is replaced by boron. These compounds are invariably isolated as adducts, e.g., C_5H_5B-pyridine. The cyclic compound borole, a structural analog of pyrrole, has not been isolated, but substituted derivatives known as boroles are known. The cyclic compound borepin is aromatic.

Boryl Compounds

Boryl anions have the formula R_2B^-. Nucleophilic anionic boryl compounds have long been elusive but a 2006 study described a boryllithium compound, which reacts as a nucleophile: Organometallic compounds with metal to boron bonds, (i.e., $M-BR_2$), are known as boryl complexes. Related ligands are borylenes (M–B(R)–M).

The absence of lithium boryl compounds is notable because in other period 2 elements lithium salts are common e.g. lithium fluoride, lithium hydroxide lithium amide and methyllithium. The gap highlights the very low electronegativity of boron. Reaction of base with a borohydride R_2BH does not result in deprotonation to the boryl anion R_2B^- but to formation of the boryl anion $R_2B^-H(base)^+$. This reaction product has a complete octet. Instead the boryl compound is prepared by reductive heterolysis of a boron-bromide bond by lithium metal. The new boryl lithium compound is very similar to and isoelectronic with N-heterocyclic carbenes. It is designed to benefit from aromatic stabilization (6-electron system counting the nitrogen lone pairs and an empty boron p-orbital) and from kinetic stabilization from the bulky 2,6-diisopropylphenyl groups. X-ray diffraction confirms sp^2 hybridization at boron and its nucleophilic addition reaction with benzaldehyde gives further proof of the proposed structure.

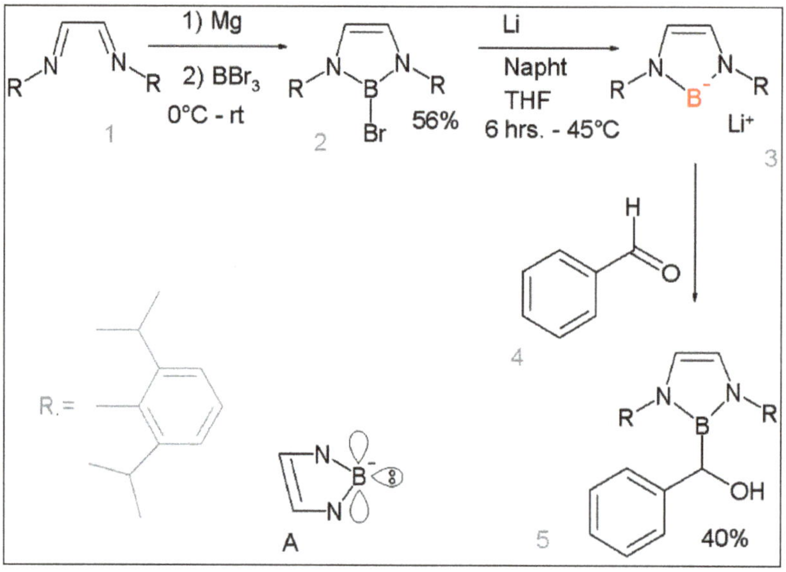

Alkylideneboranes

Alkylideneboranes of the type RB=CRR with a boron – carbon double bond are rarely encountered. An example is borabenzene. The parent compound is HB=CH2 which can be detected at low temperatures. A fairly stable derivative is $CH_3B = C(SiMe_3)_2$ but is prone to cyclodimerisation.

NHC Adducts of Boron

NHCs and boranes form stable NHC borane adducts. Triethylborane adducts can be synthesised directly from the imidazolium salt and lithium triethylborohydride. Members of this compound class are investigated for use as reagent or catalyst.

Diborenes

Chemical compounds with boron to boron double bonds are rare. In 2007 the first neutral diborene (RHB=BHR) was presented by Gregory Robinson of the University of Georgia. Each boron atom has a proton attached to it and each boron atom is coordinated to a NHC carbene. The parent structure with the additional carbene ligands is diborane(2).

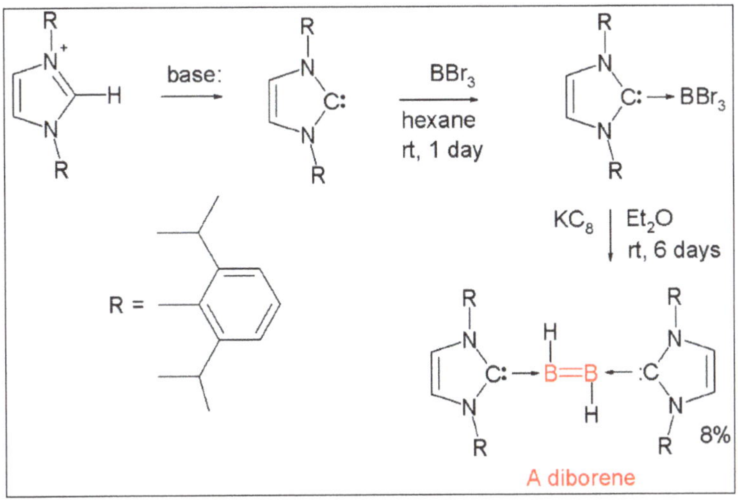

A reported diboryne is based on similar chemistry.

Other Uses

TEB – Triethylborane was used to ignite the JP-7 fuel of the Pratt & Whitney J58 variable cycle engines powering the Lockheed SR-71 Blackbird.

ORGANOLITHIUM REAGENT

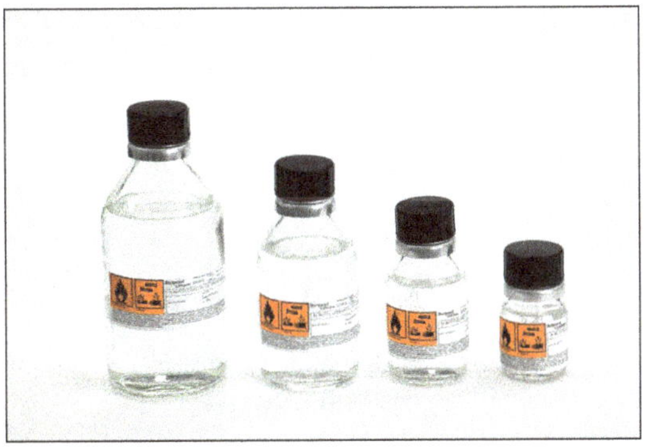

Glass bottles containing butyllithium.

Organolithium reagents are organometallic compounds that contain carbon – lithium bonds. They are important reagents in organic synthesis, and are frequently used to transfer the organic group or the lithium atom to the substrates in synthetic steps, through nucleophilic addition or simple deprotonation. Organolithium reagents are used in industry as an initiator for anionic polymerization, which leads to the production of various elastomers. They have also been applied in asymmetric synthesis in the pharmaceutical industry. Due to the large difference in electronegativity between the carbon atom and the lithium atom, the C-Li bond is highly ionic. Owing to the polar nature of the C-Li bond, organolithium reagents are good nucleophiles and strong bases. For laboratory organic synthesis, many organolithium reagents are commercially available in solution form. These reagents are highly reactive, and are sometimes pyrophoric.

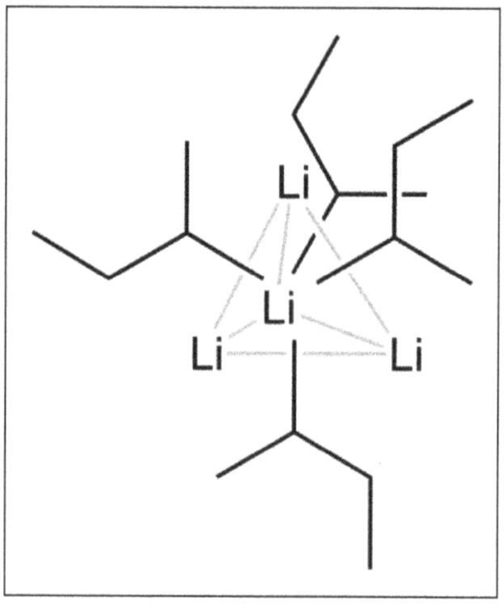

Sec-Butyllithium aggregate.

Structure

Although simple alkyllithium species are often represented as monomer RLi, they exist as aggregates (oligomers) or polymers. The degree of aggregation depends on the organic substituent and the presence of other ligands. These structures have been elucidated by a variety of methods, notably ^{6}Li, ^{7}Li, and ^{13}C NMR spectroscopy and X-ray diffraction analysis. Computational chemistry supports these assignments.

Nature of Carbon-lithium Bond

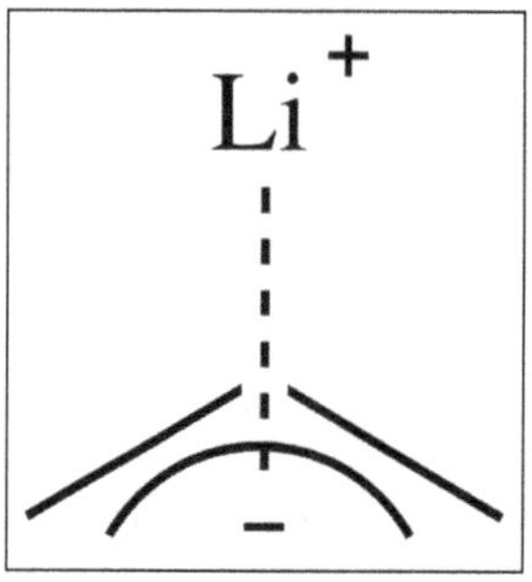

Delocalized electron density in allyllithium reagents.

The relative electronegativities of carbon and lithium suggest that the C-Li bond will be highly polar. However, certain organolithium compounds possess properties such as solubility in nonpolar solvents that complicate the issue. While most data suggest the C-Li bond to be essentially ionic, there has been debate as to whether a small covalent character exists in the C-Li bond. One estimate puts the percentage of ionic character of alkyllithium compounds at 80 to 88%.

In allyl lithium compounds, the lithium cation coordinates to the face of the carbon π bond in an η^3 fashion instead of a localized, carbanionic center, thus, allyllithiums are often less aggregated than alkyllithiums. In aryllithium complexes, the lithium cation coordinates to a single carbanion center through a Li-C σ type bond.

Solid State Structure

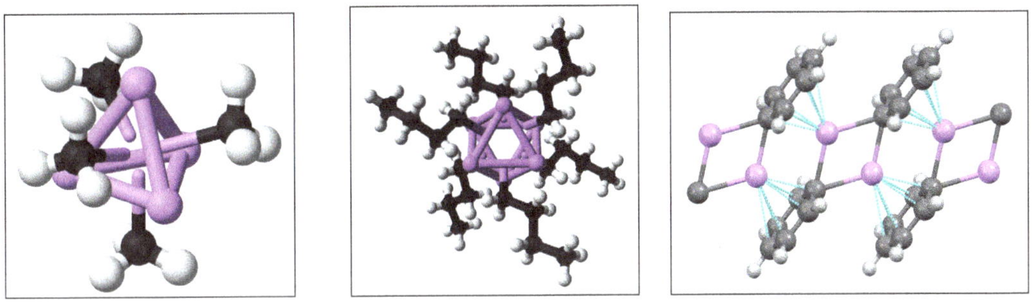

Solid state structures of methyllithium tetramers, n-butyllithium hexamers and polymeric ladder of phenyllithium.

Like other species consisting of polar subunits, organolithium species aggregate. Formation of aggregates is influenced by electrostatic interactions, the coordination between lithium and surrounding solvent molecules or polar additives, and steric effects.

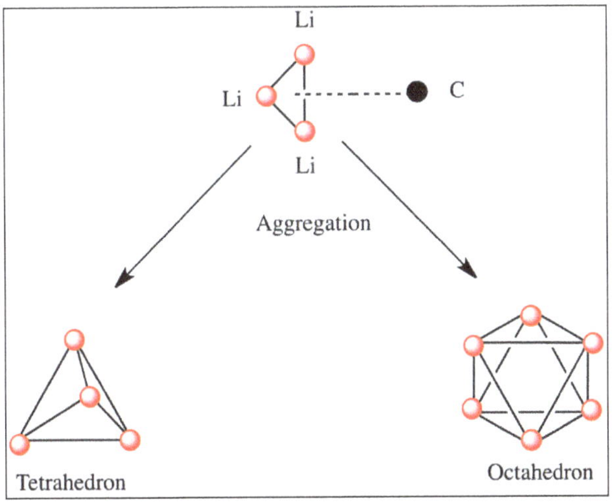

Tetrahedron and octahedron metal cores formed by aggregation of the Li_3 triangle - carbanion coordinate complex.

A basic building block toward constructing more complex structures is a carbanionic center interacting with a Li_3 triangle in an η^{-3} fashion. In simple alkyllithium reagents, these triangles aggregate to form tetrahedron or octahedron structures. For example, methyllithium, ethyllithium and tert-butyllithium all exist in the tetramer $[RLi]_4$. Methyllithium exists as tetramers in a cubane-type cluster in the solid state, with four lithium centers forming a tetrahedron. Each methanide in the tetramer in methyllithium can have agostic interaction with lithium cations in adjacent tetramers. Ethyllithium and tert-butyllithium, on the other hand, do not exhibit this interaction, and are thus soluble in non-polar hydrocarbon solvents. Another class of alkyllithium adopts hexameric structures, such as n-butyllithium, isopropyllithium, and cyclohexanyllithium.

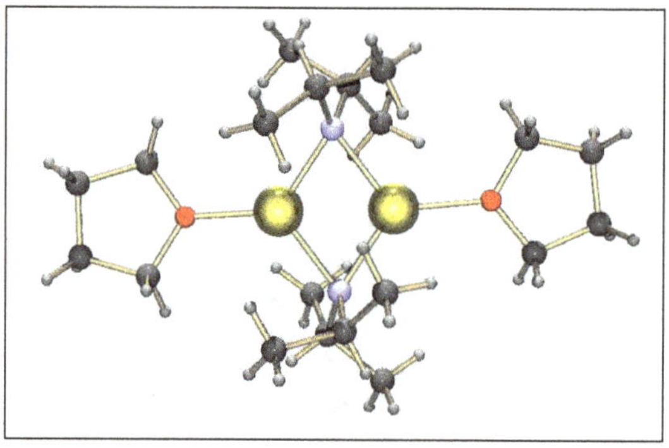

LDA dimer with THF coordinated to Li cations.

Common lithium amides, e.g. lithium bis(trimethylsilyl)amide and lithium diisopropylamide, are also subject to aggregation. Lithium amides adopt polymeric-ladder type structures in non-coordinating solvent in the solid state, and they generally exist as dimers in ethereal solvents. In the presence of strongly donating ligands, tri- or tetrameric lithium centers are formed. For example, LDA exists primarily as dimers in THF. The structures of common lithium amides, such as lithium diisopropylamide (LDA) and lithium hexamethyldisilazide (LiHMDS) have been extensively studied by Collum and coworkers using NMR spectroscopy. Another important class of reagents is silyllithiums, extensively used in the synthesis of organometallic complexes and polysilane dendrimers. In the solid state, in contrast with alkyllithium reagents, most silyllithiums tend to form monomeric structures coordinated with solvent molecules such as THF, and only a few silyllithiums have been characterized as higher aggregates. This difference can arise from the method of preparation of silyllithiums, the steric hindrance caused by the bulky alkyl substituents on silicon, and the less polarized nature of Si-Li bonds. The addition of strongly donating ligands, such as TMEDA and (-)-sparteine, can displace coordinating solvent molecules in silyllithiums.

Solution Structure

Relying solely on the structural information of organolithium aggregates obtained in the solid state from crystal structures has certain limits, as it is possible for organolithium reagents to adopt different structures in reaction solution environment. Also, in some cases the crystal structure of an organolithium species can be difficult to isolate.

Therefore, studying the structures of organolithium reagents, and the lithium-containing intermediates in solution form is extremely useful in understanding the reactivity of these reagents. NMR spectroscopy has emerged as a powerful tool for the studies of organolithium aggregates in solution. For alkyllithium species, C-Li J coupling can often use to determine the number of lithium interacting with a carbanion center, and whether these interactions are static or dynamic. Separate NMR signals can also differentiate the presence of multiple aggregates from a common monomeric unit.

The structures of organolithium compounds are affected by the presence of Lewis bases such as tetrahydrofuran (THF), diethyl ether (Et_2O), tetramethylethylene diamine (TMEDA) or hexamethylphosphoramide (HMPA). Methyllithium is a special case, in which solvation with ether, or polar additive HMPA does not deaggregate the tetrameric structure in the solid state. On the other hand, THF deaggregates hexameric butyl lithium: the tetramer is the main species, and ΔG for interconversion between tetramer and dimer is around 11 kcal/mol. TMEDA can also chelate to the lithium cations in n-butyllithium and form solvated dimers such as $[(TMEDA)\ LiBu\text{-}n)]_2$. Phenyllithium has been shown to exist as a distorted tetramer in the crystallized ether solvate, and as a mixture of dimer and tetramer in ether solution.

Table: Aggregates of some alkyllithiums in solvents.

	Solvent	Structure
Methyllithium	THF	Tetramer
Methyllithium	Ether/HMPA	Tetramer
N-butyllithium	Pentane	Hexamer
N-butyllithium	Ether	Tetramer
N-butyllithium	THF	Tetramer-dimer
Sec-butyllithium	Pentane	Hexamer-tetramer
Isopropyllithium	Pentane	Hexamer-tetramer
Tert-butyllithium	Pentane	Tetramer
Tert-butyllithium	THF	Monomer
Phenyllithium	Ether	Tetramer-dimer
Phenyllithium	Ether/HMPA	Dimer

Structure and Reactivity

As the structures of organolithium reagents change according to their chemical environment, so do their reactivity and selectivity. One question surrounding the structure-reactivity relationship is whether there exists a correlation between the degree of aggregation and the reactivity of organolithium reagents. It was originally proposed that lower aggregates such as monomers are more reactive in alkyllithiums. However, reaction pathways in which dimer or other oligomers are the reactive species have also been discovered, and for lithium amides such as LDA, dimer-based reactions are common. A series of solution kinetics studies of LDA - mediated reactions suggest that lower aggregates of enolates do not necessarily lead to higher reactivity.

Also, some Lewis bases increase reactivity of organolithium compounds. However, whether these additives function as strong chelating ligands, and how the observed increase in reactivity relates to structural changes in aggregates caused by these additives are not always clear. For example, TMEDA increases rates and efficiencies in many reactions involving organolithium reagents. Toward alkyllithium reagents, TMEDA functions as a donor ligand, reduces the degree of aggregation, and increases the nucleophilicity of these species. However, TMEDA does not always function as a donor ligand to lithium cation, especially in the presence of anionic oxygen and nitrogen centers. For example, it only weakly interacts with LDA and LiHMDS even in hydrocarbon solvents with no competing donor ligands. In imine lithiation, while THF acts as a strong donating ligand to LiHMDS, the weakly coordinating TMEDA readily dissociates from LiHMDS, leading to the formation of LiHMDS dimers that is the more reactive species. Thus, in the case of LiHMDS, TMEDA does not increase reactivity by reducing aggregation state. Also, as opposed to simple alkyllithium compounds, TMEDA does not deaggregate lithio-acetophenolate in THF solution. The addition of HMPA to lithium amides such as LiHMDS and LDA often results in a mixture of dimer/monomer aggregates in THF. However, the ratio of dimer/monomer species does not change with increased concentration of HMPA, thus, the observed increase in reactivity is not the result of deaggregation. The mechanism of how these additives increase reactivity is still being researched.

Reactivity and Applications

The C-Li bond in organolithium reagents is highly polarized. As a result, the carbon attracts most of the electron density in the bond and resembles a carbanion. Thus, organolithium reagents are strongly basic and nucleophilic. Some of the most common applications of organolithium reagents in synthesis include their use as nucleophiles, strong bases for deprotonation, initiator for polymerization, and starting material for the preparation of other organometallic compounds.

Nucleophile

Carbolithiation Reactions

As nucleophiles, organolithium reagents undergo carbolithiation reactions, whereby the carbon-lithium bond adds across a carbon-carbon double or triple bond, forming new organolithium species. This reaction is the most widely employed reaction of organolithium compounds. Carbolithiation is key in anionic polymerization processes, and n-butyllithium is used as a catalyst to initiate the polymerization of styrene, butadiene, or isoprene or mixtures thereof.

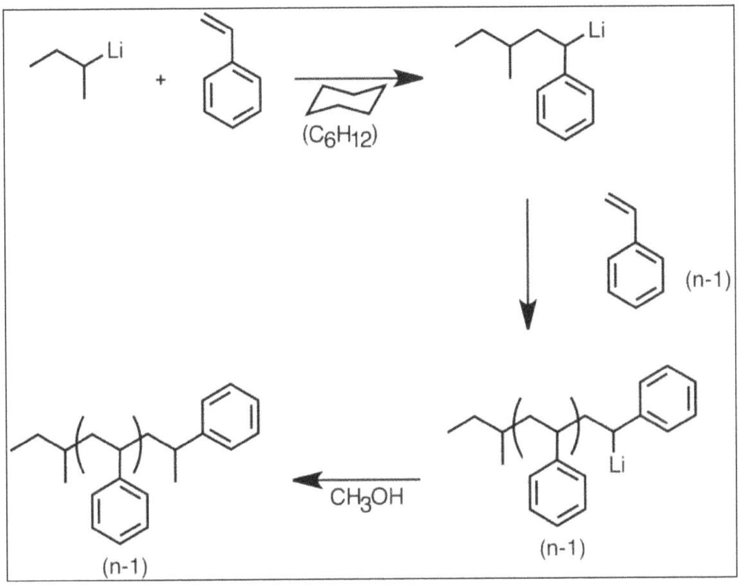

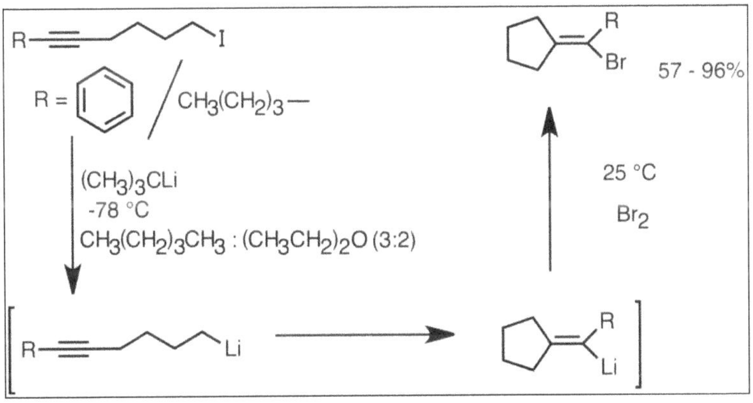

Another application that takes advantage of this reactivity is the formation of carbocyclic and heterocyclic coumpounds by intramolecular carbolithiation. As a form of anionic cyclization, intramolecular carbolithiation reactions offer several advantages over radical cyclization. First, it is possible for the product cyclic organolithium species to react with electrophiles, whereas it is often difficult to trap a radical intermediate of the corresponding structure. Secondly, anionic cyclizations are often more regio- and stereospecific than radical cyclization, particularly in the case of 5-hexenyllithiums. Intramolecular carbolithiation allows addition of the alkyl-, vinyllithium to triple bonds and mono-alkyl substituted double bonds. Aryllithiums can also undergo addition if a 5-membered ring is formed. The limitations of intramolecular carbolithiation include difficulty of forming 3 or 4-membered rings, as the intermediate cyclic organolithium species often tend to undergo ring-openings. Below is an example of intramolecular carbolithiation reaction. The lithium species derived from the lithium-halogen exchange cyclized to form the vinyllithium through 5-exo-trig ring closure. The vinyllithium species further reacts with electrophiles and produce functionalized cyclopentylidene compounds.

Addition to Carbonyl Compounds

Nucleophilic organolithium reagents can add to electrophilic carbonyl double bonds to form carbon-carbon bonds. They can react with aldehydes and ketones to produce alcohols. The addition proceeds mainly via polar addition, in which the nucleophilic organolithium species attacks from the equatorial direction, and produces the axial alcohol. Addition of lithium salts such as $LiClO_4$ can improve the stereoselectivity of the reaction.

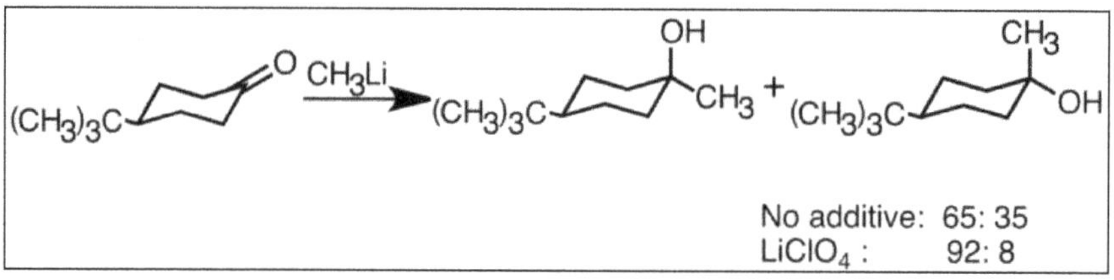

Then the ketone is sterically hindered, using Grignard reagents often leads to reduction of the carbonyl group instead of addition. However, alkyllithium reagents are less likely to reduce the ketone, and may be used to synthesize substituted alcohols. Below is an example of ethyllithium addition to adamantone to produce tertiary alcohol.

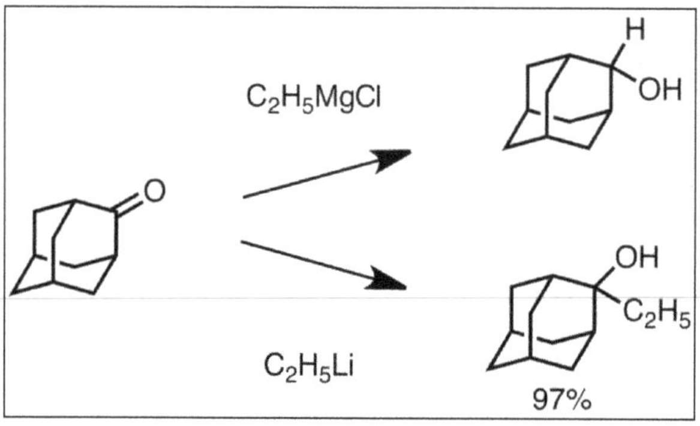

Organolithium reagents are also better than Grignard reagents in their ability to react with carboxylic acids to form ketones. This reaction can be optimized by carefully controlling the amount of organolithium reagent addition, or using trimethylsilyl chloride to quench excess lithium reagent. A more common way to synthesize ketones is through the addition of organolithium reagents to Weinreb amides (N-methoxy-N-methyl amides). This reaction provides ketones when the organolithium reagents is used in excess, due to chelation of the lithium ion between the N-methoxy oxygen and the carbonyl oxygen, which forms a tetrahedral intermediate that collapses upon acidic work up.

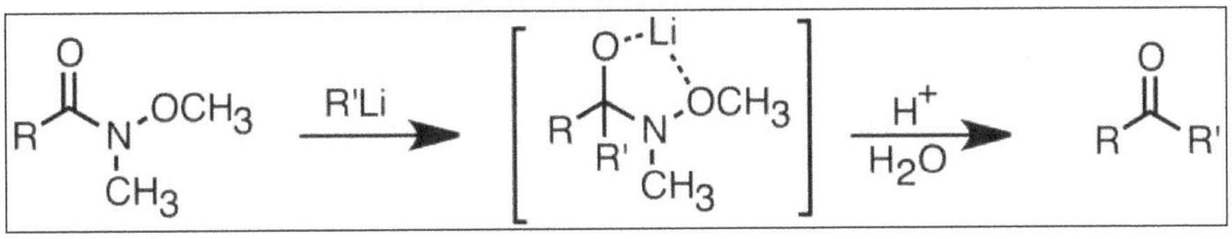

Organolithium reagents can also react with carbon dioxide to form carboxylic acids.

In the case of enone substrates, where two sites of nucleophilic addition are possible (1,2 addition to the carbonyl carbon or 1,4 conjugate addition to the β carbon), most highly reactive organolithium species favor the 1,2 addition, however, there are several ways to propel organolithium reagents to undergo conjugate addition. First, since the 1,4 adduct is the likely to be the more thermodynamically favorable species, conjugate addition can be achieved through equilibration (isomerization of the two product), especially when the lithium nucleophile is weak and 1,2 addition is reversible. Secondly, adding donor ligands to the reaction forms heteroatom-stabilized lithium species which favors 1,4 conjugate addition. In one example, addition of low-level of HMPA to the solvent favors the 1,4 addition. In the absence of donor ligand, lithium cation is closely coordinated to the oxygen atom, however, when the lithium cation is solvated by HMPA, the coordination between carbonyl oxygen and lithium ion is weakened. This method generally cannot be used to affect the regioselectivity of alkyl- and aryllithium reagents.

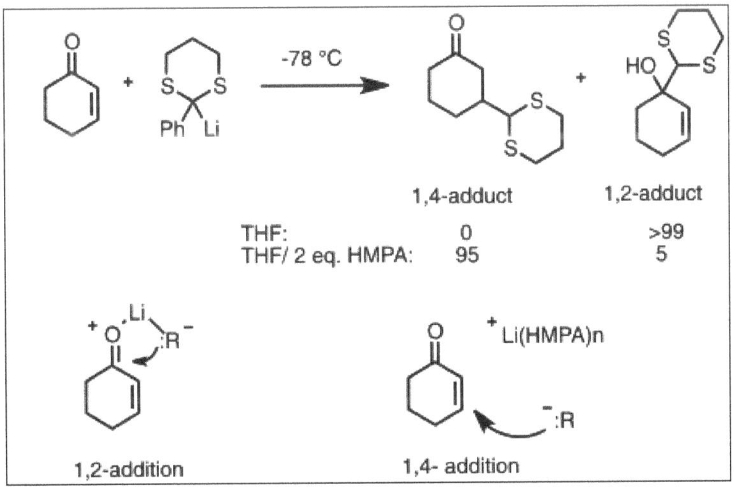

Proposed mechanism for 1,2- and 1,4-addition in thr presence of HMPA.

Organolithium reagents can also perform enantioselective nucleophilic addition to carbonyl and its derivatives, often in the presence of chiral ligands. This reactivity is widely applied in the industrial syntheses of pharmaceutical compounds. An example is the Merck and Dupont synthesis of Efavirenz, a potent HIV reverse transcriptase inhibitor. Lithium acetylide is added to a prochiral ketone to yield a chiral alcohol product. The structure of the active reaction intermediate was determined by NMR spectroscopy studies in the solution state and X-ray crystallography of the solid state to be a cubic 2:2 tetramer.

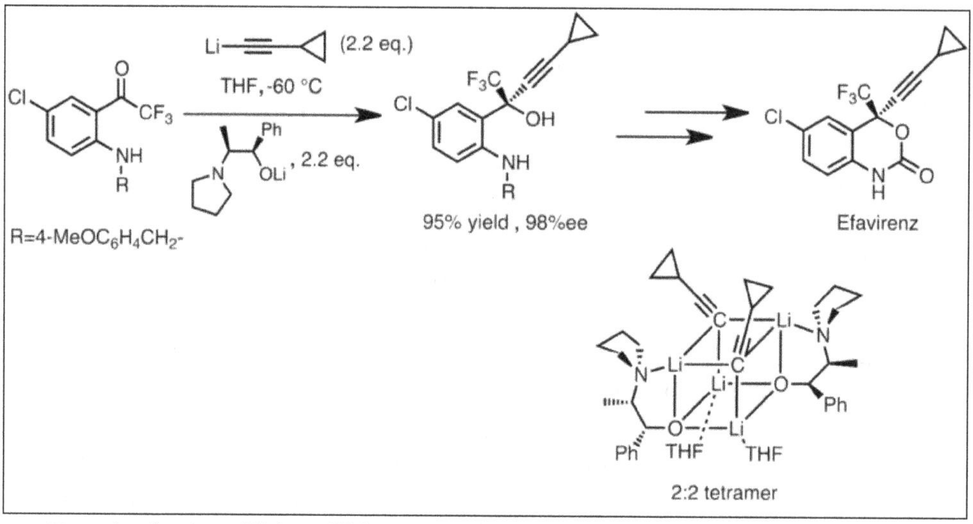

Enantioselective addition of lithium acetylide in the synthesis of HIV drug Efavirenz.

S_N2 Type Reactions

Organolithium reagents can serve as nucleophiles and carry out S_N2 type reactions with alkyl or allylic halides. Although they are considered more reactive than Grignards reactions in alkylation, their use is still limited due to competing side reactions such as radical reactions or metal-halogen exchange. Most organolithium reagents used in a alkylations are more stabilized, less basic, and less aggregated, such as heteroatom stabilized, aryl- or allyllithium reagents. HMPA has been shown to increase reaction rate and product yields, and the reactivity of aryllithium reagents is often enhanced by the addition of potassium alkoxides. Organolithium reagents can also carry out nucleophilic attacks with epoxides to form alcohols.

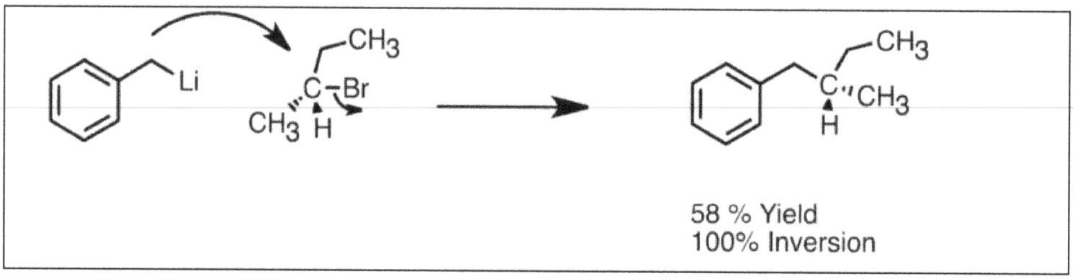

Alkylation of benzylithium reagents with secondary alkyl bromide proceed with inversion.

As Base

Organolithium reagents provide a wide range of basicity. tert-Butyllithium, with three weakly electron donating alkyl groups, is the strongest base commercially available (pKa = 53). As a result, the acidic protons on -OH, -NH and -SH are often protected in the presence of organolithium reagents. Some commonly used lithium bases are alkyllithium species such as n-butyllithium and lithium dialkylamides ($LiNR_2$). Reagents with bulky R groups such as lithium diisopropylamide (LDA) and lithium bis(trimethylsilyl)amide (LiHMDS) are often sterically hindered for nucleophilic addition, and are thus more selective toward deprotonation. Lithium dialkylamides ($LiNR_2$) are widely used in enolate formation and aldol reaction. The reactivity and selectivity of these bases are also influenced by solvents and other counter ions.

Metalation

Metalation with organolithium reagents, also known as lithiation or lithium-hydrogen exchange, is achieved when an organolithium reagent, most commonly an alkyllithium, abstracts a proton and forms a new organolithium species.

$$R - H + R'Li \rightarrow RLi + R'H$$

Common metalation reagents are the butyllithiums. Tertiary-butyllithium and secondary-butyllithium are generally more reactive and have better selectivity than n-butyllithium, however, they are also more expensive and difficult to handle. Metalation is a common way of preparing versatile organolithium reagents. The position of metalation is mostly controlled by the acidity of the C-H bond. Lithiation often occurs at a position α to electron withdrawing groups, since they are good at stabilizing the electron-density of the anion. Directing groups on aromatic compounds and heterocycles provide regioselective sites of metalation; directed ortho metalation is an important class of metalation reactions. Metalated sulfones, acyl groups and α-metalated amides are important intermediates in chemistry synthesis. Metalation of allyl ether with alkyllithium or LDA forms an anion α to the oxygen, and can proceed to 2,3-Wittig rearrangement. Addition of donor ligands such as TMEDA and HMPA can increase metalation rate and broaden substrate scope. Chiral organolithium reagents can be accessed through asymmetric metalation.

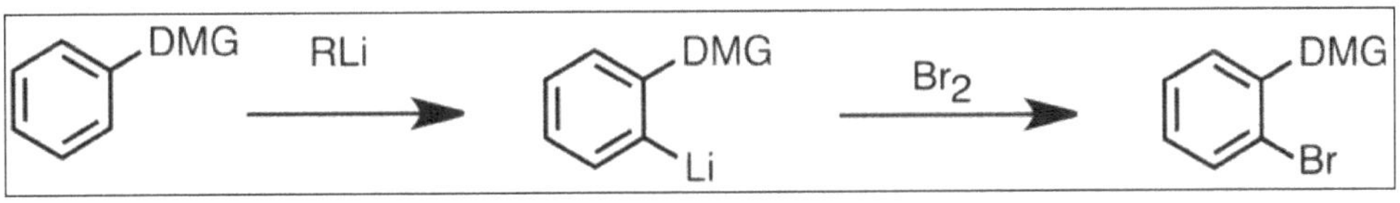

Directed ortho metalation is an important tool in the synthesis of regiospecific substituted aromatic compounds. This approach to lithiation and subsequent quenching of the intermediate lithium species with electrophile is often better than the electrophilic aromatic substitution due to its high regioselectivity. This reaction proceeds through deprotonation by organolithium reagents at the positions α to the direct metalation group (DMG) on the aromatic ring. The DMG is often a functional group containing a heteroatom that is Lewis basic, and can coordinate to the Lewis-acidic lithium cation. This generates a complex-induced proximity effect, which directs deprotonation at the α position to form an aryllithium species that can further react with electrophiles. Some of the most effective DMGs are amides, carbamates, sulfones and sulfonamides. They are strong electron-withdrawing groups that increase the acidity of alpha-protons on the aromatic ring. In the presence of two DMGs, metalation often occurs ortho to the stronger directing group, though mixed products are also observed. A number of heterocycles that contain acidic protons can also undergo ortho-metalation. However, for electron-poor heterocycles, lithium amide bases such as LDA are generally used, since alkyllithium has been observed to perform addition to the electron-poor heterocycles rather than deprotonation. In certain transition metal-arene complexes, such as ferrocene, the transition metal attracts electron-density from the arene, thus rendering the aromatic protons more acidic, and ready for ortho-metalation.

Superbases

Addition of potassium alkoxide to alkyllithium greatly increases the basicity of organolithium species. The most common "superbase" can be formed by addition of KOtBu to butyllithium, often abbreviated as "LiCKOR" reagents. These "superbases" are highly reactive and often stereoselective reagents. In the example below, the LiCKOR base generates a stereospecific crotylboronate species through metalation and subsequent lithium-metalloid exchange.

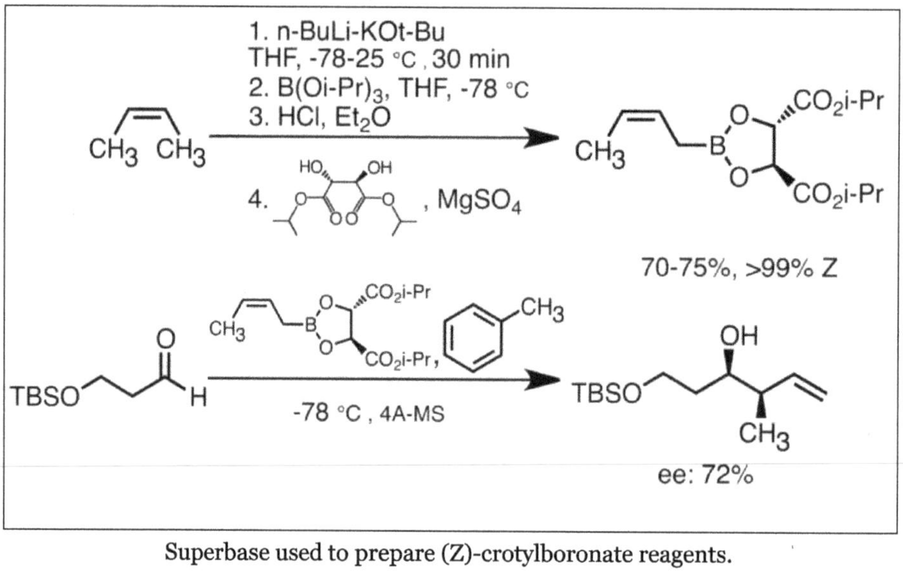

Superbase used to prepare (Z)-crotylboronate reagents.

Asymmetric Metalation

Enantioenriched organlithium species can be obtained through asymmetric metalation of prochiral substrates. Asymmetric induction requires the presence of a chiral ligand such as (-)-sparteine. The enantiomeric ratio of the chiral lithium species is often influenced by the differences in rates of deprotonation. In the example below, treatment of N-Boc-N-benzylamine with n-butyllithium in the presence of (-)-sparteine affords one enantiomer of the product with high enantiomeric excess. Transmetalation with trimethyltinchloride affords the opposite enantiomer.

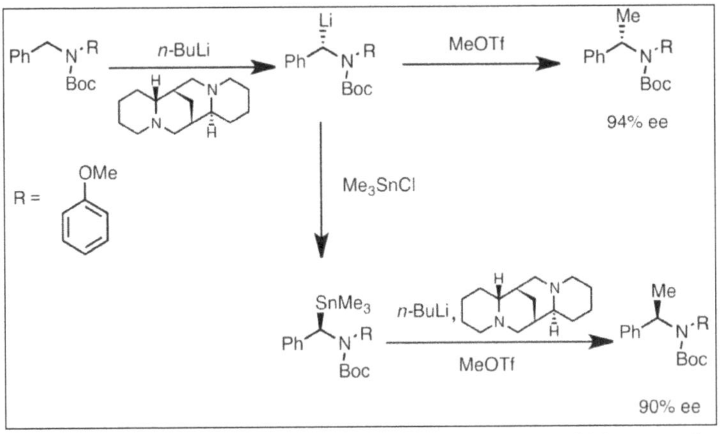

Enantioselective synthesis with n-BuLi in the presence of (-) separteine.

Enolate Formation

Lithium enolates are formed through deprotonation of a C-H bond α to the carbonyl group by an organolithium species. Lithium enolates are widely used as nucleophiles in carbon-carbon bond formation reactions such as aldol condensation and alkylation. They are also an important intermediate in the formation of silyl enol ether.

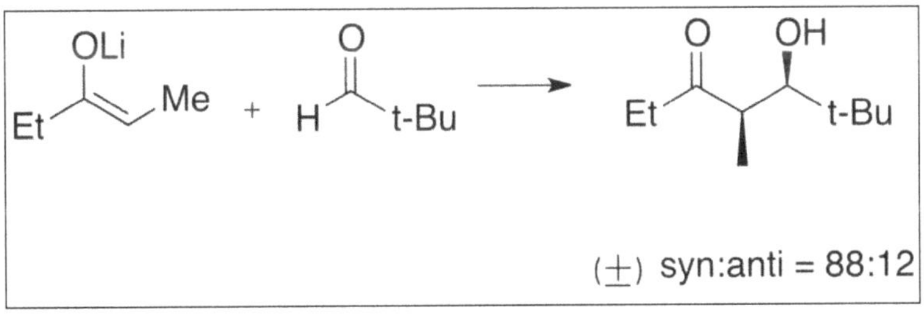

Lithium enolate formation can be generalized as an acid-base reaction, in which the relatively acidic proton α to the carbonyl group (pK =20-28 in DMSO) reacts with organolithium base. Generally, strong, non-nucleophilic bases, especially lithium amides such LDA, LiHMDS and LiTMP are used. THF and DMSO are common solvents in lithium enolate reactions.

The stereochemistry and mechanism of enolate formation have gained much interest in the chemistry community. Many factors influence the outcome of enolate stereochemistry, such as steric effects, solvent, polar additives, and types of organolithium bases. Among the many models used to explain and predict the selectivity in stereochemistry of lithium enolates is the Ireland model.

In this assumption, a monomeric LDA reacts with the carbonyl substrate and form a cyclic Zimmerman-Traxler type transition state. The (E)-enolate is favored due to an unfavorable syn-pentane interaction in the (Z)-enolate transition state.

Addition of polar additives such as HMPA or DMPU favors the formation of (Z) enolates. The Ireland model argues that these donor ligands coordinate to the lithium cations, as a result, carbonyl oxygen and lithium interaction is reduced, and the transition state is not as tightly bound as a six-membered chair. The percentage of (Z) enolates also increases when lithium bases with bulkier side chains (such as LiHMDS) are used. However, the mechanism of how these additives reverse stereoselectivity is still being debated.

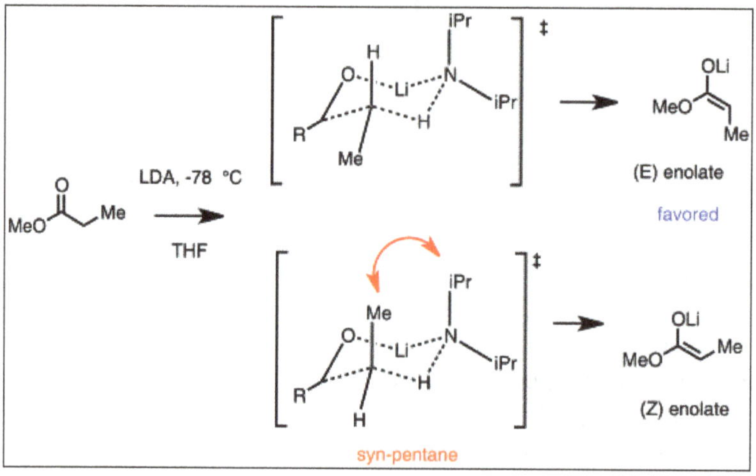

There have been some challenges to the Ireland model, as it depicts the lithium species as a monomer in the transition state. In reality, a variety of lithium aggregates are often observed in solutions of lithium enolates, and depending on specific substrate, solvent and reaction conditions, it can be difficult to determine which aggregate is the actual reactive species in solution.

Lithium-halogen Exchange

Lithium-halogen exchange is a metathesis reaction between an organohalide and organolithium species. Gilman and Wittig independently discovered this method in the late 1930s.

$$R - Li + R' - X \rightarrow R - X + R' - Li$$

The mechanism of lithium-halogen exchange is still debated. One possible pathway involves a nucleophilic mechanism that generates a reversible "ate-complex" intermediate. Farnham and Calabrese were able to isolate "ate-complex" lithium bis(pentafluorophenyl) iodinate complexed with TMEDA and obtain an X-ray crystal structure. The "ate-complex" further reacts with electrophiles and provides pentafluorophenyl iodide and C_6H_5Li. A number of kinetic studies also support a nucleophilic pathway in which the carbanion on the lithium species attacks the halogen atom on the aryl halide. Another possible mechanism involves single electron transfer and the generation of radicals. In reactions of secondary and tertiary alkyllithium and alkyl halides, radical species were detected by EPR spectroscopy. However, whether these radicals are reaction intermediates is not definitive. The mechanistic studies of lithium-halogen exchange is also complicated by the formation of aggregates of organolithium species.

The rate of lithium halogen exchange is extremely fast. It is usually faster than nucleophilic addition and can sometimes exceed the rate of proton transfer. In the example below, the exchange between lithium and primary iodide is almost instantaneous, and outcompetes proton transfer from methanol to tert-butyllithium. The major alkene product is formed in over 90% yield.

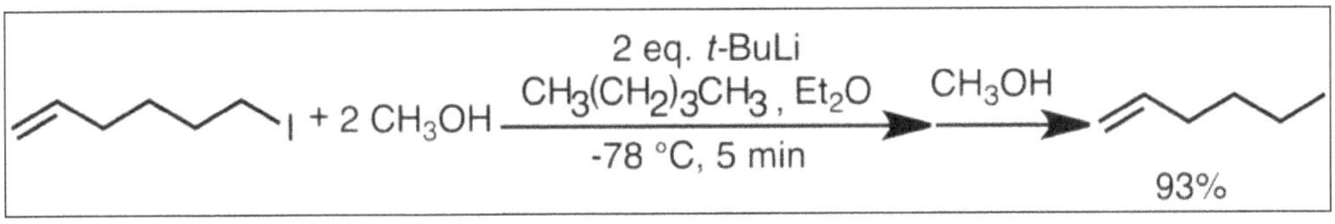

Lithium-halogen exchange is very useful in preparing new organolithium reagents. Exchange rates usually follow the trend I > Br > Cl. Alkyl- and arylfluoride are generally unreactive toward organolithium reagents. Lithium halogen exchange is kinetically controlled, and the rate of exchange is primarily influenced by the stabilities of the carbanion intermediates (sp > sp2 > sp3) of the organolithium reagents. For example, the more basic tertiary organolithium reagents (usually n-butyllithium, sec-butyllithium or tert-butyllithium) are the most reactive, and will react with primary alkyl halide (usually bromide or iodide) to form the more stable organolithium species. Therefore, lithium halogen exchange is most frequently used to prepare vinyl-, aryl- and primary alkyllithium reagents. Lithium halogen exchange is also facilitated when alkoxy groups or heteroatoms are present to stabilize the carbanion, and this method is especially useful for the preparation of functionalized lithium reagents which cannot tolerate the harsher conditions required for reduction with lithium metal. Substrates such as vinyl halides usually undergo lithium-halogen exchange with retention of the stereochemistry of the double bond.

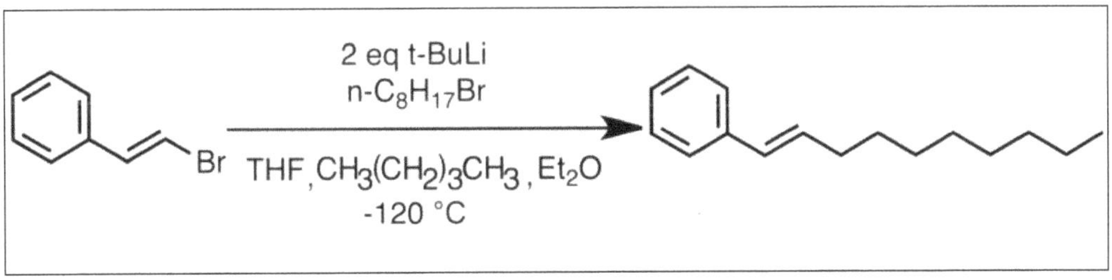

Double bond geomistry is preserved in lithium-halogen exchange.

Below is an example of the use of lithium-halogen exchange in the synthesis of morphine. Here, n-butyllithium is used to perform lithium-halogen exchange with bromide. The nucleophilic carbanion center quickly undergoes carbolithiation to the double bond, generating an anion stabilized by the adjacent sulfone group. An intramolecular S_N2 reaction by the anion forms the cyclic backbone of morphine.

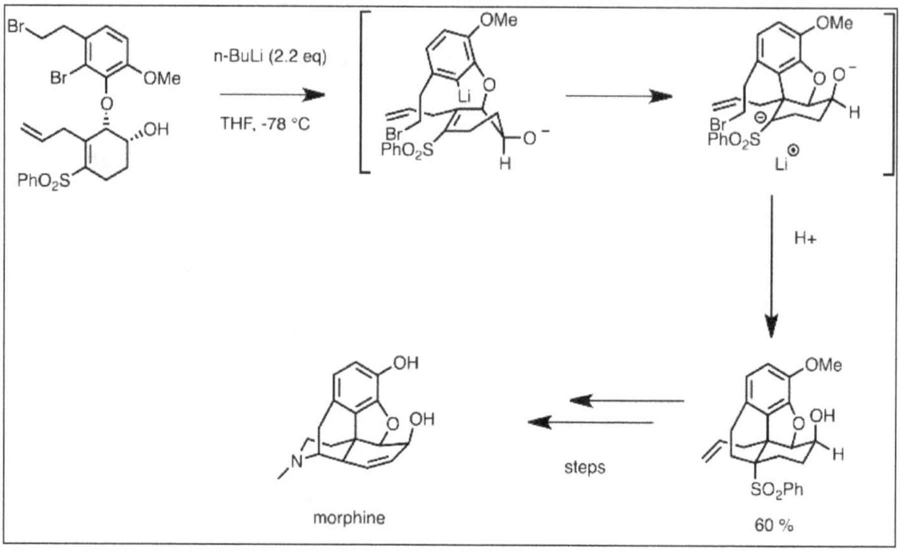

Lithium-halogen exchange in the synthesis of morphine to form tetracyclic core structure.

Lithium halogen exchange is a crucial part of Parham cyclization. In this reaction, an aryl halide (usually iodide or bromide) exchanges with organolithium to form a lithiated arene species. If the arene bears a side chain with an electrophillic moiety, the carbanion attached to the lithium will perform intramolecular nucleophilic attack and cyclize. This reaction is a useful strategy for heterocycle formation. In the example below, Parham cyclization was used to in the cyclization of an isocyanate to form isoindolinone, which was then converted to a nitrone. The nitrone species further reacts with radicals, and can be used as "spin traps" to study biological radical processes.

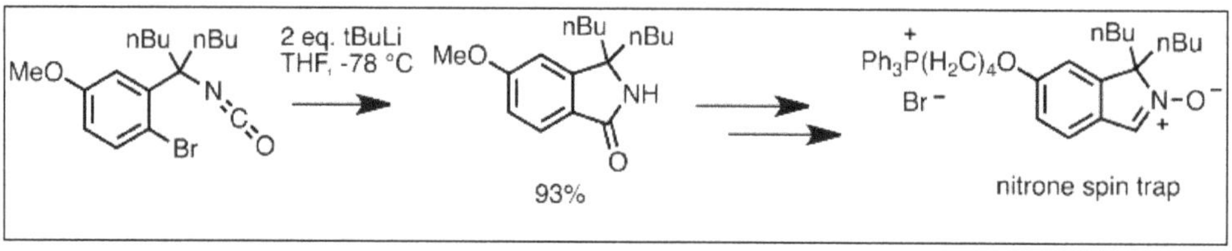

Parham cyclization in the synthesis of a nitrone "spin trap".

Transmetalation

Organolithium reagents are often used to prepare other organometallic compounds by transmetalation. Organocopper, organotin, organosilicon, organoboron, organophosphorus, organocerium and organosulfur compounds are frequently prepared by reacting organolithium reagents with appropriate electrophiles.

$$R - M + n - BuLi \rightarrow R - Li + n - BuM$$

Common types of transmetalation include Li/Sn, Li/Hg, and Li/Te exchange, which are fast at low temperature. The advantage of Li/Sn exchange is that the tri-alkylstannane precursors undergo few side reactions, as the resulting n-Bu_3Sn byproducts are unreactive toward alkyllithium reagents. In the following example, vinylstannane, obtained by hydrostannylation of a terminal alkyne, forms vinyllithium through transmetalation with n-BuLi.

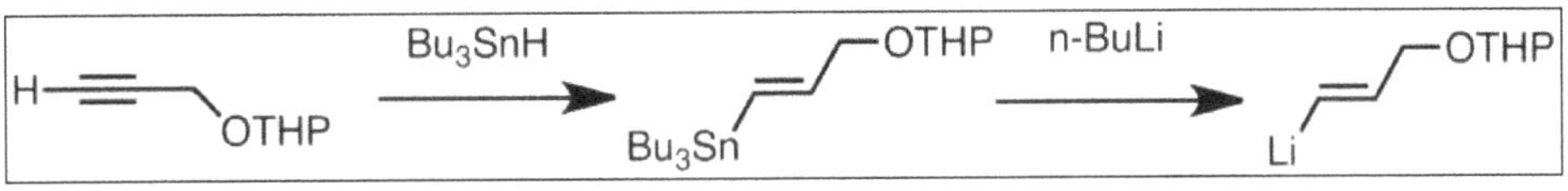

Organolithium can also be used in to prepare organozinc compounds through transmetalation with zinc salts.

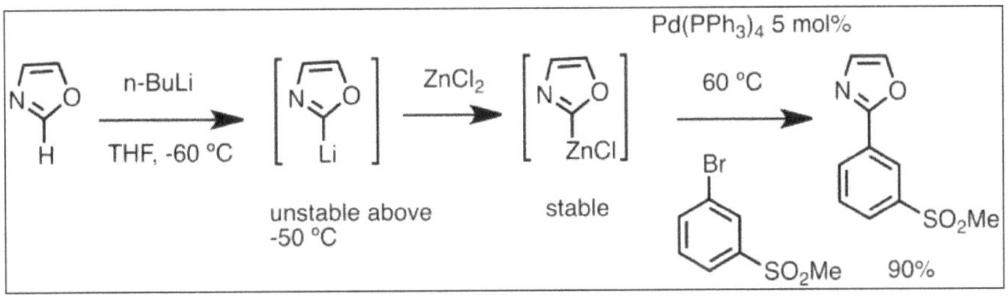

Lithium diorganocuprates can be formed by reacting alkyl lithium species with copper(I) halide. The resulting organocuprates are generally less reactive toward aldehydes and ketones than organolithium reagents or Grignard reagents.

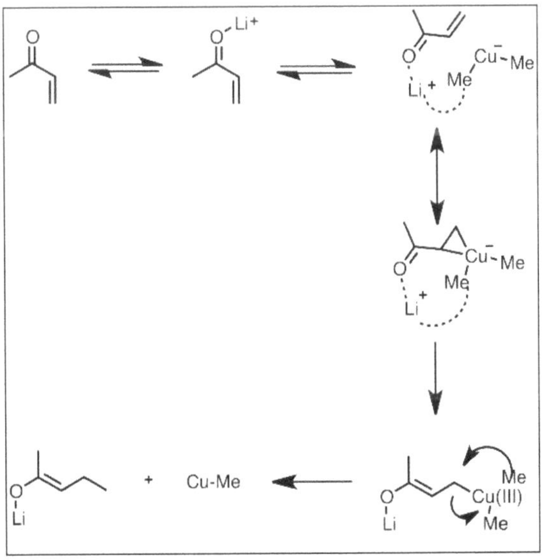

Reaction pathway of 1,4 conjugate addition of organocuprate.

Preparation

Most simple alkyllithium reagents and common lithium amides are commercially available in a variety of solvents and concentrations. Organolithium reagents can also be prepared in the laboratory. Below are some common methods for preparing organolithium reagents.

Reaction with Lithium Metal

Reduction of alkyl halide with metallic lithium can afford simple alkyl and aryl organolithium reagents.

$$R - X + 2Li \rightarrow R - Li + Li - X$$

Industrial preparation of organolithium reagents is achieved using this method by treating the alkyl chloride with metal lithium containing 0.5-2% sodium. The conversion is highly exothermic. The sodium initiates the radical pathway and increases the rate. The reduction proceeds via a radical pathway. Below is an example of the preparation of a functionalized lithium reagent using reduction with lithium metal? Sometimes, lithium metal in the form of fine powders are used in the reaction with certain catalysts such as naphthalene or 4,4'-di-t-butylbiphenyl (DTBB). Another substrate that can be reduced with lithium metal to generate alkyllithium reagents is sulfides. Reduction of sulfides is useful in the formation of functionalized organolithium reagents such as alpha-lithio ethers, sulfides, and silanes.

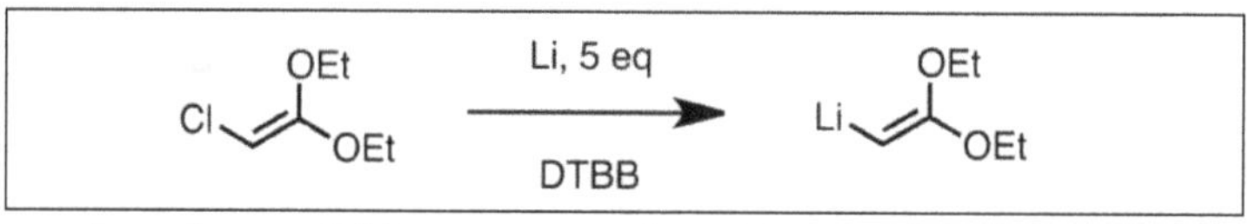

Preparation of organolithium reagents through reduction with lithium metal.

Metalation

A second method of preparing organolithium reagents is a metalation (lithium hydrogen exchange). The relative acidity of hydrogen atoms controls the position of lithiation.

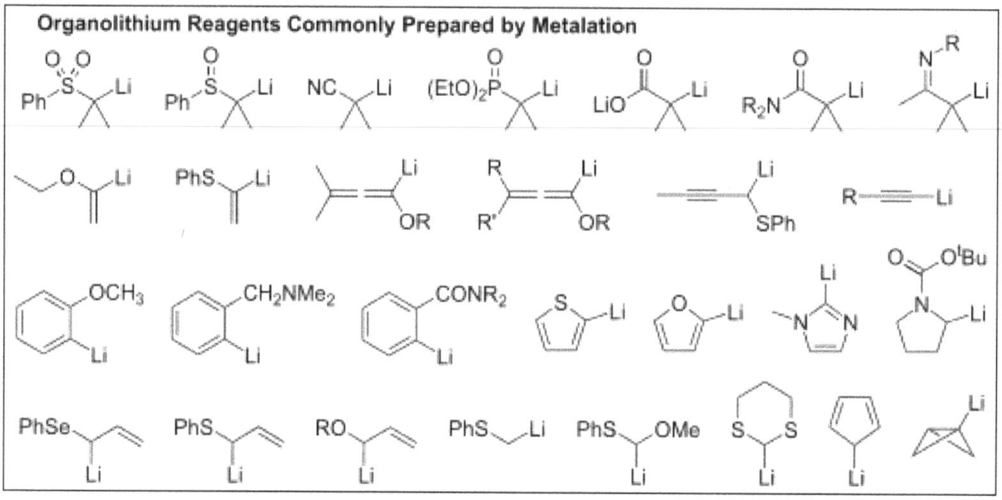

This is the most common method for preparing alkynyllithium reagents, because the terminal hydrogen bound to the sp carbon is very acidic and easily deprotonated. For aromatic compounds, the position of lithiation is also determined by the directing effect of substituent groups. Some of the most effective directing substituent groups are alkoxy, amido, sulfoxide, sulfonyl. Metalation often occurs at the position ortho to these substituents. In heteroaromatic compounds, metalation usually occurs at the position ortho to the heteroatom.

Lithium Halogen Exchange

A third method to prepare organolithium reagents is through lithium halogen exchange.

Tert-butyllithium or n-butyllithium are the most commonly used reagents for generating new organolithium species through lithium halogen exchange. Lithium-halogen exchange is mostly used to convert aryl and alkenyl iodides and bromides with sp2 carbons to the corresponding organolithium compounds. The reaction is extremely fast, and often proceed at -60 to -120 °C.

Transmetalation

The fourth method to prepare organolithium reagents is through transmetalation. This method can be used for preparing vinyllithium.

Shapiro Reaction

In the Shapiro reaction, two equivalents of strong alkyllithium base react with p-tosylhydrazone compounds to produce the vinyllithium, or upon quenching, the olefin product.

Handling

Organolithium compounds are highly reactive species and require specialized handling techniques. They are often corrosive, flammable, and sometimes pyrophoric (spontaneous ignition when exposed to oxygen or moisture). Alkyllithium reagents can also undergo thermal decomposition to form the corresponding alkyl species and lithium hydride. Organolithium reagents are typically stored below 10 °C. Reactions are conducted using air-free techniques. The concentration of alkyllithium reagents is often determined by titration.

Organolithium reagents react, often slowly, with ethers, which nonetheless are often used as solvents.

Table: Approximate half-lives of common lithium reagents in typical solvents.

Solvent	Temp	n-BuLi	s-BuLi	t-BuLi	MeLi	CH_2= C(OEt)-Li	CH_2= C(-$SiMe_3$)-Li
THF	-40 °C			338 min			

THF	-20 °C			42 min			
THF	0 °C	17 h					
THF	20 °C	107 min				>15 h	17 h
THF	35 °C	10 min					
THF/TMEDA	-20 °C	55 h					
THF/TMEDA	0 °C	340 min					
THF/TMEDA	20 °C	40 min					
Ether	-20 °C			480 min			
Ether	0 °C			61 min			
Ether	20 °C	153 h		<30 min			17 d
Ether	35 °C	31 h					
Ether/TMEDA	20 °C	603 min					
DME	-70 °C		120 min	11 min			
DME	-20 °C	110 min	2 min	≪2 min			
DME	0 °C	6 min					

ZEISE'S SALT

This is the first metal complex identified as an organometallic compound $KPtCl_3(C_2H_4)$ obtained from reaction of ethylene with platinum (II) chloride by William Zeise in 1825. The anion of this air-stable, yellow, coordination complex contains an η^2-ethylene ligand. The anion features a platinum atom with a square planargeometry. The salt is of historical importance in the area of organometallic chemistry as one of the first examples of a transition metal alkene complex and is named for its discoverer, William Christopher Zeise. This compound is commercially available as a hydrate. The hydrate is commonly prepared from $K_2[PtCl_4]$ and ethylene in the presence of a catalytic amount of $SnCl_2$. The water of hydration can be removed in vacuo. The alkene C = C bond is approximately perpendicular to the $PtCl_3$ plane. In Zeise's salt and related compounds, the alkene rotates about the metal-alkene bond with modest activation energy. Analysis of the barrier heights indicates that the π-bonding between most metals and the alkene is weaker than the σ-bonding. In Zeise's anion, this rotational barrier has not been assessed. Zeise's salt was one of the first organometallic compounds to be reported. It was discovered by William Christopher Zeise, a professor at the University of Copenhagen, who prepared this compound in 1830 while investigating the reaction of $PtCl_4$ with boiling ethanol. Following careful analysis he proposed that the resulting compound contained ethylene. Justus von Liebig, a highly influential chemist of that era, often criticised Zeise's proposal, but Zeise's proposal was decisively supported in 1868 when Birnbaum prepared the complex using ethylene. Zeise's salt received a great deal of attention during the second half of the 19th century because chemists could not explain its molecular structure. This question remained unanswered until the determination of its X-ray crystal structure in the 20th century. Zeise's salt stimulated much scientific research in the field of organometallic chemistry and would be key in defining new concepts in chemistry. The Dewar–Chatt–Duncanson model explains how the metal is coordinated to the C=C double bond.

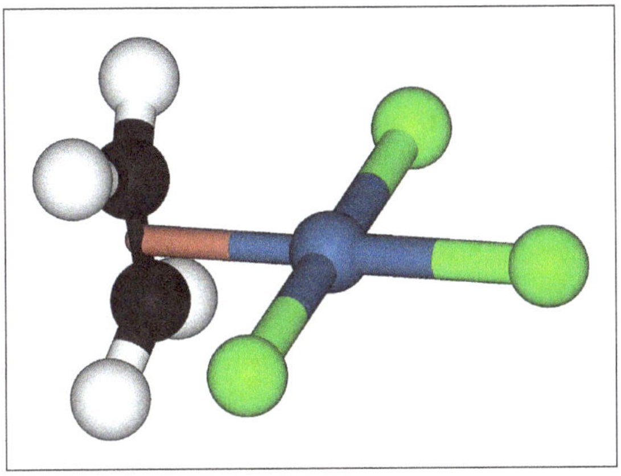

Zeise's Salt Preparation

W. C. Zeise, a professor at the University of Copenhagen was the first person to prepare zeise's salt; he prepared this compound in 1820s while investigating the reaction of PtCl4 with boiling ethanol, and proposed that the resulting compound contained ethylene. In 1868 Birnbaum prepared the complex using ethylene. Zeise's salt compound is now commercially available as a hydrate. Hydrates are inorganic salts "containing water molecules combined in a definite ratio as an integral part of the crystal that are either bound to a metal center or that have crystallized with the metal complex. Such hydrates are also said to contain water of crystallization or water of hydration. If the water is heavy water, where the hydrogen involved is the isotope deuterium, then the term deuterate may be used in place of hydrate. The hydrate is commonly prepared from $K_2[PtCl_4]$ and ethylene in the presence of a catalytic amount of $SnCl_2$. The water of hydration can be removed in vacuo.

Zeise's Salt Properties

The first olefin complex of a platinum group metal to be discovered was Zeise's salt, K[C,H,PtCl,]. These olefin complexes have, however, been the subject of intensive study during the past decade, with the result that not only are large numbers of these compounds known but the structures of many are also understood and applications for them are now being discovered.

In a homogeneous catalytic reaction where the net process is the conversion of anolefin to some other species, the original platinum metal salt or complex may be regarded as the catalyst and the olefm-metal complex as the catalyst-substrate combination.

There is much evidence to suggest that an olefin molecule co-ordinated to a metal atom has a reactivity which is different in kind from the normal reactivity of olefins. The act of co-ordination decreases the electron density between the olefinic carbon atoms, and hence renders the olefin liable to attack by nucleophilic reagents such as hydroxyl or acetate ions, or indeed any electron-rich species which is seeking an electron deficient site. It is not possible to judge at the present how far this hypothesis accounts for the reactions undergone by co-ordinated olefins, although it almost certainly explains the occurrence of oxidation and hydrogenation processes. Study of olefin oxidation, it was observed that if the reaction was stopped before completion the olefin had isomerised and that an equilibrium mixture of isomers was present. This led to the discovery that palladous chloride and its olefin and nitrile complexes catalysed olefin.

Zeise's Salt Structure and Characterization

Zeise's salt $K[PtC_{13}.C_2H_4].H_{20}$ forms monoclinic crystals, a = 10.750_+ 0"006, b = 8.405 _+ 0-003, c 4.836+0-002 Å, fl= 97-73 + 0.06 °, space group P21, with two formula units in the unit cell. In Zeise's salt and related compounds, the alkene rotates about the metal-alkene bond with modest activation energy. Analysis of the barrier heights indicates that the π-bonding between most metals and the alkene is weaker than the σ-bonding. In Zeise's anion, this rotational barrier cannot be assessed by NMR spectroscopy because all four protons are equivalent. Lower symmetry complexes of ethylene, e.g. $CpRh(C_2H_4)_2$, are, however, suitable for analysis of the rotational barriers associated with the metal-ethylene bond. CC-bond of ethylene in Zeise's salt still possesses double bond character but to a lesser degree than free ethylene. This is because the metal donates electrons to the antibonding p*-orbital of the olefin thus reducing the bond order and accordingly, the CC stretching frequency. With increasing strength of the olefin-metal interaction, the metal-carbon bond distance will decrease and the CC-bond distances will increase.

In zeise's salt Pt(II) is the central metal ion which is a square planar complex and the complex is diamagnetic. Central Pt^{2+} ion coordinated with 3 Cl^- ions and 1 ethene molecule in a square planar geometry while the 4 coplanar H's are not coplanar with the 2 C's. Pt—Cl (1) bond length is longer than the Pt—Cl (2) and Pt—Cl (3) bond length. Which can be termed as the Trans effect is and is kinetically controlled. The intensity of the trans effect (as measured by the increase in rate of substitution of the trans ligand) follows this sequence: F^-, H_{20}, OH^- < NH_3 < py < Cl– < Br– < I–, SCN–, NO^{2-}, $SC(NH_2)_2$, Ph^- < $SO^{2-}{}_3$ < PR_3, AsR_3, SR_2, CH^{3-} < H^-, NO, CO, CN^-,C_2H_4. This can be proved with X-Ray Crystallography, and neutron diffraction. The bond stretching frequency for: C = C is 18^{27} cm^{-1} (B.O. = 2), C—C is 16^{50} cm^{-1} (B.O. = 1) and Ethene present in the Zeise's salt 17^{38} cm^{-1} (B.O. = 1.54). The pi-acid alkene donates electron density into a metal d-orbital from π-symmetry bonding orbital between the carbon atoms (II). The metal donates electrons back from (a different) filled d-orbital into the empty π^* anti bonding orbital (I). Both of these effects tend to reduce the carbon-carbon bond order, leading to an elongated C-C distance and a lowering its vibration frequency.

The coordination round the platinumatom is essentially square planar. The deviations from the coordination plane (i. e. the best plane throughthe platinum and three chlorine atoms) are Pt, + 0-002;CI(1), 0.000; C1(2),0-0.015; CI(3), +0-005; centre of C-C bond, -0.15 Å. The ethylene group is bound bya re-type bond to the platinum atom, with the C-C bond approximately perpendicular to the coordination plane. The deviations from ideal symmetry are small, though significant: the C-C bond is inclined at 86° to the coordination plane instead of 90 o, and its mid-point is 0.15 Å below the plane; in addition, C(1) is displaced from the plane perpendicular to the coordination plane and passing through Pt and CI(1), by 0.12 Å. The Pt-C distances are effectively equal, 2.15(2) Å, and are rather longer than 2.08 Å, the distance predicted for a single Pt-C a-bond. The C-C bond-length shows that coordination to the metal atom reduces the bond-order in ethylene from 2 in the uncoordinated molecule to about 1½ in this complex. Similar results were obtained in the substituted acetylene complex [Pt Clz (But C = C But) (ptoluidine)], where coordination reduces the C - C bond order by about ½, and the Pt-C bond [2.19(2) Å] is also longer than expected for a single bond. It may be that olefins bond more strongly to platinum than to palladium.

References

- Organomagnesium-Compounds: encyclopedia2.thefreedictionary.com, Retrieved 29 June, 2020
- Garst, J.F.; Soriaga, M.P. "Grignard reagent Formation", Coord. Chem. Rev. 2004, 248, 623 - 652. doi:10.1016/j.ccr.2004.02.018
- Zeises-salt: azecs.az, Retrieved 02 August, 2020
- Negishi, Ei-ichi (2002). Organometallics In Synthesis: A Manual (Ed. M. Schlosser). Chichester, West Sussex, UK: Wiley. pp. 963–975. ISBN 978-0471984160
- Almennin, A.; Halvorse, S.; Haaland, A. (2005). "Gas Phase Electron Diffraction Investigation of Molecular Structures of Trimethylaluminum Monomer and Dimer". Acta Chemica Scandinavica. 44 (15): 2232–2234. doi:10.3891/acta.chem.scand.25-1937

4

Organometallic Compounds of Transition Metals

Organometallic compounds of transition metals belong to the d-block and f-block and follow similar rules as the main-group organometallics. Organotin compounds, organoiron compounds, organocobalt compounds, organotitanium compounds, etc. come under it. All the organometallic compounds of transition metals have been carefully analyzed in this chapter.

ORGANOTIN COMPOUNDS

Organotin compounds are chemicals with at least one covalent Sn–C bond. They are widely used as polyvinyl chloride (PVC) stabilizers, biocides, or antifouling paints and have given rise to ubiquitous environmental contamination. They are known to be immunotoxic, to induce thymus atrophy in experimental animals, and to decrease proliferation of human lymphocytes. Apoptosis by organotin compounds is apparently mediated by caspase activation triggered via the mitochondrial pathway and the death receptor pathway. Like the pleiotropic effects of PHAHs the effects of organotin differ depending on the cell type and on the exposure regimen. In natural killer cells, for instance, the MAP kinases p38 and p44/42 are activated by the organotin tributyltin. Consideration of the dose effect is highlighted in organotins, as different doses have strikingly different outcomes, ranging from necrosis and apoptosis to subtle changes in cytokine secretion by T helper cells.

Organotin compounds are widely applied as pesticides; as preservatives of wood, paper, textiles, leather, and glass; in heat/light protection of PVC plastics; and in antifouling paints. The use of dialkyltin compounds can cause hepatotoxicity, whereas trialkyltin compounds can cause neurotoxicity. Toxicity is greater for substances with alkyl groups of shorter chain length.

The effect of organotins on the immune system has been evaluated for di- or trialkyltin compounds, either in the form of oxides or chlorides. Rats treated daily with oral tri-n-butyltinoxide (TBTO) for 4–6 weeks or longer develop lymphopenia, and have reductions in weights of thymus and spleen. This immunosuppression is accompanied by reduced T-cell-mediated immune responses, including delayed-type hypersensitivity reactions, graft-versus-host reactivity, and T_H activity in antibody formation after immunization with immunogens like ovalbumin. In vivo, resistance to the nematode Trichinella spiralis is decreased. Serum IgG concentrations are decreased, but increased serum IgM levels have been noted. Natural killer cell activity is decreased, and macrophage phagocytosis is also affected by di-n-octyltindichloride (DOTC), dipropyltinchloride, and TBTO.

Other toxic effects of organotin compounds, primarily on the developing immune system, indicate a primary effect on the thymus. For both short-term and long-term exposures, 1-year-old rats require higher doses of TBTO than 4- to 6-week-old animals to manifest similar systemic effects, such as decreased resistance to Trichinella spiralis and Listeria monocytogenes.

Rats are more susceptible than mice and guinea pigs to the toxic action of organotin compounds on the thymus. Adrenalectomy has no effect on organotin-induced thymic involution, and growth hormone treatment after hypophysectomy has no effect on recovery after organotin-induced thymic atrophy. These data provide additional evidence for the absence of a glucocorticosteroid-mediated mechanism. Thymic atrophy is observed for both tri- and diorganotin compounds. Since TBTO and tri-n-butyltinchloride (TBTC) are rapidly dealkylated after oral administration, the atrophy may in fact be induced by the dialkyl metabolite.

In vitro and in vivo studies have both shown that tri-organotin compounds directly inhibit thymocyte viability and proliferation. Bone marrow seems unaffected. The antiproliferative effect is associated with a reduction in the capacity to form rosettes with sheep erythrocytes. This rosetting capacity is mediated by the CD2 molecule on the thymocyte surface, which has a transmembrane signaling function in early thymocyte activation (with leukocyte function-associated molecule LFA-3 on the stroma as the ligand). Another factor involved in early thymocyte activation is Il-2, the synthesis of which is down-regulated at the mRNA level by DOTC. These data indicate that a blockade of intrathymic T-cell maturation is induced by organotin compounds.

The presence of organotin compounds as pollutants in the aquatic environment has initiated some toxicity studies in fish species. In Poecilia reticulata (guppy), the thymus is among the organs most sensitive to TBTO (exposure to water levels above 0.03 μg/liter during 1–3 month exposure studies) and DBTC (at levels of 320 μg/liter or higher). The thymic atrophy is species-dependent; it is present in the guppy, but not in Oryzias latipes (medaka). Additional studies show that TBTO also affects the thymus of Platichthys flesus (flounder), as shown by morphometric analysis.

A number of mechanisms have been suggested to explain the effects just mentioned. The toxic effects of dialkyltin substances can be reversed or prevented by dithiol compounds, which points to the possible involvement of sulfhydryl groups in toxicity. These groups are pivotal in membrane-related functions, such as secretion, phagocytosis, transport, cytoskeletal function, cell-to-cell contact, and transmembrane signaling. It is, therefore, not surprising that the immune system, especially the thymus, is very susceptible to sulfhydryl-reactive compounds, which include p-benzoquinone, cytochalasin A, heavy metals, and organotin compounds. Other proposed mechanisms of toxicity include interference with mitochondrial metabolism and glycolytic pathway, and decreased cAMP production. Synthesis of both RNA and protein is also inhibited by organotin compounds. The increased sensitivity of thymocytes over hepatocytes may be related to the unfavorable pattern of purine metabolic enzymes in immature thymocytes, making these cells prone to "suicide" after activation.

Organotins with four organic groups, e.g. tetrabutyltin, tetraoctyltin and tetraphenyltin, can be used as starting materials or catalysts. Organotins with three organic groups can be powerful fungicides and bactericides, depending on the organic group R. Tributyltins (e.g. TBTO, Incidin) are industrial biocides used in antifouling paints and in wood treatment and preservation. They find use as disinfectants and agents for destroying molluscs such as snails (molluscicides). Other uses include antifungal action in textiles and industrial water systems such as cooling tower and refrigeration water systems, wood pulp and paper mill systems, and breweries. They can also be used for the control of schistosomiasis (a chronic tropical disease). Triphenyltins (e.g. Fentin, Tinmate, Brestanol) are used as agricultural fungicides and in antifouling paints. Other organotins in this class include Plictran, Bay Bue 1452, Vendex and Torque which are used as miticides/acaricides (substances used for killing mites or ticks). Organotins with two organic groups, e.g. dimethyltin, dibutyltin and diestertins, can be used as PVC heat stabilisers, as catalysts, in polyurethane formation and silicon curing, and as heat stabilisers for rigid food-packaging PVC. Organotins with one organic group, e.g. methyltin, butyltin, octyltin and monoestertins, are employed as PVC heat stabilisers. Common uses for selected organotin compounds follow. Tributyltin oxide is used as a fungicide and pesticide in timber treatments. It has been used in antifouling paints but this application is now banned by international protocol. Triphenyltin acetate is used as an insect control agent and a fungicide. Triphenyltin chloride is used as a biocide and an intermediate. Triphenyltin hydroxide is used as a chemical to sterilise insects and as a fungicide. Fenbutatin oxide is used for the control of a wide range of mites which feed on plants. Azocyclotin is a long-acting acaricide for the control of spider mites on fruits (including citrus), vines, hops, cotton, vegetables and ornamentals. Cyhexatin is used as an acaricide/miticide.

Physical Properties

Organotin compounds vary in their physical properties. Properties of selected organotins follow:

- Tributyltin oxide is a thin colourless to yellow liquid with a weak odour. Its specific gravity is 1.17.
- Triphenyltin acetate is an off-white crystalline solid. It melts at 124-126 °C.
- Triphenyltin chloride is a white crystalline solid which melts at 103-106 °C.
- Triphenyltin hydroxide is an off-white powder which melts at 124-126 °C.

- Fenbutatin oxide is a white crystalline solid which is extremely stable to heat, light and atmospheric oxygen. It melts at 138-139°C and its specific gravity is 1.29-1.33.
- Azocyclotin is a colourless crystalline solid which melts at 219 °C.
- Cyhexatin is a white crystalline solid that melts at 195-198 °C.

Chemical Properties

Organotin compounds vary in their chemical and biological properties. Most commercial organotins have very low water solubility. Some general properties for different classes of organotins follow:

- Tetraorganotins are very stable. They are ineffective as biocides (substances that are capable of killing living organisms) and are relatively non-toxic. However, they can be slowly decomposed or metabolised to more toxic triorganotin compounds.
- Triorganotin compounds are generally more toxic than organotins in other classes. Trialkyltins with linear organic groups cannot be used as agricultural biocides due to their high toxicity to plants (phytotoxicity). Diorganotins show no antifungal activity. Their antibacterial and toxic activity is low, except for the diphenyl derivatives.
- Monoorganotins show no biocidal activity and have very low toxicity to mammals.

Major environmental implications are related to triorganotins used as agricultural and industrial biocides and to a lesser extent to the application of some diorganotin and monoorganotin compounds in rigid PVC. Properties for selected organotin compounds follow:

- Solubility of tributyltin oxide in water is less than 20 ppm. It mixes with organic solvents.
- Triphenyltin acetate does not mix with water. It is soluble in ether.
- Triphenyltin chloride reacts with water.
- Triphenyltin hydroxide does not mix with water.
- Fenbutatin oxide is a triorganotin compound and is almost insoluble in water (0.005 milligram/L at 23 °C). It is soluble in acetone, benzene and dichloromethane.
- Azocyclotin is a triorganotin compound. Its solubility in water is less than 1 ppm.
- Cyhexatin is a triorganotin compound. Its solubility in water is less than 1 ppm at 25 °C. It is slightly soluble in acetone and soluble in methanol and chloroform.

ORGANOIRON COMPOUNDS

Organoiron chemistry is the chemistry of iron compounds containing a carbon-to-iron chemical bond. Organoiron compounds are relevant in organic synthesis as reagents such as iron pentacarbonyl, diiron nonacarbonyl and disodium tetracarbonylferrate. Iron adopts oxidation states from Fe(−II) through to Fe(VII). Although iron is generally less active in many catalytic applications, it is less expensive and "greener" than other metals. Organoiron compounds feature a wide range of ligands that support the Fe-C bond; as with other organometals, these supporting ligands prominently include phosphines, carbon monoxide, and cyclopentadienyl, but hard ligands such as amines are employed as well.

Iron(0) and more Reduced States

Carbonyl Complexes

Important iron carbonyls are the three neutral binary carbonyls, iron pentacarbonyl, diiron nonacarbonyl, and triiron dodecacarbonyl. One or more carbonyl ligands in these compounds can be replaced by a variety of other ligands including alkenes and phosphines. An iron(-II) complex, disodium tetracarbonylferrate ($Na_2[Fe(CO)_4]$), also known as "Collman's Reagent," is prepared by reducing iron pentacarbonyl with metallic sodium. The highly nucleophilic anionic reagent can be alkylated and carbonylated to give the acyl derivatives that undergo protonolysis to afford aldehydes:

$$\mathrm{LiFe(CO)_4(C(O)R) + H^+ \rightarrow RCHO\ (+\ iron\ containing\ products)}$$

Similar iron acyls can be accessed by treating iron pentacarbonyl with organolithium compounds:

$$\mathrm{ArLi + Fe(CO)_5 \rightarrow LiFe(CO)_4C(O)R}$$

In this case, the carbanion attacks a CO ligand. In a complementary reaction, Collman's reagent can be used to convert acyl chlorides to aldehydes. Similar reactions can be achieved with $[HFe(CO)_4]^-$ salts.

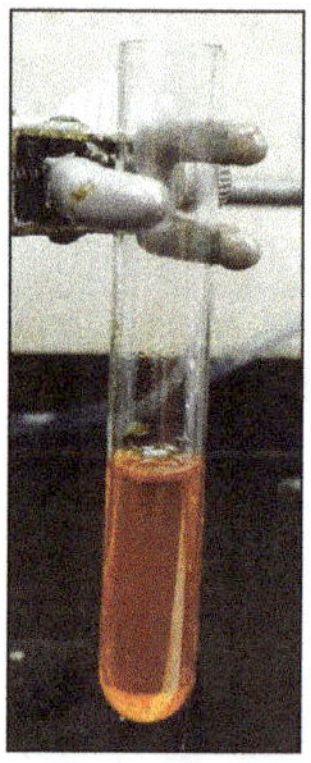

Iron pentacarbonyl.

Alkene-Fe(o)-CO Derivatives

Monoalkenes

Iron pentacarbonyl reacts photochemically with alkenes to give $Fe(CO)_4$(alkene).

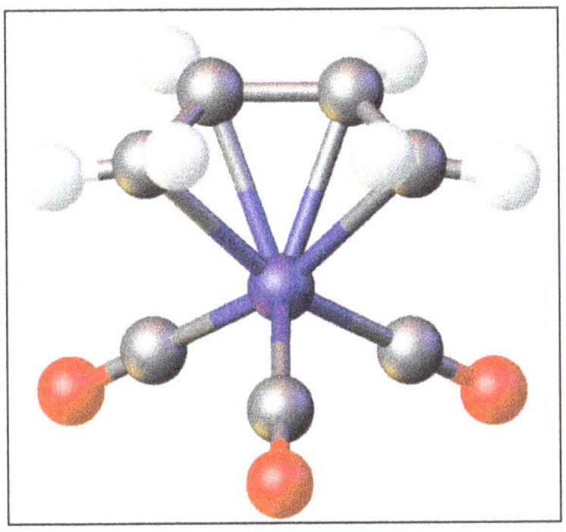

(Butadiene) iron tricarbonyl.

Diene-Fe(o)-CO Derivatives

Iron diene complexes are usually prepared from $Fe(CO)_5$ or $Fe_2(CO)_9$. Derivatives are known for common dienes like cyclohexadiene, norbornadiene and cyclooctadiene, but even cyclobutadiene can be stabilized. In the complex with butadiene, the diene adopts a cis-conformation. Iron carbonyls are used as a protective group for dienes in hydrogenations and Diels-Alder reactions. Cyclobutadieneiron tricarbonyl is prepared from 3,4-dichlorocyclobutene and $Fe_2(CO)_9$.

Cyclohexadienes, many derived from Birch reduction of aromatic compounds, form derivatives (diene)$Fe(CO)_3$. The affinity of the $Fe(CO)_3$ unit for conjugated dienes is manifested in the ability of iron carbonyls catalyse the isomerisations of 1,5-cyclooctadiene to 1,3-cyclooctadiene. Cyclohexadiene complexes undergo hydride abstraction to give cyclohexadienyl cations, which add nucleophiles. Hydride abstraction from cyclohexadiene iron(o) complexes gives ferrous derivatives.

The enone complex (benzylideneacetone)iron tricarbonyl serves as a source of the Fe(-CO)$_3$ subunit and is employed to prepare other derivatives. It is used similarly to $Fe_2(CO)_9$.

Alkyne-Fe(o)-CO Derivatives

Alkynes react with iron carbonyls to give a large variety of derivatives. Derivatives include ferroles ($Fe_2(C_4R_4)(CO)_6$), (p-quinone)$Fe(CO)_3$, (cyclobutadiene)$Fe(CO)_3$ and many others.

Tri- and Polyene Fe(o) Complexes

Stable iron-containing complexes with and without CO ligands are known for a wide variety of polyunsaturated hydrocarbons, e.g. cycloheptatriene, azulene, and bullvalene.

In the case of cyclooctatetraene (COT), derivatives include $Fe(COT)_2$, $Fe_3(COT)_3$, and several mixed COT-carbonyls (e.g. $Fe(COT)(CO)_3$ and $Fe_2(COT)(CO)_6$).

Iron(I) and Iron(II)

As Fe(II) is a common oxidation state for Fe, many organoiron(II) compounds are known. Fe(I) compounds often feature Fe-Fe bonds, but exceptions occur, such as $[Fe(anthracene)_2]^-$.

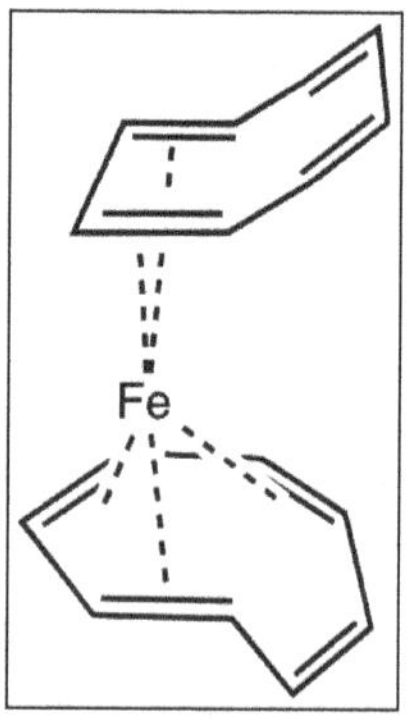

Bis(cyclooctatetraene)iron is an Fe(o) complex lacking CO ligands.

Fp_2, Fp^- and Fp^+ and Derivatives

$Fe(CO)_5$ reacts with cyclopentadiene to give the cyclopentadienyliron dicarbonyl dimer ($[FeCp(CO)_2]_2$), called Fp_2. Pyrolysis of Fp_2 gives the cuboidal cluster $[FeCp(CO)]_4$.

Reduction of Fp_2 with sodium gives "NaFp", containing a potent nucleophile and precursor to many derivatives of the type $CpFe(CO)_2R$. The derivative $[FpCH_2S(CH_3)_2]^+$ has been used in cyclopropanations. The complex $Cp(CO^2)Fe^+(\eta^2\text{-vinyl ether}]^+$ is a masked vinyl cation.

Fp-R compounds are prochiral, and studies have exploited the chiral derivatives $CpFe(PPh_3)(CO)acyl$.

Alkyl, Allyl and Aryl Compounds

The simple peralkyl and peraryl complexes of iron are less numerous than are the Cp and CO derivatives. One example is tetramesityldiiron.

Compounds of the type $[(\eta^3\text{-allyl})Fe(CO)_4]^+X^-$ are allyl cation synthons in allylic substitution. In contrast, compounds of the type $[(\eta^5\text{-}C_5H_5)Fe(CO)_2(CH_2CH = CHR)]$ possessing η^1-allyl groups are analogous to main group allylmetal species (M = B, Si, Sn, etc.) and react with carbon electrophiles to give allylation products with S_E2' selectivity. Similarly, allenyl(cyclopentadienyliron) dicarbonyl complexes exhibit reactivity analogous to main group allenylmetal species and serve as nucleophilic propargyl synthons.

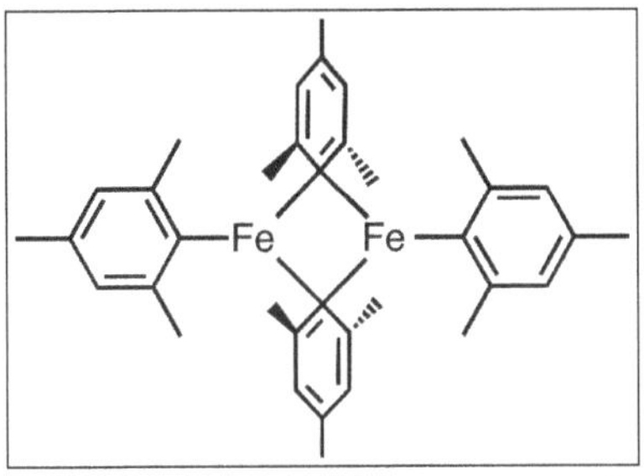

Tetramesityldiiron is a rare example of a neutral per-organo complex of iron.

Sulfur and Phosphorus Derivatives

Complexes of the type $Fe_2(SR)_2(CO)_6$ and $Fe_2(PR_2)_2(CO)_6$ form, usually by the reaction of thiols and secondary phosphines with iron carbonyls. The thiolates can also be obtained from the tetrahedrane $Fe_2S_2(CO)_6$.

Iron(III)

Some organoiron(III) compounds are prepared by oxidation of organoiron(II) compounds. A long-known example being ferrocenium $[(C_5H_5)_2Fe]^+$. Organoiron(III) porphyrin complexes are numerous.

Iron(IV)

In Fe(norbornyl)$_4$, Fe(IV) is stabilized by an alkyl ligand that resists beta-hydride elimination. Two-electron oxidation of decamethylferrocene gives the dication $[Fe(C_5Me_5)_2]^{2+}$, which forms a carbonyl complex, $[Fe(C_5Me_5)_2(CO)](SbF_6)_2$.

Organoiron Compounds in Organic Synthesis and Homogeneous Catalysis

In industrial catalysis, iron complexes are seldom used in contrast to cobalt and nickel. Because of low cost and low toxicity of its salts, iron is attractive as a stoichiometric reagent. Some areas of investigation include:

- Hydrogenation and reduction, example catalyst Knölker complex.
- Cross-coupling reactions. Iron compounds such as Fe(acac)$_3$ catalyze a wide range of cross-coupling reactions with one substrate an aryl or alkyl Grignard and the other substrate an aryl, alkenyl (vinyl), or acyl organohalide. In the related Kumada coupling the catalysts are based on palladium and nickel.
- Complexes derived from Schiff bases are active catalysts for olefin polymerization.

Biochemistry

In the area of bioorganometallic chemistry, organoiron species are found at the active sites of the three hydrogenase enzymes as well as carbon monoxide dehydrogenase.

Ferrocene

Ferrocene is an organometallic compound with the formula $Fe(C_5H_5)_2$. The molecule consists of two cyclopentadienyl rings bound on opposite sides of a central iron atom. It is an orange solid with a camphor-like odor that sublimes above room temperature, and is soluble in most organic solvents. It is remarkable for its stability: it is unaffected by air, water, strong bases, and can be heated to 400 °C without decomposition. In oxidizing conditions it can reversibly react with strong acids to form the ferrocenium cation $Fe(C_5H_5)^+_2$.

The rapid growth of organometallic chemistry is often attributed to the excitement arising from the discovery of ferrocene and its many analogues (metallocenes).

Structure and Bonding

Mössbauer spectroscopy indicates that the iron center in ferrocene should be assigned the +2 oxidation state. Each cyclopentadienyl (Cp) ring should then be allocated a single negative charge. Thus ferrocene could be described as iron(II) bis(cyclopentadienide), $Fe^{2+}[C_5H^-_5]_2$.

The number of π-electrons on each ring is then six, which makes it aromatic according to Hückel's rule. These twelve π-electrons are then shared with the metal via covalent bonding. Since Fe^{2+} has six d-electrons, the complex attains an 18-electron configuration, which accounts for its stability. In modern notation, this sandwich structural model of the ferrocene molecule is denoted as $Fe(\eta^5\text{-}C_5H_5)_2$.

The carbon–carbon bond distances around each five-membered ring are all 1.40 Å, and the Fe–C bond distances are all 2.04 Å. From room temperature down to 164K, X-ray crystallography yields the monoclinic space group; the cyclopentadienide rings are a staggered conformation, resulting in a centrosymmetric molecule, with symmetry group D_{5d}. However, Below 110 K, ferrocene crystallizes in an orthorhombic crystal lattice in which the Cp rings are ordered and eclipsed, so that the molecule has symmetry group D_{5h}. In the gas phase, electron diffraction and computational studies show that the Cp rings are eclipsed.

The Cp rings rotate with a low barrier about the $Cp_{(centroid)}\text{–}Fe\text{–}Cp_{(centroid)}$ axis, as observed by measurements on substituted derivatives of ferrocene using ^1H and ^{13}C nuclear magnetic resonance spectroscopy. For example, methylferrocene ($CH_3C_5H_4FeC_5H_5$) exhibits a singlet for the C_5H_5 ring.

Synthesis

Grignard Reagent

The first reported syntheses of ferrocene were nearly simultaneous. Pauson and Kealy synthesised ferrocene using iron(III) chloride and a Grignard reagent, cyclopentadienyl magnesium bromide. Iron(III) chloride is suspended in anhydrous diethyl ether and added to the Grignard reagent. A redox reaction occurs, forming the cyclopentadienyl radical and iron(II) ions. Dihydrofulvalene is produced by radical-radical recombination while the iron(II) reacts with the Grignard reagent to form ferrocene. Oxidation of dihydrofulvalene to fulvalene with iron(III), the outcome sought by Kealy and Pauson, does not occur.

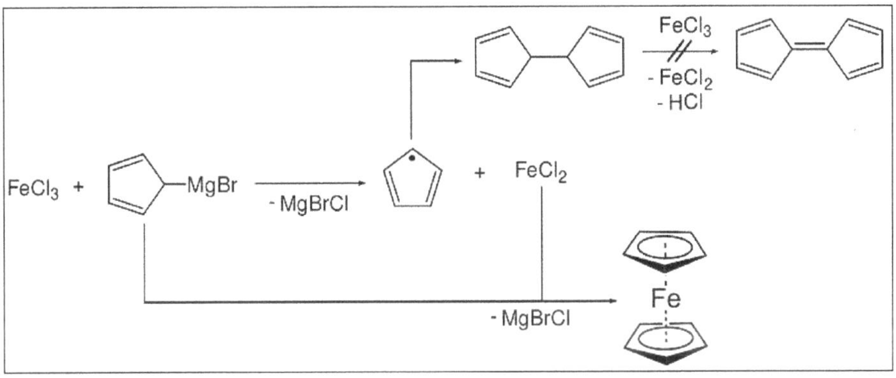

Gas-metal Reaction

The other early synthesis of ferrocene was by Miller, who reacted metallic iron directly with gas-phase cyclopentadiene at elevated temperature. An approach using iron pentacarbonyl was also reported.

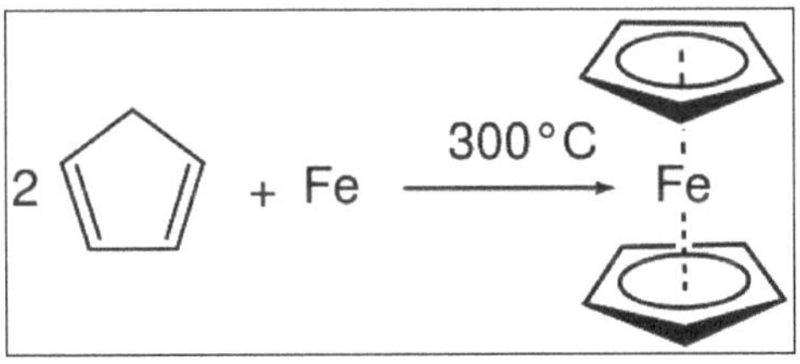

$$Fe(CO)_5 + 2C_5H_6(g) \rightarrow Fe(C_5H_5)_2 + 5CO(g) + H_2(g)$$

Alkali Cyclopentadienide

More efficient preparative methods are generally a modification of the original transmetalation sequence using either commercially available sodium cyclopentadienide or freshly cracked cyclopentadiene deprotonated with potassium hydroxide and reacted with anhydrous iron(II) chloride in ethereal solvents.

Modern modifications of Pauson and Kealy's original Grignard approach are known:

- Using sodium cyclopentadienide: $2\,NaC_5H_5 + FeCl_2 \rightarrow Fe(C_5H_5)_2 + 2\,NaCl$.
- Using freshly-cracked cyclopentadiene: $FeCl_2 \cdot 4H_2O + 2\,C_5H_6 + 2\,KOH \rightarrow Fe(C_5H_5)_2 + 2\,KCl + 6\,H_2O$.
- Using an iron(II) salt with a Grignard reagent: $2\,C_5H_5MgBr + FeCl_2 \rightarrow Fe(C_5H_5)_2 + 2\,MgBrCl$.

Even some amine bases (such as diethylamine) can be used for the deprotonation, though the reaction proceeds more slowly than when using stronger bases:

$$2\,C_5H_6 + 2\,(CH_3CH_2)_2NH + FeCl_2 \rightarrow Fe(C_5H_5)_2 + 2\,(CH_3CH_2)_2NH_2Cl$$

Direct transmetalation can also be used to prepare ferrocene from other metallocenes, such as manganocene:

$$FeCl_2 + Mn(C_5H_5)_2 \rightarrow MnCl_2 + Fe(C_5H_5)_2$$

Properties

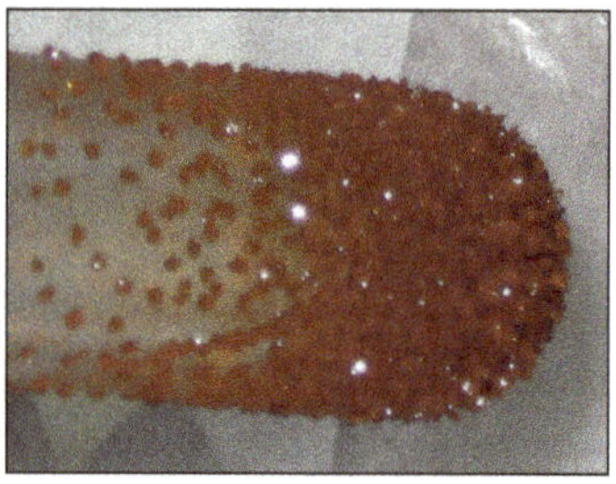

Crystals of ferrocene after purification by vacuum sublimation.

Ferrocene is an air-stable orange solid with a camphor-like odor. As expected for a symmetric, uncharged species, ferrocene is soluble in normal organic solvents, such as benzene, but is insoluble in water. It is stable to temperatures as high as 400 °C.

Ferrocene readily sublimes, especially upon heating in a vacuum. Its vapor pressure is about 1 Pa at 25 °C, 10 Pa at 50 °C, 100 Pa at at 80 °C, 1000 Pa at 116 °C, and 10,000 Pa (nearly 0.1 atm) at 162 °C.

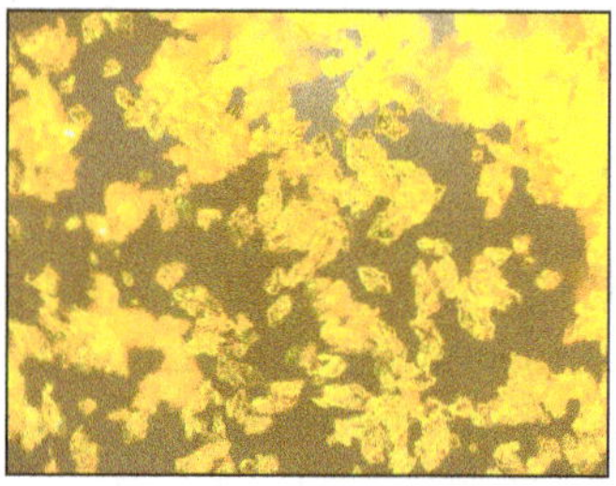

Close-up of ferrocene crystals.

Reactions

Electrophiles

Ferrocene undergoes many reactions characteristic of aromatic compounds, enabling the preparation of substituted derivatives. A common undergraduate experiment is the Friedel–Crafts reaction of ferrocene with acetic anhydride (or acetyl chloride) in the presence of phosphoric acid as a catalyst.

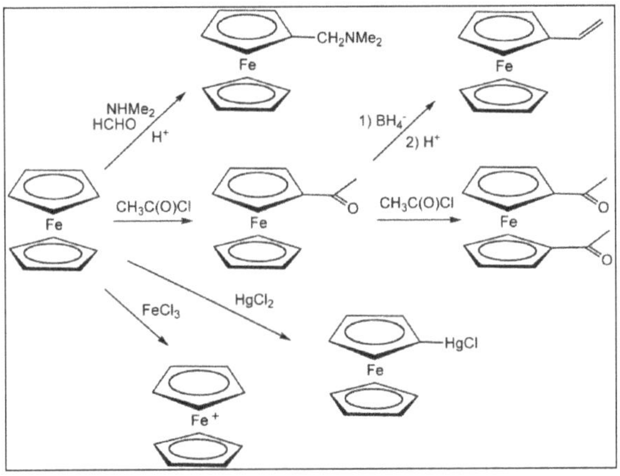

Important reactions of ferrocene with electrophiles and other reagents.

Protonation of ferrocene allows isolation of $[Cp_2FeH]PF_6$.

In the presence of aluminium chloride Me_2NPCl_2 and ferrocene react to give ferrocenyl dichlorophosphine, whereas treatment with phenyldichlorophosphine under similar conditions forms P,P-diferrocenyl-P-phenyl phosphine.

Ferrocene reacts with P4S10 forms a diferrocenyl-dithiadiphosphetane disulfide.

Lithiation

Ferrocene reacts with butyllithium to give 1,1′-dilithioferrocene, which is a versatile nucleophile. Tert-Butyllithium produces monolithioferrocene. Dilithioferrocene reacts with S_8, chlorophosphines, and chlorosilanes. The strained compounds undergo ring-opening polymerization.

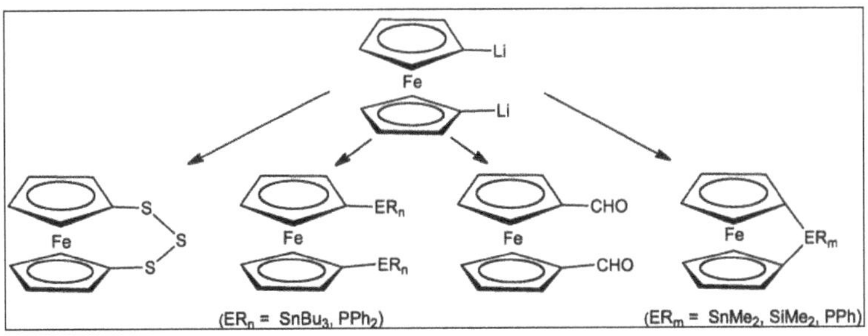

Some transformations of dilithioferrocene.

The phosphine ligand 1,1'-bis(diphenylphosphino)ferrocene (dppf) is prepared from dilithioferrocene.

Redox Chemistry – The Ferrocenium Ion

Ferrocene undergoes a one-electron oxidation at around 0.5 V versus a saturated calomel electrode (SCE). This reversible oxidation has been used as standard in electrochemistry as Fc+/Fc = 0.40 V versus the standard hydrogen electrode. Ferrocenium tetrafluoroborate is a common reagent.

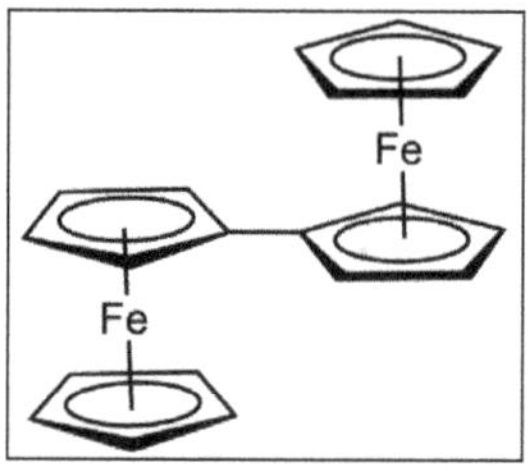

The one-electron oxidized derivative of biferrocene has attracted much research attention.

Substituent on the cyclopentadienyl ligands alters the redox potential in the expected way: electron-withdrawing groups such as a carboxylic acid shift the potential in the anodic direction (i.e. made more positive), whereas electron-releasing groups such as methyl groups shift the potential in the cathodic direction (more negative). Thus, decamethylferrocene is much more easily oxidised than ferrocene and can even be oxidised to the corresponding dication. Ferrocene is often used as an internal standard for calibrating redox potentials in non-aqueous electrochemistry.

Stereochemistry of Substituted Ferrocenes

Disubstituted ferrocenes can exist as either 1,2-, 1,3- or 1,1'- isomers, none of which are interconvertible. Ferrocenes that are asymmetrically disubstituted on one ring are chiral – for example [CpFe(EtC$_5$H$_3$Me)]. This planar chirality arises despite no single atom being a stereogenic centre. The substituted ferrocene shown at right (a 4-(dimethylamino)pyridine derivative) has been shown to be effective when used for the kinetic resolution of racemic secondary alcohols.

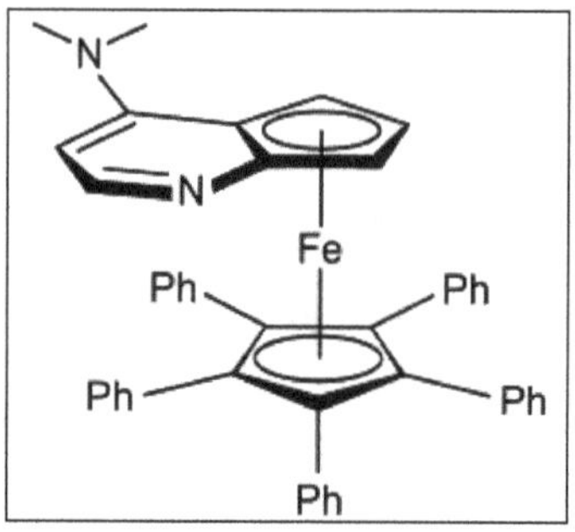

A planar chiral ferrocene derivative.

Applications of Ferrocene and its Derivatives

Ferrocene and its numerous derivatives have no large-scale applications, but have many niche uses that exploit the unusual structure (ligand scaffolds, pharmaceutical candidates), robustness (anti-knock formulations, precursors to materials), and redox (reagents and redox standards).

Ligand Scaffold

Chiral ferrocenyl phosphines are employed as ligands for transition-metal catalyzed reactions. Some of them have found industrial applications in the synthesis of pharmaceuticals and agrochemicals. For example, the diphosphine 1,1′-bis(diphenylphosphino)ferrocene (dppf) is a valued for palladium-coupling reactions and Josiphos ligand is useful for hydrogenation catalysis. They are named after the technician who made the first one, Josi Puleo.

Fuel Additives

Ferrocene and its derivatives are antiknock agents used in the fuel for petrol engines; they are safer than tetraethyllead, previously used. Petrol additive solutions containing ferrocene can be added to unleaded petrol to enable its use in vintage cars designed to run on leaded petrol. The iron-containing deposits formed from ferrocene can form a conductive coating on the spark plug surfaces. What is more, ferrocene polyglycol copolymer, prepared by effecting a polycondensation reaction between a ferrocene derivative and a substituted dihydroxy alcohol, has especially promising applications as a component of rocket propellants. In particular, these copolymers provide the rocket propellants with heat stability, serving as a propellant binder and controlling the burn rate of the propellant.

In a similar light, ferrocene also has been found to be effective at reducing the smoke and sulfur trioxide produced when burning coal. The addition by any practical means, impregnating the coal or simply adding ferrocene to the combustion chamber, can significantly cut down the amount of these undesirable byproducts, even with a small amount of the metal cyclopentadienyl compound.

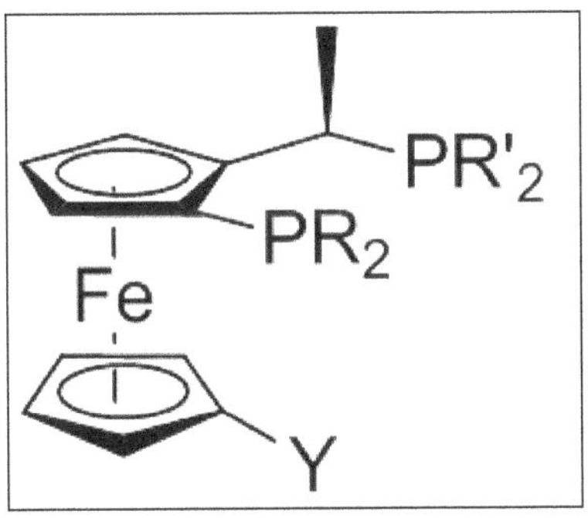

Josiphos ligand.

Pharmaceutical

Ferrocene derivatives have been investigated as drugs. Only one drug has entered clinic trials, Ferroquine (7-chloro-N-(2-((dimethylamino)methyl)ferrocenyl)quinolin-4-amine), an antimalarial. Ferrocene-containing polymer-based drug delivery systems have been investigated.

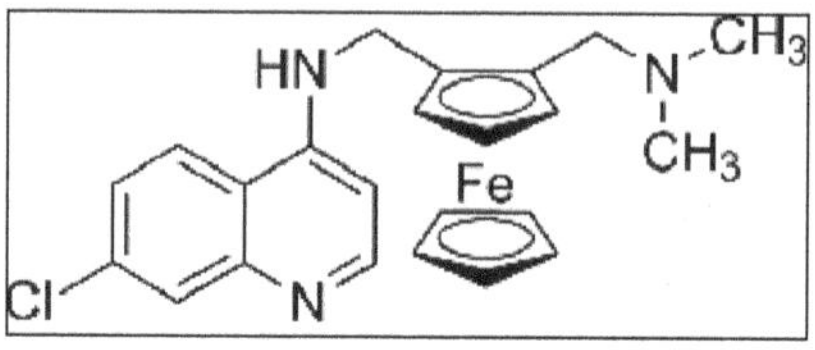

Ferroquine.

The anticancer activity of ferrocene derivatives was first investigated in the late 1970s, when derivatives bearing amine or amide groups were tested against lymphocytic leukemia. Some ferrocenium salts exhibit anticancer activity, but no compound has seen evaluation in the clinic. In particular, ferrocene derivatives have strong inhibitory activity against human lung cancer cell line A549, colorectal cancer cell line HCT116, and breast cancer cell line MCF-7. An experimental drug was reported which is a ferrocenyl version of tamoxifen. The idea is that the tamoxifen will bind to the estrogen binding sites, resulting in cytotoxicity.

Solid Rocket Propellant

Ferrocene and related derivatives are used as powerful burn rate catalysts in Ammonium perchlorate composite propellant.

Derivatives and Variations

Ferrocene analogues can be prepared with variants of cyclopentadienyl. For example, bisindenyliron and bisfluorenyliron.

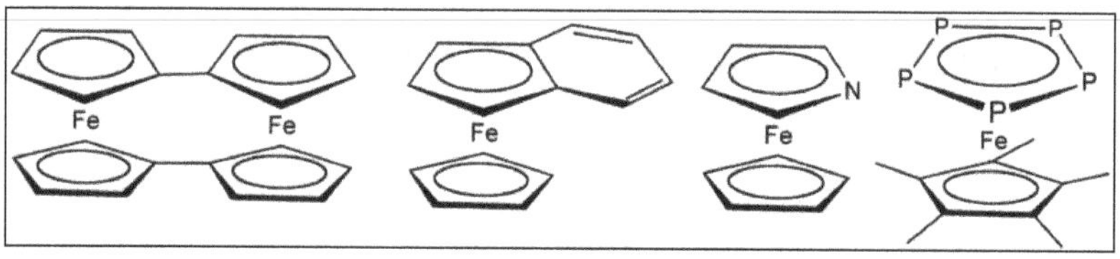

Carbon atoms can be replaced by heteroatoms as illustrated by Fe(η^5-C_5Me_5)(η^5-P_5) and Fe(η^5-C_5H_5)(η^5-C_4H_4N) ("azaferrocene"). Azaferrocene arises from decarbonylation of Fe(η^5-C_5H_5)$(CO)_2$(η^1-pyrrole) in cyclohexane. This compound on boiling under reflux in benzene is converted to ferrocene.

Because of the ease of substitution, many structurally unusual ferrocene derivatives have been prepared. For example, the penta(ferrocenyl)cyclopentadienyl ligand, features a cyclopentadienyl anion derivatized with five ferrocene substituents.

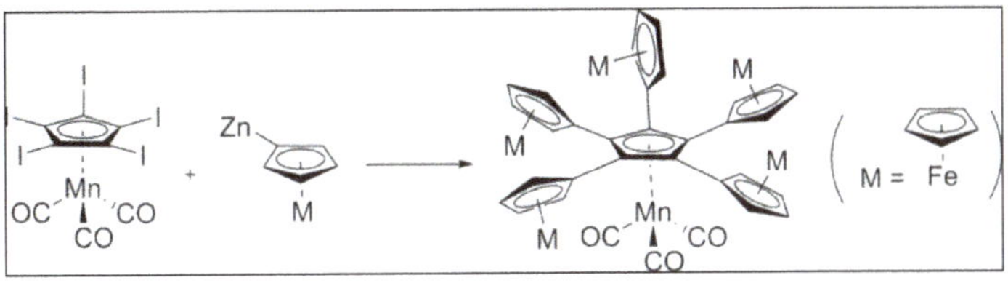

In hexaferrocenylbenzene, $C_6[(\eta^5\text{-}C_5H_4)Fe(\eta^5\text{-}C_5H_5)]_6$, all six positions on a benzene molecule have ferrocenyl substituents (R). X-ray diffraction analysis of this compound confirms that the cyclopentadienyl ligands are not co-planar with the benzene core but have alternating dihedral angles of +30° and −80°. Due to steric crowding the ferrocenyls are slightly bent with angles of 177° and have elongated C-Fe bonds. The quaternary cyclopentadienyl carbon atoms are also pyramidalized. Also, the benzene core has a chair conformation with dihedral angles of 14° and displays bond length alternation between 142.7 pm and 141.1 pm, both indications of steric crowding of the substituents.

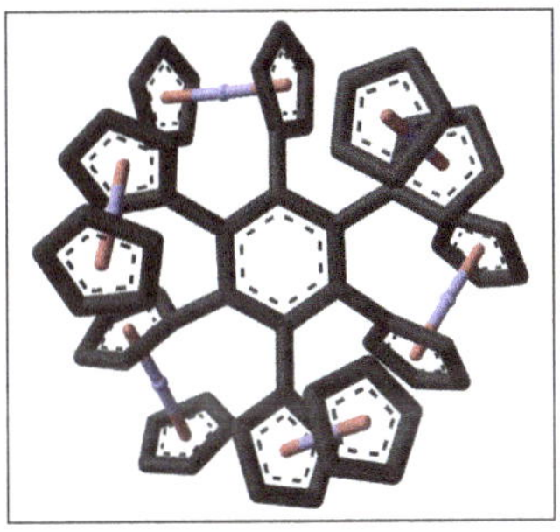

Structure of hexaferrocenylbenzene.

The synthesis of hexaferrocenylbenzene has been reported using Negishi coupling of hexaiodidobenzene and diferrocenylzinc, using tris(dibenzylideneacetone)dipalladium(0) as catalyst, in tetrahydrofuran:

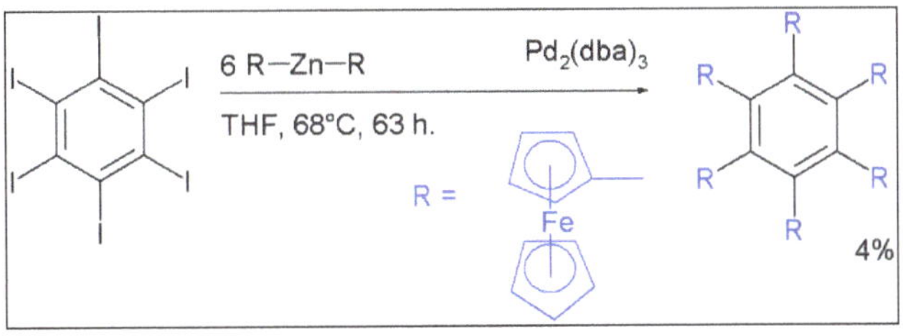

The yield is only 4%, which is further evidence consistent with substantial steric crowding around the arene core.

Materials Chemistry

Ferrocene, a precursor to iron nanoparticles, can be used as a catalyst for the production of carbon nanotubes. The vinylferrocene can be made by a Wittig reaction of the aldehyde, a phosphonium salt, and sodium hydroxide. The vinyl ferrocene can be converted into a polymer (polyvinylferrocene, PVFc), a ferrocenyl version of polystyrene (the phenyl groups are replaced with ferrocenyl groups). Another polyferrocene which can be formed is poly(2-(methacryloyloxy)ethyl ferrocenecarboxylate), PFcMA. In addition to using organic polymer backbones, these pendant ferrocene units have been attached to inorganic backbones such as polysiloxanes, polyphosphazenes, and polyphosphino-boranes, $(-PH(R)-BH_2-)_n$, and the resulting materials exhibit unusual physical and electronic properties relating to the ferrocene/ferrocinium redox couple. Both PVFc and PFcMA have been tethered onto silica wafers and the wettability measured when the polymer chains are uncharged and when the ferrocene moieties are oxidised to produce positively charged groups. The contact angle with water on the PFcMA-coated wafers was 70° smaller following oxidation, while in the case of PVFc the decrease was 30°, and the switching of wettability is reversible. In the PFcMA case, the effect of lengthening the chains and hence introducing more ferrocene groups is significantly larger reductions in the contact angle upon oxidation.

Iron Pentacarbonyl

Iron pentacarbonyl, also known as iron carbonyl, is the compound with formula $Fe(CO)_5$. Under standard conditions $Fe(CO)_5$ is a free-flowing, straw-colored liquid with a pungent odour. Older samples appear darker. This compound is a common precursor to diverse iron compounds, including many that are useful in small scale organic synthesis.

Properties

Iron pentacarbonyl is a homoleptic metal carbonyl, where carbon monoxide is the only ligand complexed with a metal. Other examples include octahedral $Cr(CO)_6$ and tetrahedral $Ni(CO)_4$. Most metal carbonyls have 18 valence electrons, and $Fe(CO)_5$ fits this pattern with 8 valence electrons on Fe and five pairs of electrons provided by the CO ligands. Reflecting its symmetrical structure and charge neutrality, $Fe(CO)_5$ is volatile; it is one of the most frequently encountered liquid metal complexes. $Fe(CO)_5$ adopts a trigonal bipyramidal structure with the Fe atom surrounded by five CO ligands: three in equatorial positions and two axially bound. The Fe–C–O linkages are each linear.

$Fe(CO)_5$ exhibits a relatively low rate of interchange between the axial and equatorial CO groups via the Berry mechanism. It is characterized by two intense ν_{CO} bands in the IR spectrum at 2034 and 2014 cm^{-1} (gas phase).

Synthesis and other Iron Carbonyls

$Fe(CO)_5$ is produced by the reaction of fine iron particles with carbon monoxide. The compound was described in a journal by Mond and Langer in 1891 as “a somewhat viscous liquid of a pale-yellow colour.” Samples were prepared by treatment of finely divided, oxide-free iron powder with carbon monoxide at room temperature.

Industrial synthesis of the compound requires relatively high temperatures and pressures (e.g. 175 atm at 150 °C) as well as special, chemically resistant equipment (e.g. composed of copper-silver alloys). Preparation of the compound at the laboratory scale avoids these complications by using an iodide intermediate:

$$FeI_2 + 4CO \rightarrow Fe(CO)_4I_2$$

$$5Fe(CO)_4I_2 + 10Cu \rightarrow 10CuI + 4Fe(CO)_5 + Fe$$

$Fe(CO)_5$ is sensitive to light. Photodissociation of $Fe(CO)_5$ produces $Fe_2(CO)_9$, a yellow-orange solid, also described by Mond. When heated, $Fe(CO)_5$ converts to small amounts of the metal cluster $Fe_3(CO)_{12}$, a green solid. Simple thermolysis, however, is not a useful synthesis, and each iron carbonyl complex exhibits distinct reactivity.

Industrial Production and Use

The industrial production of this compound is somewhat similar to the Mond process in that the metal is treated with carbon monoxide to give a volatile gas. In the case of iron pentacarbonyl, the reaction is more sluggish. It is necessary to use iron sponge as the starting material, and harsher reaction conditions of 5–30 MPa of carbon monoxide and 150–200 °C. Similar to the Mond process, sulfur acts as a catalyst. The crude iron pentacarbonyl is purified by distillation. Ullmann's Encyclopedia of Industrial Chemistry reports that there are only three plants manufacturing pentacarbonyliron; BASF in Germany and GAF in Alabama have capacities of 9000 and 1500–2000 tonnes/year respectively.

Most iron pentacarbonyl produced is decomposed on site to give pure carbonyl iron in analogy to carbonyl nickel. Some iron pentacarbonyl is burned to give pure iron oxide. Other uses of pentacarbonyliron are small in comparison.

Reactions

Decarbonylation

Purified iron pentacarbonyl can be decomposed to produce carbonyl iron, a high-purity preparation of iron metal.

CO Substitution Reactions

Many compounds are derived from $Fe(CO)_5$ by substitution of CO by Lewis bases, L, to give derivatives $Fe(CO)_{5-x}L_x$. Common Lewis bases include isocyanides, tertiary phosphines and arsines, and alkenes. Usually these ligands displace only one or two CO ligands, but certain acceptor ligands such as PF_3 and isocyanides can proceed to tetra- and pentasubstitution. These reactions are often induced with a catalyst or light. Illustrative is the synthesis of the bis(triphenylphosphine) complex $Fe(CO)_3(P(C_6H_5)_3)_2$. This transformation can be accomplished photochemically, but it is also induced by the addition of NaOH or $NaBH_4$. The catalyst attacks a CO ligand, which labilizes another CO ligand toward substitution. The electrophilicity of $Fe(CO)_4L$ is less than that of $Fe(CO)_5$, so the nucleophilic catalyst, disengages and attacks another molecule of $Fe(CO)_5$.

Oxidation and Reduction

Most metal carbonyls can be halogenated. Thus, treatment of $Fe(CO)_5$ with halogens gives the ferrous halides $Fe(CO)_4X_2$ for X = I, Br, Cl. These species, upon heating lose CO to give the ferrous halides, such as iron(II) chloride.

Reduction of $Fe(CO)_5$ with Na gives $Na_2Fe(CO)_4$, "tetracarbonylferrate" also called Collman's reagent. The dianion is isoelectronic with $Ni(CO)_4$ but highly nucleophilic.

Acid-base Reactions

$Fe(CO)_5$ is not readily protonated, but it is attacked by hydroxide. Treatment of $Fe(CO)_5$ with aqueous base produces $[HFe(CO)_4]^-$, via the Metallacarboxylate intermediate. The oxidation of this monoanion gives triiron dodecarbonyl, $Fe_3(CO)_{12}$. Acidification of solutions of $[HFe(CO)_4]^-$ gives iron tetracarbonyl hydride, $H_2Fe(CO)_4$, the first metal hydride ever reported.

Diene Adducts

Dienes react with $Fe(CO)_5$ to give (diene)$Fe(CO)_3$, wherein two CO ligands have been replaced by two olefins. Many dienes undergo this reaction, notably norbornadiene and 1,3-butadiene. One of the more historically significant derivatives is cyclobutadieneiron tricarbonyl (C_4H_4)$Fe(CO)_3$, where C_4H_4 is the otherwise unstable cyclobutadiene. Receiving the greatest attention are complexes of the cyclohexadienes, the parent organic 1,4-dienes being available through the Birch reductions. 1,4-Dienes isomerize to the 1,3-dienes upon complexation.

$Fe(CO)_5$ reacts in dicyclopentadiene to form $[Fe(C_5H_5)(CO)_2]_2$, cyclopentadienyliron dicarbonyl dimer. This compound, called "Fp dimer" can be considered a hybrid of ferrocene and $Fe(CO)_5$, although in terms of its reactivity, it resembles neither.

Other Uses

In Europe, iron pentacarbonyl was once used as an anti-knock agent in petrol in place of tetraethyllead. Two more modern alternative fuel additives are ferrocene and methylcyclopentadienyl manganese tricarbonyl. $Fe(CO)_5$ is used in the production of "carbonyl iron", a finely divided form of Fe, a material used in magnetic cores of high-frequency coils for radios and televisions and for manufacture of the active ingredients of some radar absorbent materials (e.g. iron ball paint). It is famous as a chemical precursor for the synthesis of various iron-based nanoparticles.

Iron pentacarbonyl has been found to be a strong flame speed inhibitor in oxygen based flames. A few hundred ppm of iron pentacarbonyl are known to reduce the flame speed of stoichiometric methane–air flame by almost 50%. However due to its toxic nature it has not been used widely as a flame retardant.

Toxicity and Hazards

$Fe(CO)_5$ is toxic, which is of concern because of its volatility (vapour pressure: 21 millimetres of mercury (2.8 kPa) at 20 °C). If inhaled, iron pentacarbonyl may cause lung irritation, toxic pneumonitis, or pulmonary edema. Like other metal carbonyls, $Fe(CO)_5$ is flammable. It is, however, considerably less toxic than nickel tetracarbonyl.

In regards to occupational exposures to iron pentacarbonyl, the National Institute for Occupational Safety and Health has set a recommended exposure limit at 0.1 ppm (0.23 mg/m^3) over an eight-hour time-weighted average, and a short-term exposure limit at 0.2 ppm (0.45 mg/m^3).

ORGANOCOBALT COMPOUNDS

Organocobalt chemistry is the chemistry of organometallic compounds containing a carbon to cobalt chemical bond. Organocobalt compounds are involved in several organic reactions and the important biomolecule vitamin B_{12} has a cobalt-carbon bond. Many organocobalt compounds exhibit useful catalytic properties, the preeminent example being dicobalt octacarbonyl.

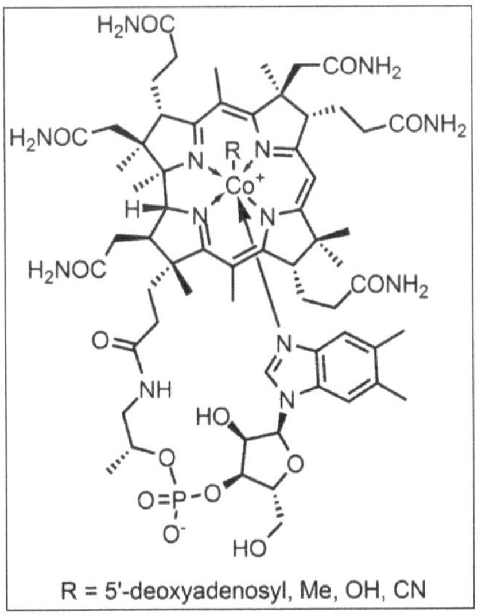

Carbonyl Complexes

Dicobalt octacarbonyl reacts with hydrogen and alkenes to give aldehydes. This reaction is the basis of hydroformylation, the formation of aldehydes from an alkene, CO and hydrogen. A key intermediate is cobalt tetracarbonyl hydride ($HCo(CO)_4$). The original Ruhrchemie process produced propanal from ethene and syngas using cobalt carbonyl has been displaced by rhodium-based catalysts. Processes involving cobalt are practiced by BASF, EXXON, and Shell mainly for the production of C7-C14 alcohols used for the production of surfactants.

In hydrocarboxylations hydrogen is replaced by water or an alcohol and the reaction product is a carboxylic acid or an ester. An example of this reaction type is the conversion of butadiene to adipic acid. Cobalt catalysts (together with iron) are relevant in the Fischer-Tropsch process in which it is assumed that organocobalt intermediates form.

Alkyne Derivatives of $Co_2(CO)_8$

Dicobalt octacarbonyl also reacts with alkynes to give "tetrahedranes" of the formula $Co_2(CO)_6(C_2R_2)$. Because the cobalt carbonyl centers can be removed later, it functions as a protective group for the alkyne. In the Nicholas reaction an alkyne group is also protected and at the same time the alpha-carbon position is activated for nucleophilic substitution.

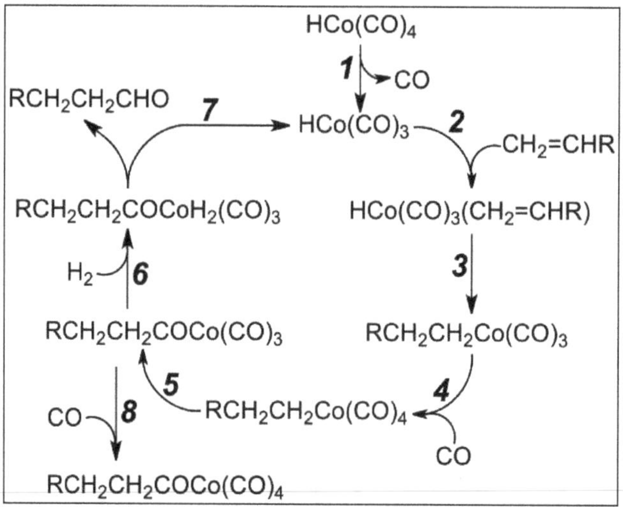

Figure shows mechanism of cobalt-catalyzed hydroformylation. The process begins with dissociation of CO from cobalt tetracarbonyl hydride to give the 16-electron species (step 1). Subsequent binding of alkene gives an 18e species (step 2). In step 3, the olefin inserts to give the 16e alkyl tricarbonyl. Coordination of another equivalent of CO give alkyl tetracarbonyl (step 4). Migratory insertion of CO gives the 16e acyl in step 5. In step 6, oxidative addition of hydrogen gives a dihydrido complex, which in step 7 releases aldehyde by reductive elimination. Step 8 is unproductive and reversible.

Cyclization Reactions

Cobalt compounds react with dialkynes and dienes to cyclic intermediates in cyclometalation. Other alkynes, alkenes, nitriles or carbon monoxide can then insert themselves into the Co-C bond. Reaction types based on this concept are the Pauson–Khand reaction (CO insertion) and alkyne trimerization (notably with cyclopentadienylcobalt dicarbonyl).

Cp, Allyl and Alkene Compounds

Sandwich Compounds

Organocobalt compounds are known with alkene, allyl, diene, and Cp ligands. A famous sandwich compound is cobaltocene, a 19-electron metallocene that is used as a reducing agent and a source of CpCo. Other sandwich compounds are $CoCp(C_6Me_6)$ and $Co(C_6Me_6)_2$, with 20 electrons and 21 electrons, respectively. Reduction of anhydrous cobalt(II) chloride with sodium in the presence of cyclooctadiene gives Co(cyclooctadiene)(cyclooctenyl), a synthetically versatile reagent.

$CpCo(CO)_2$ and Derivatives

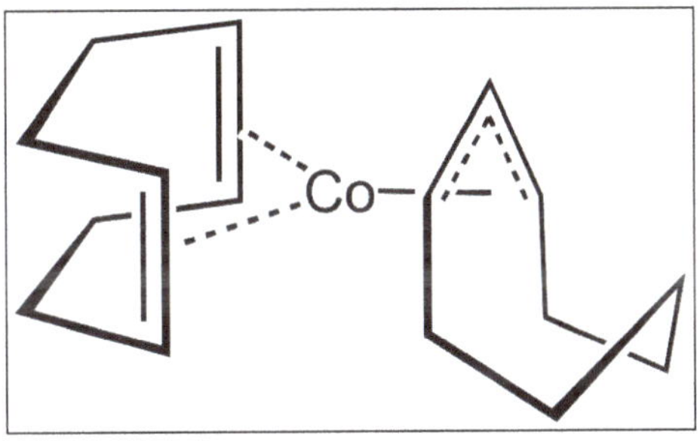

Co(1,5-cyclooctadiene)(cyclooctenyl).

The half-sandwich compounds are particularly versatile catalysts. These include cyclopentadienylcobalt dicarbonyl ($CpCo(CO)_2$) as well as the olefin complexes $CpCo(C_2H_4)_2$ and CpCo(cod). They catalyze alkyne trimerisation. which has been applied to the synthesis of a variety of complex structures.

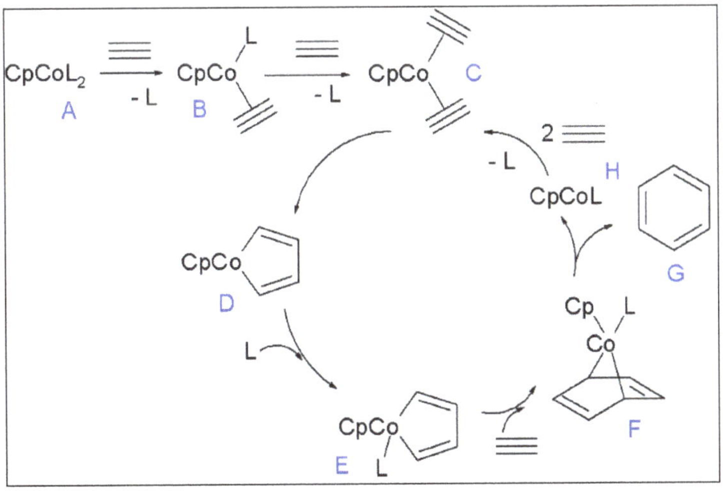

Vitamin B_{12}-type Compounds

Cobalt is found in vitamin B_{12} and related enzymes. These cofactors catalyze unusual reactions involving the intermediacy of Co-C bonds. In these reactions, the oxidation state of cobalt can vary from Co(III) to Co(I). In methylcobalamin the ligand is a methyl group, which is electrophilic. In vitamin B_{12}, the alkyl ligand is an adenosyl group. Related to vitamin B_{12} are cobalt porphyrins, dimethylglyoximates, and related complexes of Schiff base ligands. These synthetic compounds also form alkyl derivatives that undergo diverse reactions reminiscent of the biological processes. The weak cobalt(III)-carbon bond in vitamin B_{12} analogues can be exploited in a type of Cobalt mediated radical polymerization of acrylic and vinyl esters (e.g. vinyl acetate), acrylic acid and acrylonitrile.

Cobaltocene

Cobaltocene, known also as bis(cyclopentadienyl)cobalt(II) or even "bis Cp cobalt", is an organocobalt compound with the formula $Co(C_5H_5)_2$. It is a dark purple solid that sublimes readily slightly above room temperature. Cobaltocene was discovered shortly after ferrocene, the first metallocene. Due to the ease with which it reacts with oxygen, the compound must be handled and stored using air-free techniques.

Synthesis

Vacuum sublimated cobaltocene under nitrogen atmosphere.

Cobaltocene is prepared by the reaction of sodium cyclopentadienide (NaC_5H_5) with anhydrous cobalt(II) chloride in THF solution. Sodium chloride is cogenerated, and the organometallic product is usually purified by vacuum sublimation.

Structure and Bonding

In $Co(C_5H_5)_2$ the Co centre is "sandwiched" between two cyclopentadienyl (Cp) rings. The Co–C bond lengths are about 2.1 Å, slightly longer than the Fe–C bond in ferrocene.

$Co(C_5H_5)_2$ belongs to a group of organometallic compounds called metallocenes or sandwich compounds. Cobaltocene has 19 valence electrons, one more than usually found in organotransition metal complexes such as its very stable relative ferrocene. This additional electron occupies an orbital that is antibonding with respect to the

Co–C bonds. Consequently, the Co–C distances are slightly longer than the Fe–C bonds in ferrocene. Many chemical reactions of $Co(C_5H_5)_2$ are characterized by its tendency to lose this "extra" electron, yielding an 18-electron cation known as cobaltocenium:

$$\underbrace{2Co(C_5H_5)_2}_{19e^-} + I_2 \rightarrow \underbrace{2Co(C_5H_5)_2^+}_{18e^-} + 2I^-$$

The otherwise close relative of cobaltocene, rhodocene does not exist as a monomer, but spontaneously dimerizes by formation of a C–C bond between Cp rings.

Reactions

Redox properties

$Co(C_5H_5)_2$ is a common one-electron reducing agent in the laboratory. In fact, the reversibility of the $Co(C_5H_5)_2$ redox couple is so well-behaved that $Co(C_5H_5)_2$ may be used in cyclic voltammetry as an internal standard. Its permethylated analogue decamethylcobaltocene ($Co(C_5Me_5)_2$) is an especially powerful reducing agent, due to inductive donation of electron density from the 10 methyl groups, prompting the cobalt to give up its "extra" electron even more so. These two compounds are rare examples of reductants that dissolve in non-polar organic solvents. The reduction potentials of these compounds follow, using the ferrocene-ferrocenium couple as the reference:

Half-reaction	E° (V)
$Fe(C_5H_5)_2^+ + e^- \rightleftharpoons Fe(C_5H_5)_2$	0.00 (by definition)
$Fe(C_5Me_5)_2^+ + e^- \rightleftharpoons Fe(C_5Me_5)_2$	−0.59
$Co(C_5H_5)_2^+ + e^- \rightleftharpoons Co(C_5H_5)_2$	−1.33
$Co(C_5Me_5)_2^+ + e^- \rightleftharpoons Co(C_5Me_5)_2$	−1.94

The data show that the decamethyl compounds are around 600 mV more reducing than the parent metallocenes. This substituent effect is, however, overshadowed by the influence of the metal: changing from Fe to Co renders the reduction more favorable by over 1.3 volts.

Carbonylation

Treatment of $Co(C_5H_5)_2$ with carbon monoxide gives the cobalt(I) derivative $Co(C_5H_5)(CO)_2$, concomitant with loss of one Cp ligand.

Cyclopentadienylcobalt Dicarbonyl

Cyclopentadienylcobalt dicarbonyl is an organocobalt compound with formula $(C_5H_5)Co(CO)_2$, abbreviated $CpCo(CO)_2$. It is an example of a half-sandwich complex. It is a dark red air sensitive liquid. This compound features one cyclopentadienyl ring that is bound in an η^5-manner and two carbonyl ligands. The compound is soluble in common organic solvents and it is best stored under nitrogen.

Preparation

$CpCo(CO)_2$ was first reported in 1954 by Piper, Cotton, and Wilkinson who produced it by the reaction of cobalt carbonyl with cyclopentadiene. It is prepared commercially by the same method:

$$Co_2(CO)_8 + 2\ C_5H_6 \rightarrow 2\ C_5H_5Co(CO)_2 + H_2 + 4\ CO$$

Alternatively, it is generated by the high pressure carbonylation of bis(cyclopentadienyl)cobalt (cobaltocene) at elevated temperature and pressures:

$$Co(C_5H_5)_2 + 2\ CO \rightarrow C_5H_5Co(CO)_2 + "C_5H_5"$$

Reactions

$CpCo(CO)_2$ catalyzes the cyclotrimerization of alkynes. The catalytic cycle begins with dissociation of one CO ligand forming bis(alkyne) intermediate.

$$CpCo(CO)_2 + 2\ R_2C_2 \rightarrow CpCo(R_2C_2)_2 + 2\ CO$$

This reaction proceeds by formation of metal-alkyne complexes by dissociation of CO. Although monoalkyne complexes $CpCo(CO)(R^1C_2R^2)$ have not been isolated, their analogues, $CpCo(PPh_3)(R^1C_2R^2)$ are made by the following reactions:

$$CpCo(CO)_2 + PR_3 \rightarrow CO + CpCo(CO)(PR_3)$$
$$CpCoL(PR_3) + R_2C_2 \rightarrow L + CpCo(PR_3)(R_2C_2) \quad (\text{where, } L = CO \text{ or } PR_3)$$

$CpCo(CO)_2$ catalyzes the formation of pyridines from a mixture of alkynes and nitriles. Reduction of $CpCo(CO)_2$ with sodium yields the dinuclear radical $[Cp_2Co_2(CO)_2]^-$, which reacts with alkyl halides to give the dialkyl complexes $[Cp_2Co_2(CO)_2R_2]$. Ketones are produced by carbonylation of these dialkyl complexes, regenerating $CpCo(CO)_2$.

Related Compounds

The pentamethylcyclopentadienyl analogue $Cp^*Co(CO)_2$ is well studied. The Rh and Ir analogues, $CpRh(CO)_2$ and $CpIr(CO)_2$ are also well known.

ORGANONICKEL COMPOUND

Organonickel chemistry is a branch of organometallic chemistry that deals with organic compounds featuring nickel-carbon bonds. They are used as a catalyst, as a building block in organic chemistry and in chemical vapor deposition. Organonickel compounds are also short-lived intermediates in organic reactions. The first organonickel compound was nickel tetracarbonyl $Ni(CO)_4$, reported in 1890 and quickly applied in the Mond process for nickel purification. Organonickel complexes are prominent in numerous industrial processes including carbonylations, hydrocyanation, and the Shell higher olefin process.

Organonickel compound.

Many complexes exist of nickel coordinated to an alkene. In these compounds nickel is formally zerovalent Ni^0 and the bonding is described with the Dewar-Chatt-Duncanson model. One common representative is Bis(cyclooctadiene)nickel(0) ($Ni(COD)_2$), which contains two cyclooctadiene ligands. It is a 18VE compound with 10 electrons provided by nickel itself and 4x2 electrons more by the double bonds. This solid, which melts at 60 °C, is used as a catalyst and as a precursor for many other nickel compounds.

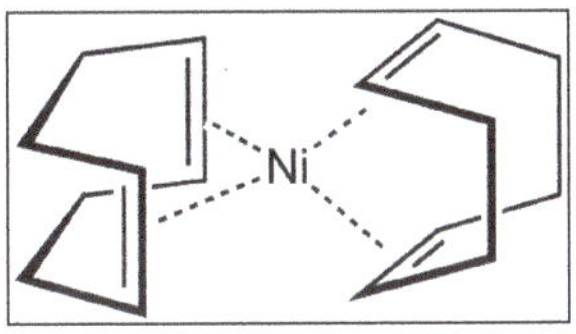

Ni Allyl Complexes

Allyl halides react with $Ni(CO)_4$ to form pi-allyl complexes, $(allyl)_2Ni_2Cl_2$. These compounds in turn are sources of allyl nucleophiles. In $(allyl)_2Ni_2Br_2$ and $(allyl)Ni(C_5H_5)$, nickel is assigned to oxidation number +2, and the electron counts are 16 and 18, respectively.

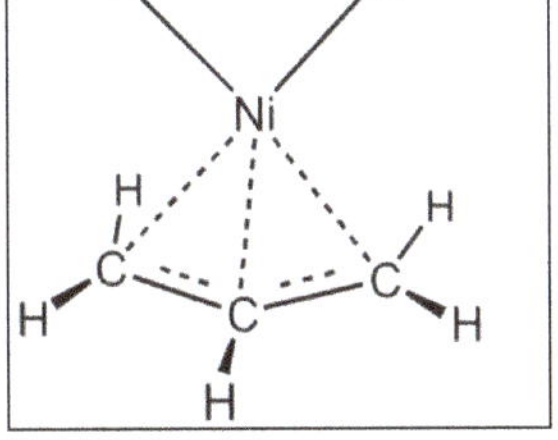

Cyclopentadienyl Complexes

Nickelocene $NiCp_2$ with +2 Ni oxidation state and 20 valence electrons is the main metallocene of nickel. It can be oxidized by one electron. The corresponding palladocene and platinocene are unknown. From nickelocene, many derivatives are generated, e.g. CpNiLCl, CpNiNO, and $Cp_2Ni_2(CO)_3$.

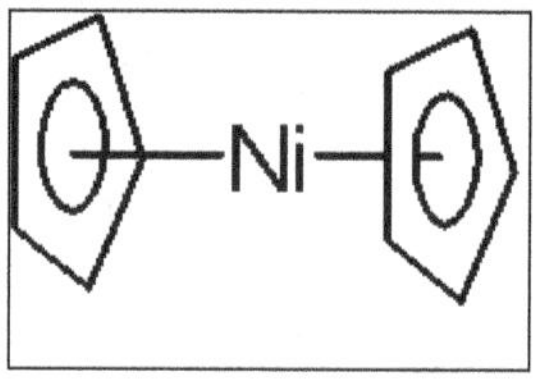

Nickel Carbene Complexes

Nickel forms carbene complexes, formally featuring C=Ni double bonds.

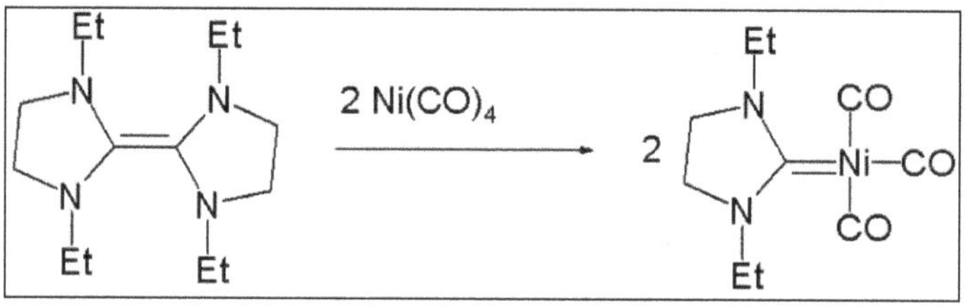

Alkyl and Aryl Precatalysts

A variety of Ni(II) alkyl and aryl complexes are known. Examples include [(dppf)Ni(-cinnamyl)Cl)], trans-$(PCy_2Ph)_2$Ni(o-tolyl)Cl]], (dppf)Ni(o-tolyl)Cl]], (TMEDA)Ni(o-tolyl)Cl, and (TMEDA)$NiMe_2$.

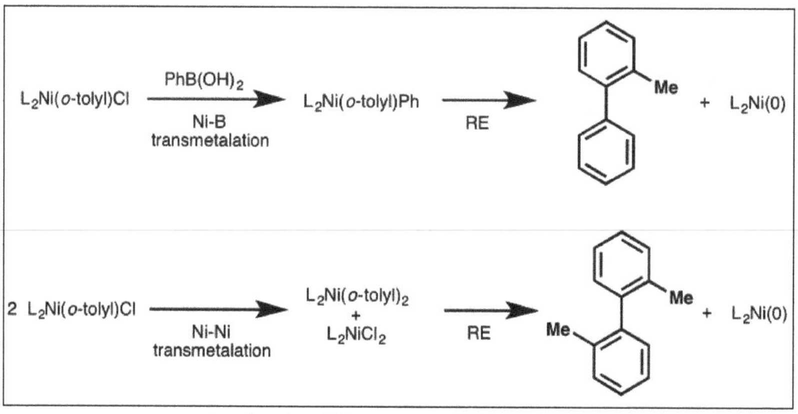

Synthesis of [(TMEDA)Ni(o-tolyl)Cl].

Nickel compounds of the type NiR_2 also exist with just 12 valence electrons. In solution however solvent always interact with the metal atom increasing the electron count. One 12 VE compound is di(mesityl)nickel prepared from $(allyl)_2Ni_2Br_2$ and the corresponding Grignard reagent.

$$(\text{allyl})_2\,\text{Ni}_2\text{Br}_2 + 4\ \text{C}_6\text{H}_2\text{Me}_3\text{MgBr} \rightarrow 2\ (\text{allyl})\text{MgBr} + 2\ \text{MgBr}_2 + 2\ (\text{C}_6\text{H}_2\text{Me}_3)_2\text{Ni}$$

Reactions

Alkene/Alkyne Oligomerizations

Nickel compounds catalyze the oligomerization of alkenes and alkynes. This property validated the research and development of Ziegler–Natta catalysts in the 1950s. That discovery shown by nickel impurities originating from an autoclave which killed the propagation reaction (Aufbau) in favor of termination reaction to a terminal alkene: the polymerization of ethylene suddenly stopped at 1-butene. This so-called nickel effect prompted the search for other catalysts capable of this reaction, with results in the finding of new catalysts that technically produced high molar mass polymers, like the modern Ziegler–Natta catalysts.

One practical implementation of alkyne oligomerization is the Reppe synthesis; for example in the synthesis of cyclooctatetraene:

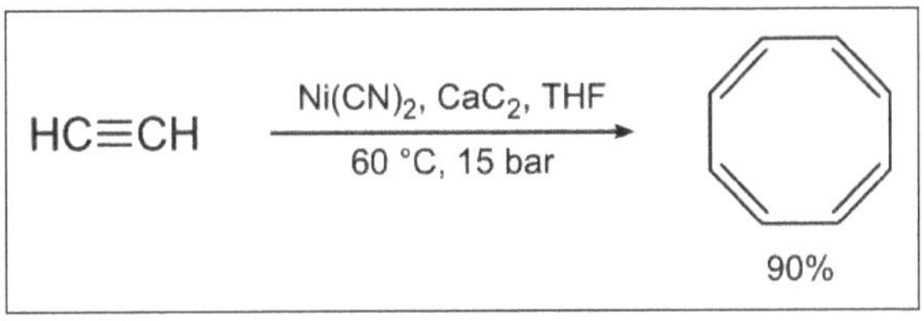

This is a formal [2+2+2+2]cycloaddition. The oligomerization of butadiene with ethylene to trans-1,4-hexadiene was an industrial process at one time.

Formal [2+2+2]cycloadditions also take place in alkyne trimerisation. This extensible trimerisation can generally include benzyne. Benzyne is generated in situ from a benzene compound attached to a triflate and a trimethylsilyl substituent in the ortho- positions and reacts with a di-yne such as 1,7-octadiyne along with a nickel(II) bromide/zinc catalyst system ($NiBr_2$ bis(diphenylphosphino) ethane/Zn) to synthesize the corresponding naphthalene derivative.

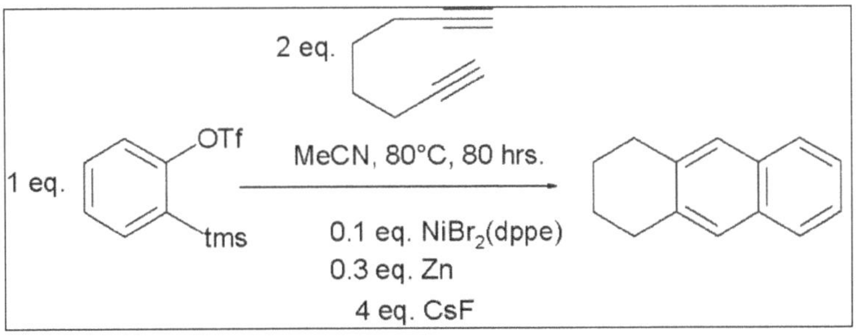

In the catalytic cycle elementary zinc serves to reduce nickel(II) to nickel(0) to which can then coordinate two alkyne bonds. A cyclometalation step follows to the nickelcyclopentadiene intermediate and then coordination of the benzyne which gives a C-H insertion reaction to the nickelcycloheptatriene compound. Reductive elimination liberates the tetrahydroanthracene compound.

The formation of organonickel compounds in this type of reaction is not always obvious but in a carefully designed experiment two such intermediates are formed quantitatively:

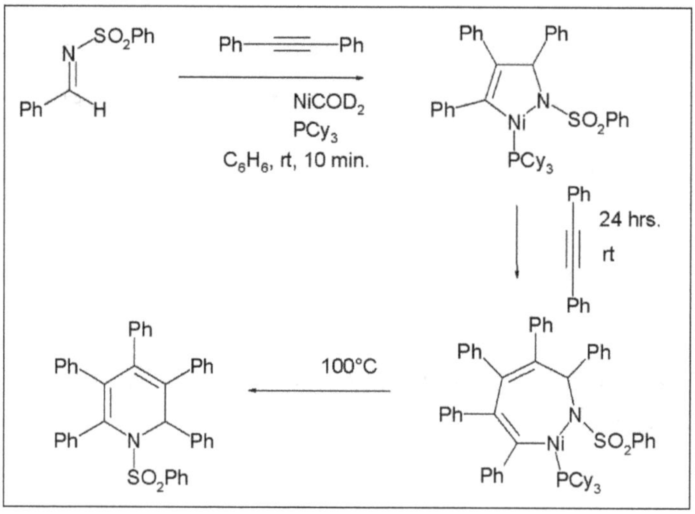

It is noted in one study that this reaction only works with acetylene itself or with simple alkynes due to poor regioselectivity. From a terminal alkyne 7 isomers are possibly differing in the position of the substituents or the double bond positions. One strategy to remedy this problem employs certain diynes:

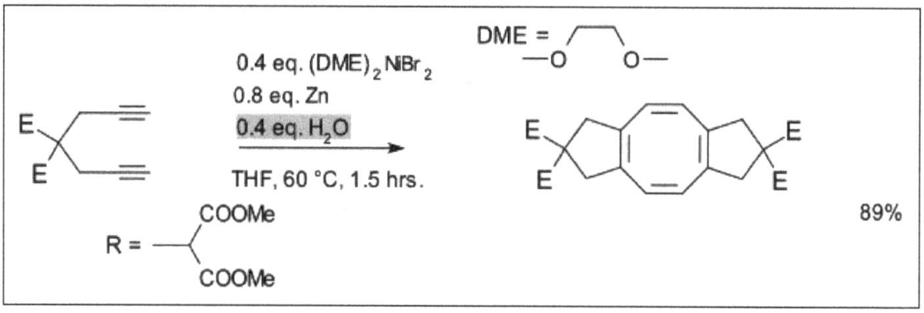

The selected reaction conditions also minimize the amount formed of competing [2+2+2]cycloaddition product to the corresponding substituted arene.

Coupling Reactions

Nickel compounds cause the coupling reaction between allyl and aryl halides. Other coupling reactions involving nickel in catalytic amounts are the Kumada coupling and the Negishi coupling.

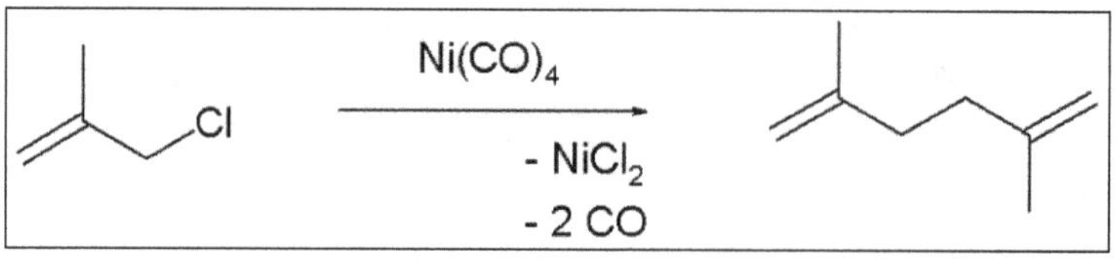

Ni Carbonylation

Ni catalyzes the addition of carbon monoxide to alkenes and alkynes. The industrial production of acrylic acid at one time consisted of combining acetylene, carbon monoxide and water at 40-55 atm and 160-200 °C with nickel(II) bromide and a copper halide.

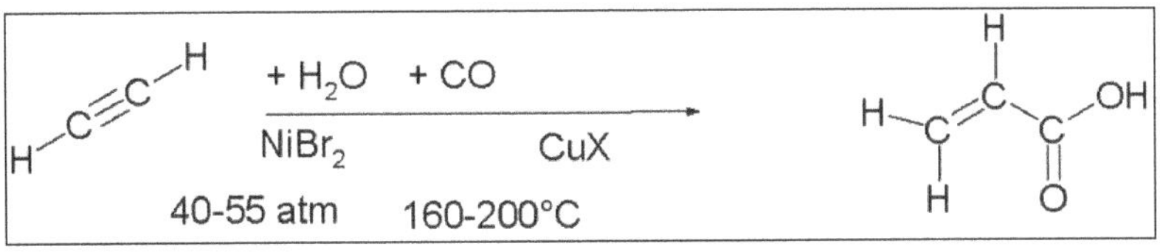

Nickel Tetracarbonyl

Nickel carbonyl (IUPAC name: tetracarbonylnickel) is the organonickel compound with the formula $Ni(CO)_4$. This colorless liquid is the principal carbonyl of nickel. It is an intermediate in the Mond process for producing very high-purity nickel and a reagent in organometallic chemistry, although the Mond Process has fallen out of common usage due to the health hazards in working with the compound. Nickel carbonyl is one of the most dangerous substances yet encountered in nickel chemistry due to its very high toxicity, compounded with high volatility and rapid skin absorption.

Structure and Bonding

In nickel tetracarbonyl, the oxidation state for nickel is assigned as zero. The formula conforms to 18-electron rule. The molecule is tetrahedral, with four carbonyl (carbon monoxide) ligands. Electron diffraction studies have been performed on this molecule, and the Ni–C and C–O distances have been calculated to be 1.838(2) and 1.141(2) angstroms respectively.

Preparation

$Ni(CO)_4$ was first synthesised in 1890 by Ludwig Mond by the direct reaction of nickel metal with CO. This pioneering work foreshadowed the existence of many other metal carbonyl compounds, including those of V, Cr, Mn, Fe, and Co. It was also applied industrially to the purification of nickel by the end of the 19th century.

At 323 K (50 °C; 122 °F), carbon monoxide is passed over impure nickel. The optimal rate occurs at 130 °C.

Laboratory Routes

$Ni(CO)_4$ is not readily available commercially. It is conveniently generated in the laboratory by carbonylation of commercially available bis(cyclooctadiene)nickel(0). It can also be prepared by reduction of ammoniacal solutions of nickel sulfate with sodium dithionite under an atmosphere of CO.

Reactions

Thermal Decarbonylation

On moderate heating, $Ni(CO)_4$ decomposes to carbon monoxide and nickel metal. Combined with the easy formation from CO and even very impure nickel, this decomposition is the basis for the Mond process for the purification of nickel or plating onto surfaces. Thermal decomposition commences near 180 °C and increases at higher temperature.

Spheres of nickel made by the Mond process.

Reactions with Nucleophiles and Reducing Agents

Like other low-valent metal carbonyls, $Ni(CO)_4$ is susceptible to attack by nucleophiles. Attack can occur at nickel center, resulting in displacement of CO ligands, or at CO. Thus, donor ligands such as triphenylphosphine react to give $Ni(CO)_3(PPh_3)$ and $Ni(CO)_2(PPh_3)_2$. Bipyridine and related ligands behave similarly. The monosubstitution of nickel tetracarbonyl with other ligands can be used to determine the Tolman electronic parameter, a measure of the electron donating or withdrawing ability of a given ligand.

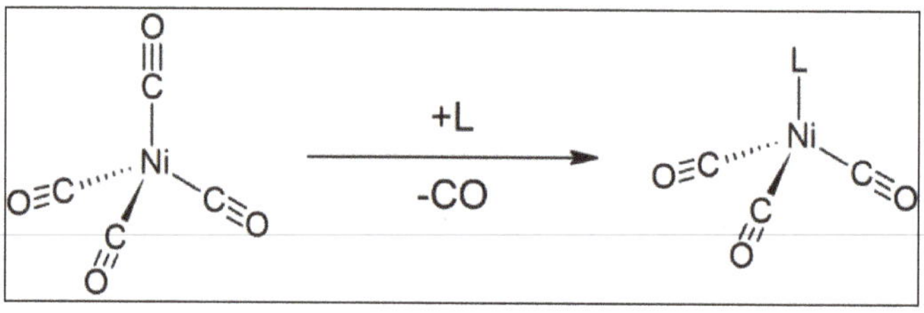

Treatment with hydroxides gives clusters such as $[Ni_5(CO)_{12}]^{2-}$ and $[Ni_6(CO)_{12}]^{2-}$. These compounds can also be obtained by reduction of nickel carbonyl.

Thus, treatment of $Ni(CO)_4$ with carbon nucleophiles (Nu^-) results in acyl derivatives such as $[Ni(CO)_3C(O)Nu]^-$.

Reactions with Electrophiles and Oxidizing Agents

Nickel carbonyl can be oxidized. Chlorine oxidizes nickel carbonyl into $NiCl_2$, releasing CO gas. Other halogens behave analogously. This reaction provides a convenient method for precipitating the nickel portion of the toxic compound.

Reactions of $Ni(CO)_4$ with alkyl and aryl halides often result in carbonylated organic products. Vinylic halides, such as PhCH = CHBr, are converted to the unsaturated esters upon treatment with $Ni(CO)_4$ followed by sodium methoxide. Such reactions also probably proceed via oxidative addition. Allylic halides give the π-allylnickel compounds, such as $(allyl)_2Ni_2Cl_2$:

$$2\,Ni(CO)_4 + 2\,ClCH_2CH{=}CH_2 \rightarrow Ni_2(\mu-Cl)_2(\eta^3-C_3H_5)_2 + 8\,CO$$

Toxicology and Safety Considerations

The hazards of $Ni(CO)_4$ are far greater than that implied by its CO content, reflecting the effects of the nickel if released in the body. Nickel carbonyl may be fatal if absorbed through the skin or more likely, inhaled due to its high volatility. Its LC_{50} for a 30-minute exposure has been estimated at 3 ppm, and the concentration that is immediately fatal to humans would be 30 ppm. Some subjects exposed to puffs up to 5 ppm described the odour as musty or sooty, but because the compound is so exceedingly toxic, its smell provides no reliable warning against a potentially fatal exposure.

The vapours of $Ni(CO)_4$ can autoignite. The vapor decomposes quickly in air, with a half-life of about 40 seconds.

Nickel carbonyl poisoning is characterized by a two-stage illness. The first consists of headaches and chest pain lasting a few hours, usually followed by a short remission. The second phase is a chemical pneumonitis which starts after typically 16 hours with symptoms of cough, breathlessness and extreme fatigue. These reach greatest severity after four days, possibly resulting in death from cardiorespiratory or acute kidney injury. Convalescence is often extremely protracted, often complicated by exhaustion, depression and dyspnea on exertion. Permanent respiratory damage is unusual. The carcinogenicity of $Ni(CO)_4$ is a matter of debate, but is presumed to be significant.

It is classified as an extremely hazardous substance in the United States as defined in Section 302 of the U.S. Emergency Planning and Community Right-to-Know Act (42 U.S.C. 11002), and is subject to strict reporting requirements by facilities which produce, store, or use it in significant quantities.

Nickelocene

Nickelocene is the organonickel compound with the formula $Ni(\eta^5-C_5H_5)_2$. Also known as bis(cyclopentadienyl)nickel or $NiCp_2$, this bright green paramagnetic solid is of enduring academic interest, although it does not yet have any known practical applications.

Structure and Bonding

$Ni(C_5H_5)_2$ belongs to a group of organometallic compounds called metallocenes. Metallocenes usually adopt structures in which a metal ion is sandwiched between two parallel cyclopentadienyl (Cp) rings. In the solid-state, the molecule has D_{5d} symmetry, wherein the two rings are staggered.

The Ni center has a formal +2 charge, and the Cp rings are usually assigned as cyclopentadienyl anions (Cp^-), related to cyclopentadiene by deprotonation. The structure is similar to ferrocene. In terms of its electronic structure, three pairs of d electrons on nickel are allocated to the three d orbitals involved in Ni–Cp bonding: d_{xy}, $d_{x^2-y^2}$, d_{z^2}. The two remaining d-electrons each reside in the d_{yz} and d_{xz} orbitals, giving rise to the molecule's paramagnetism, as manifested in the unusually high field chemical shift observed in its ^1H NMR spectrum. With 20 valence electrons, nickelocene has the highest electron count of the transition metal metallocenes. Cobaltocene, $Co(C_5H_5)_2$, with only 19 valence electrons is, however, a stronger reducing agent, illustrating the fact that electron energy, not electron count determines redox potential.

Preparation

Nickelocene was first prepared by E. O. Fischer in 1953, shortly after the discovery of ferrocene, the first metallocene compound. It has been prepared in a one-pot reaction, by deprotonating cyclopentadiene with ethylmagnesium bromide, and adding anhydrous nickel(II) acetylacetonate. A modern synthesis entails treatment of anhydrous sources of $NiCl_2$ (such as hexaamminenickel chloride) with sodium cyclopentadienyl:

$$[Ni(NH_3)_6]Cl_2 + 2\ NaC_5H_5 \rightarrow Ni(C_5H_5)_2 + 2\ NaCl + 6\ NH_3$$

Properties

Like many organometallic compounds, $Ni(C_5H_5)_2$ does not tolerate extended exposure to air before noticeable decomposition. Samples are typically handled with air-free techniques.

Most chemical reactions of nickelocene are characterized by its tendency to yield 18-electron products with loss or modification of one Cp ring.

$$Ni(C_5H_5)_2 + 4PF_3 \rightarrow Ni(PF_3)_4 + \text{organic products}$$

The reaction with secondary phosphines follows a similar pattern:

$$2Ni(C_5H_5)_2 + 2PPh_2H \rightarrow [Ni_2(PPh_2)_2(C_5H_5)_2] + 2C_5H_6$$

Nickelocene can be oxidized to the corresponding cation, which contains Ni(III).

Gaseous $Ni(C_5H_5)_2$ decomposes to a nickel mirror upon contact with a hot surface, releasing the hydrocarbon ligands as gaseous coproducts. This process has been considered as a means of preparing nickel films.

Nickelocene reacts with nitric acid to produce cyclopentadienyl nickel nitrosyl, a highly toxic organonickel compound.

ORGANOCOPPER COMPOUND

Organocopper compounds in organometallic chemistry contain carbon to copper chemical bonds. Organocopper chemistry is the science of organocopper compounds describing their physical properties, synthesis and reactions. They are reagents in organic chemistry.

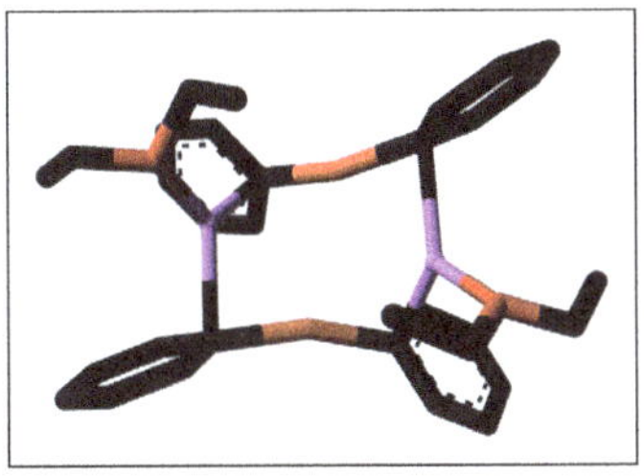

Lithium diphenylcuprate etherate dimer from crystal structure.

The first organocopper compound, the explosive copper(I) acetylide Cu_2C_2 (Cu – C ≡ C - Cu), was synthesized by Rudolf Christian Böttger in 1859 by passing acetylene gas through copper(I) chloride solution:

$$C_2H_2 + 2\ CuCl \rightarrow Cu_2C_2 + 2\ HCl$$

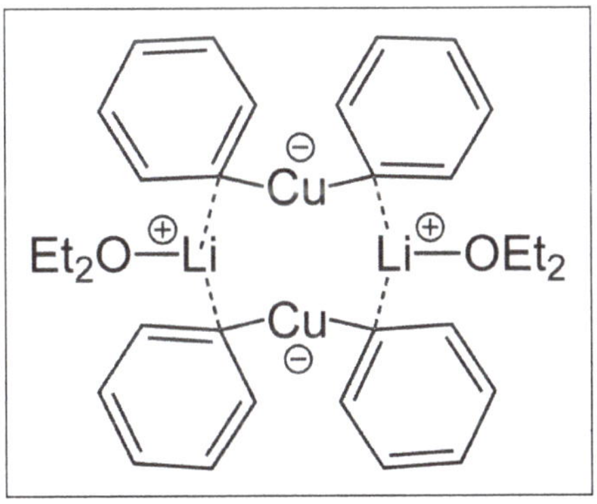

Skeletal formula of lithium diphenylcuprate etherate dimer.

Structure and Bonding

Organocopper compounds are diverse in structure and reactivity, but organocopper compounds are largely limited in oxidation states to copper(I), sometimes denoted Cu^+. As a d^{10} metal center, it is related to Ni(0), but owing to its higher oxidation state, it engages in less pi-backbonding. Organic derivatives of Cu(II) and Cu(III) are invoked as intermediates but rarely isolated or even observed. In terms of geometry, copper(I) adopts symmetrical structures, in keeping with its spherical electronic shell. Typically one of three coordination geometries is adopted: linear 2-coordinate, trigonal 3-coordinate, and tetrahedral 4-coordinate. Organocopper compounds form complexes with a variety of soft ligands such as alkylphosphines (R_3P), thioethers (R_2S), and cyanide (CN^-).

Simple Complexes with CO, Alkene and Cp Ligands

Copper(I) salts have long been known to bind CO, albeit weakly. A representative complex is CuCl(CO), which is polymeric. In contrast to classical metal carbonyls, pi-backbonding is not strong in these compounds.

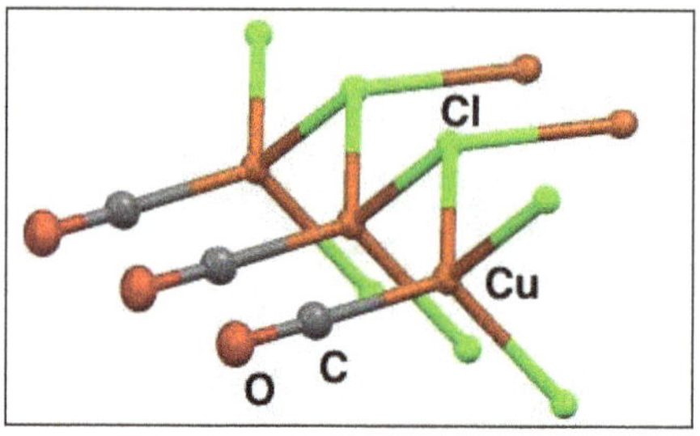

Part of the framework of CuCl(CO). In this coordination polymer, the Cu centers are tetrahedral linked by triply bridging chloride ligands.

Alkenes bind to copper(I), although again generally weakly. The binding of ethylene to Cu in proteins is of broad significance in plant biology so much so that ethylene is classified as a plant hormone. Its presence, detected by the Cu-protein, affects ripening and many other developments.

Although copper does not form a metallocene, half-sandwich complexes can be produced. One such derivative is (η-cyclopentadienyl triethylphosphine) copper.

Alkyl and Aryl Copper Compounds

Alkyl and Aryl Copper(I) Compounds

Copper halides react with organolithium reagents to give organocopper compounds. The area was pioneered by Henry Gilman, who reported methylcopper in 1936. Thus, phenylcopper is prepared by reaction of phenyllithium with copper(I) bromide in diethyl ether. Grignard reagents can be used in place of organolithium compounds. Gilman also investigated the dialkylcuprates. These are obtained by combining two equivalent of RLi with Cu(I) salts. Alternatively, these cuprates are prepared from oligomeric neutral organocopper compounds by treatment with one equivalent of organolithium reagent.

Compounds of the type $[CuR_n]^{(n-1)-}$ are reactive towards oxygen and water, forming copper(I) oxide. They also tend to be thermally unstable, which can be useful in certain coupling reactions. Despite or because of these difficulties, organocopper reagents are frequently generated and consumed in situ with no attempt to isolate them. They are used in organic synthesis as alkylating reagents because they exhibit greater functional group tolerance than corresponding Grignard and organolithium reagents. The electronegativity of copper is much higher than its next-door neighbor in the group 12 elements, zinc, suggesting diminished nucleophilicity for its carbon ligands.

Copper salts react with terminal alkynes to form the acetylides.

Alkyl halides react with organocopper compounds with inversion of configuration. On the other hand, reactions of organocopper compound with alkenyl halides proceed with retention of subtrate's configuration.

Organocopper compounds couple with aryl halides:

$$ArX + (Ar')_2CuLi \rightleftharpoons ArAr'CuLi + Ar'X$$
$$2ArAr'CuLi \rightleftharpoons (Ar)_2CuLi + (Ar')_2CuLi$$
$$ArAr'CuLi + O_2 \rightarrow Ar - Ar'$$

Structures

Alkyl and aryl copper complexes aggregate both in crystalline form and in solution. Aggregation is especially evident for charge-neutral organocopper compounds, i.e. species with the empirical formula (RCu), which adopt cyclic structures. Since each copper center requires at least two ligands, the organic group is a bridging ligand. This effect is illustrated by the structure of mesitylcopper, which is a pentamer. A cyclic structure is also seen for $CuCH_2SiMe_3$, first 1:1 organocopper compound to be analyzed by X-ray crystallography. This compound is relatively stable because the bulky trimethylsilyl groups provide steric protection. It is a tetramer, forming an 8-membered ring with alternating Cu-C bonds. In addition the four copper atoms form a planar Cu_4 ring based on three-center two-electron bonds. The copper to copper bond length is 242 pm compared to 256 pm in bulk copper. In pentamesitylpentacopper a 5-membered copper ring is formed, similar to (2,4,6-trimethylphenyl)gold, and pentafluorophenylcopper is a tetramer.

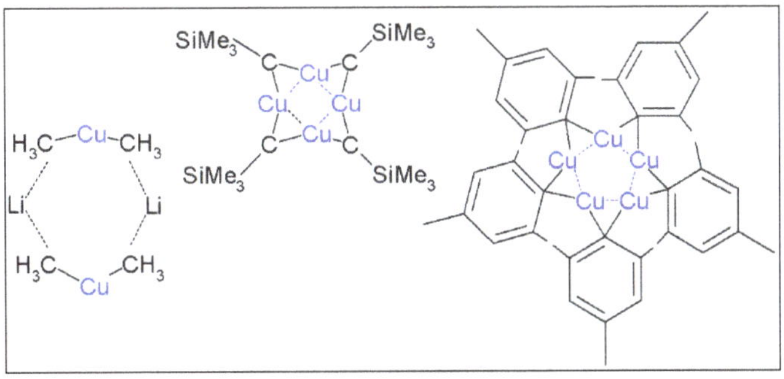

Lithium dimethylcuprate is a dimer in diethyl ether, forming an 8-membered ring with two lithium atoms linking two methyl groups. Similarly, lithium diphenylcuprate forms a dimeric etherate, $[\{Li(OEt_2)\}(CuPh_2)]_2$, in the solid state.

Alkyl and Aryl Copper(III) Compounds

The involvement of the otherwise rare Cu(III) oxidation state has been demonstrated in the conjugate addition of the Gilman reagent to an enone: In a so-called rapid-injection NMR experiment at -100 °C, the Gilman reagent Me_2CuLi (stabilized by lithium iodide) was introduced to cyclohexenone 1. enabling the detection of the copper — alkene pi complex 2. On subsequent addition of trimethylsilyl cyanide the Cu(III) species 3 is formed (indefinitely stable at that temperature) and on increasing the temperature to -80 °C the conjugate addition product 4. According to an accompanying in silico experiments the Cu(III) intermediate has a square planar molecular geometry with the cyano group in cis orientation with respect to the cyclohexenyl methine group and anti-parallel to the methine proton. With other ligands than the cyano group this study predicts room temperature stable Cu(III) compounds.

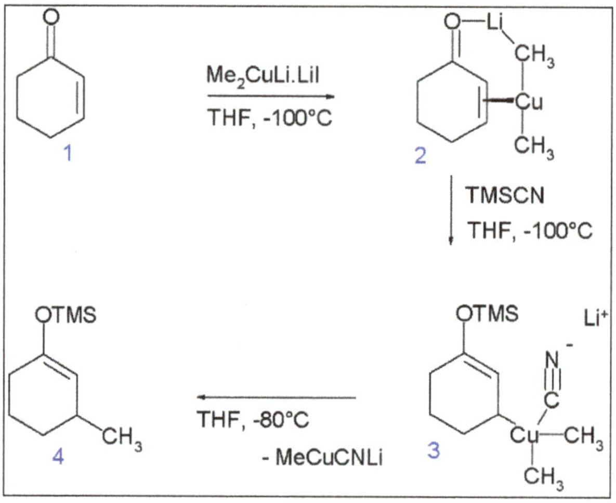

Reactions of Organocuprates

Cross-coupling Reactions

Prior to the development of palladium-catalyzed cross coupling reactions, copper was the preferred catalyst for almost a century. Palladium offers a faster, more selective reaction. However, in recent years copper has reemerged as a synthetically useful metal, because of its lower cost and because it is an eco-friendly metal.

Reactions of R_2CuLi with alkyl halides R'-X give the coupling product:

$$R_2CuLi + R'X \rightarrow R - R' + CuR + LiX$$

The reaction mechanism involves oxidative addition (OA) of the alkyl halide to Cu(I), forming a planar Cu(III) intermediate, followed by reductive elimination (RE). The nucleophilic attack is the rate-determining step. In the substitution of iodide, a single-electron transfer mechanism is proposed.

$$[R - Cu - R]^{-}\ Li^{+} \xrightarrow{R'-X} \left[\begin{array}{c} R' \\ | \\ R - Cu - R \\ | \\ X \end{array}\right]^{-} Li^{+} \rightarrow R - Cu + R - R' + Li - X$$

Many electrophiles participate in this reaction. The approximate order of reactivity, beginning with the most reactive, is as follows: acid chlorides > aldehydes > tosylates ~ epoxides > iodides > bromides > chlorides > ketones > esters > nitriles >> alkenes.

Generally the OA-RE mechanism is analogous to that of palladium-catalyzed cross coupling reactions. One difference between copper and palladium is that copper can undergo single-electron transfer processes.

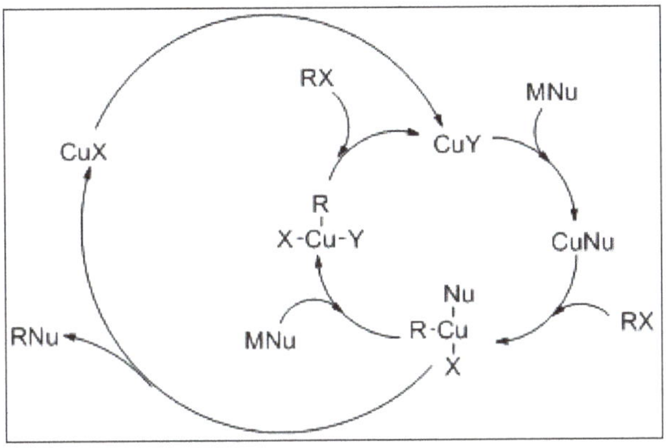

Coupling Reactions

Oxidative coupling is the coupling of copper acetylides to conjugated alkynes in the Glaser coupling (for example in the synthesis of cyclooctadecanonaene) or to aryl halides in the Castro-Stephens Coupling.

Reductive coupling is a coupling reaction of aryl halides with a stoichiometric equivalent of copper metal that occurs in the Ullmann reaction. In an example of a present-day cross coupling reaction called decarboxylative coupling, a catalytic amount of Cu(I) displaces a carboxyl group forming the arylcopper (ArCu) intermediate. Simultaneously, a palladium catalyst converts an aryl bromide to the organopalladium intermediate (Ar'PdBr), and on transmetallation the biaryl is formed from ArPdAr'.

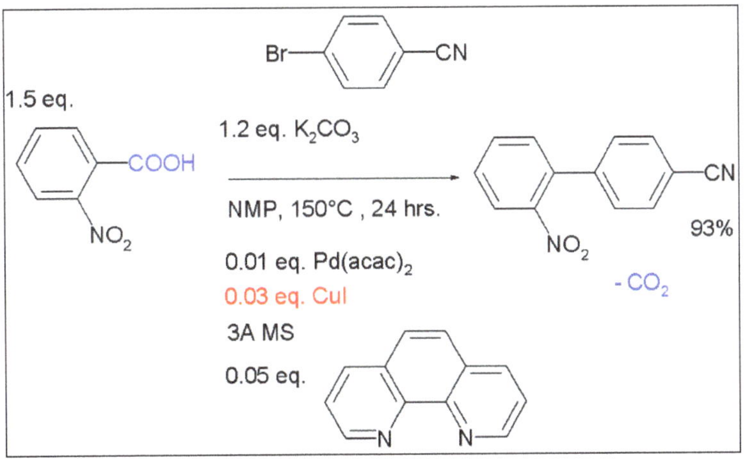

Redox neutral coupling is the coupling of terminal alkynes with halo-alkynes with a copper(I) salt in the Cadiot-Chodkiewicz coupling. Thermal coupling of two organocopper compounds is also possible.

Carbocupration

Carbocupration is a nucleophilic addition of organocopper reagents (R-Cu) to acetylene or terminal alkynes resulting in an alkenylcopper compound (RC = C - Cu). It is a special case of carbometalation and also called the Normant reaction.

Figure show catalytic cycle for carbocupration for the Synthesis of Aldol, Baylis-Hillman Type Products.

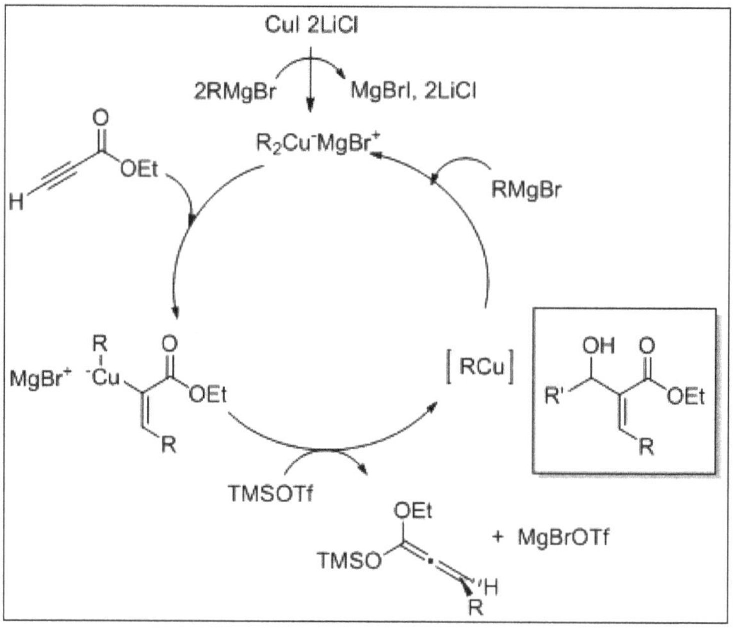

Synthetic Applications

- The Chan-Lam coupling enables the formation of aryl carbon-hetoroatom bonds. It involves coupling of boronic acids, stannanes, or siloxanes with NH- or OH-containing substrates.
- Ullmann reaction involves copper-mediated reactions of aryl halides. Two types of Ullmann reaction are recognized.
- Classic copper-promoted synthesis of symmetric biaryl compounds.
- Copper-promoted nucleophilic aromatic substitution.
- Sonogashira coupling reaction, which utilizes both copper and palladium, entails the coupling of aryl and/or vinyl halides with terminal alkynes.

Reducing Agents

Copper hydrides are specialized reagents used occasionally as reducing agent. The best known copper hydride is called Stryker's reagent, a cluster compound with the formula $[(PPh_3)CuH]_6$. It reduces the alkene α,β-unsaturated carbonyl compounds.

The Buchwald reaction is a copper-catalyzed asymmetric reduction of activated alkenes. The reagent is generated in situ from copper(I) NHC complex. The hydride equivalents are provided by a silane.

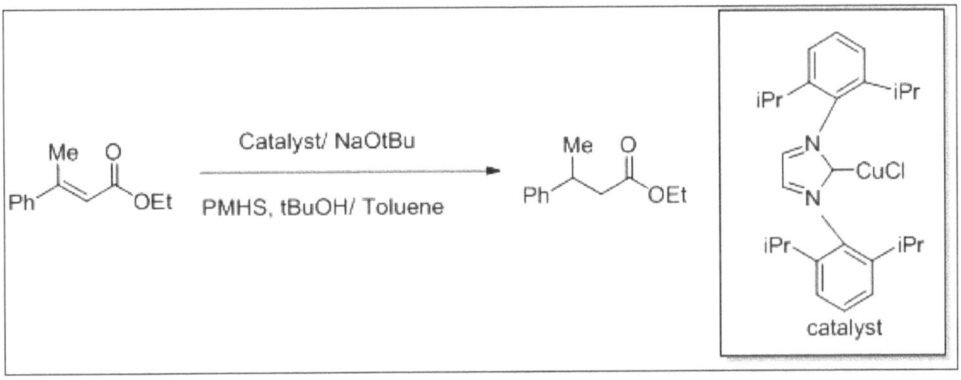

Synthesis of Z-Fluoro alkene dipeptide isosteres. Other effort to make this a more selective reactions includes the use of oxidation reduction condition for the reaction. Fluoride acts as a leaving group and it enhances regioselectivity in the transformation the Z- fluoroalkene.

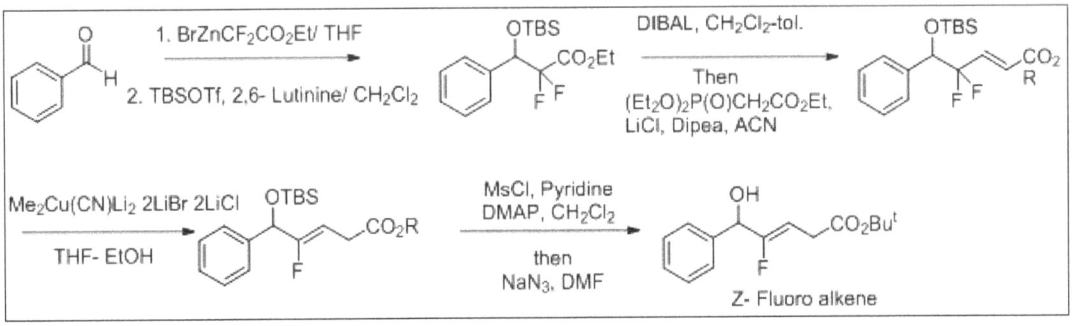

Cu Alkylation Reaction

Generally, the alkylation reaction of organocopper reagents proceeds via gamma-alkylation. Cis-gamma attack occurs better in cyclohexyl carbamate due to sterics. The reaction is reported to be favorable in ethereal solvents. This method was proved to be very effective for the oxidative coupling of amines and alkyl, including tertbutyl, and aryl halides.

Vicinal Functionalization Reactions

Vicinal functionalization using a Carbocupration-Mukaiyama aldol reaction sequence.

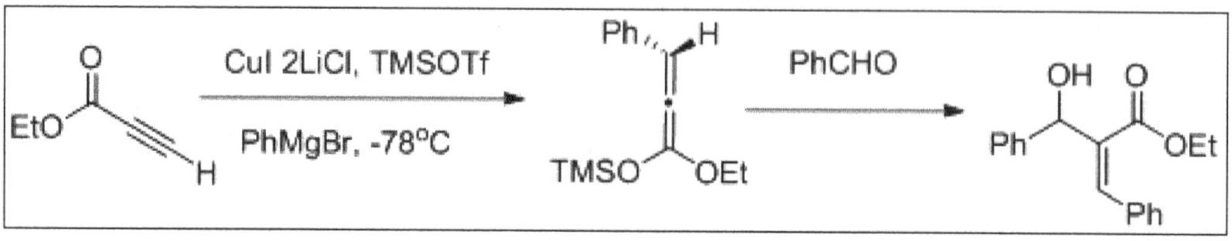

Muller and collaborators reported a vicinal functionalization of α,β-acetylenic esters using a Carbocupration/Mukaiyama aldol reaction sequence carbocupration favors the formation of the Z- aldol.

Copper(I) Acetylide

Copper(I) acetylide, or cuprous acetylide, is a chemical compound with the formula Cu_2C_2. Although never characterized by X-ray crystallography, the material has been claimed at least since 1856. One form is claimed to be a monohydrate with formula $Cu_2C_2.H_2O$). It is a reddish solid, that easily explodes when dry.

Synthesis

Materials purported to be copper acetylide can be prepared by treating acetylene with a solution of copper(I) chloride and ammonia:

$$C_2H_2 + 2\,CuCl \rightarrow Cu_2C_2 + 2\,HCl$$

This reaction produces a reddish solid precipitate.

Properties

When dry, copper acetylide is a heat and shock sensitive high explosive, more thermally sensitive than silver acetylide.

Copper acetylide is thought to form inside pipes made of copper or an alloy with high copper content, which may result in violent explosion. This was found to be the cause of explosions in acetylene plants, and led to abandonment of copper as a construction material in such plants. Copper catalysts used in petrochemistry can also possess a degree of risk under certain conditions.

Reactions

Copper acetylide is the substrate of Glaser coupling for the formation of polyynes. In a typical reaction, a suspension of $Cu_2C_2.H_2O$ in an amoniacal solution is treated with air. The copper is oxidized to Cu^{2+} and forms a blue soluble complex with the ammonia, leaving behind a black solid residue. The latter has been claimed to consist of carbyne, an elusive allotrope of carbon:

$$Cu^+\ {}^-C(\equiv C-C\equiv)_n\,C^-\ Cu^+$$

This interpretation has been disputed.

Freshly prepared copper acetylide reacts with hydrochloric acid to form acetylene and copper(I) chloride. Samples that have been aged with exposure to air or to copper(II) ions liberate also higher polyynes $H(-C\equiv C-)_nH$, with n from 2 to 6, when decomposed by hydrochloric acid. A "carbonaceous" residue of this decomposition also has the spectral signature of $(-C\equiv C-)_n$ chains. It has been conjectured that oxidation causes polymerization of the acetylide anions C_2^{2-} in the solid into carbyne-type anions. $C(\equiv C-C\equiv)_nC^{2-}$ or polycumulene-type anions $C(=C=C=)_mC^{4-}$.

Thermal decomposition of copper acetylide in vacuum is not explosive and leaves copper as a fine powder at the bottom of the flask, while depositing a fluffy very fine carbon powder on the walls. On the basis of spectral data, this powder was claimed to be carbyne $C(-C\equiv C-)_nC$ rather than graphite as expected.

Applications

Though not practically useful as an explosive due to high sensitivity and reactivity towards water, it is interesting as a curiosity because it is one of the very few explosives that do not liberate any gaseous products upon detonation.

The formation of copper acetylide when a gas is passed through a solution of copper(I) chloride is used as a test for the presence of acetylene.

Reactions between Cu^+ and alkynes occur only if a terminal hydrogen is present (as it is slightly acidic in nature). Thus, this reaction is used for identification of terminal alkynes.

ORGANOTITANIUM COMPOUND

Organotitanium compounds in organometallic chemistry contain carbon-to-titanium chemical bonds. Organotitanium chemistry is the science of organotitanium compounds describing their physical properties, synthesis and reactions. They are reagents in organic chemistry and are involved in major industrial processes.

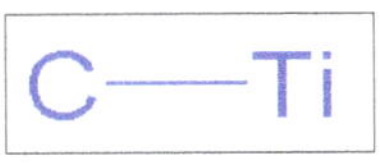

Organotitanium compounds.

Properties

The titanium electron configuration ($[Ar]3d^24s^2$) vaguely resembles that of carbon and like carbon, the +4 oxidation state dominates. Titanium is however a much larger element than carbon, reflected by the Ti-C bond lengths being about 30% longer, e.g. 210 pm in tetrabenzyltitanium vs a typical C-C bond of 155 pm. Simple tetraalkyltitanium compounds however are not typically isolable, owing to the large size of titanium and the electron-deficient nature of its tetrahedral complexes. More abundant and more useful than the simple tetraalkyl compounds are mixed ligand complexes with alkoxide and cyclopentadienyl coligands. Titanium is capable of forming complexes with high coordination numbers.

In terms of oxidation states, most organotitanium chemistry, in solution at least, focuses on derivatives of Ti(IV) and Ti(III). Ti(II) compounds are rarer, examples being titanocene dicarbonyl and $Ti(CH_3)_2(dmpe)_2$. $[Ti(CO)_6]^{2-}$ is formally a complex of Ti(-II). Although Ti(III) is involved in Ziegler–Natta catalysis, the organic derivatives of Ti(III) are uncommon. One example is the dimer $[Cp_2Ti^{III}Cl]_2$.

The oxidation states −1, 0, +1 are also known in organotitanium compounds.

Due to the low electronegativity of titanium, Ti-C bonds are polarized toward carbon. Consequently, alkyl ligands in many titanium compounds are nucleophilic. Titanium is characteristically oxophilic, which presents challenges to handling these compounds, which require air-free techniques. On the other hand, high oxophilicity means that titanium alkyls are effective for abstracting or exchanging organyl ligands for oxo groups.

Compounds

Alkyl Titanium Chlorides and Alkoxides

Organotitanium compounds are somewhat useful reagents in organic chemistry. Many of these reagents and catalysts are incompletely defined. At least from the commercial perspective, the most useful organotitanium compounds are generated by combining titanium(III) chloride and diethylaluminium chloride. As Ziegler–Natta catalysts, such species efficiently catalyze the polymerization of ethylene. The process is heterogeneous and no organotitanium intermediates have been well characterized for this process.

Numerous organotitanium reagents are produced by combining titanium tetrachloride, titanium tetraalkoxides, or mixtures thereof with organolithium, organomagnesium, and organozinc compounds. Such compounds find occasional use as stoichiometric reagents in organic synthesis. "Methyltitanium trichloride", nominally CH_3TiCl_3, can be prepared by treating titanium(IV) chloride with dimethylzinc in dichloromethane at -78 °C. It delivers a methyl groups to carbonyl compounds and alkyl halides. "Methyltriisopropoxytitanium" is a related reagent. A dialkyltitanium species is implicated for Ti-promoted cyclopropanations starting from a Grignard reagent and an ester. This reaction is the basis of the Kulinkovich reaction. The "Lombardo's reagent" is a methylenation reagent. It is functionally related to the Dibromomethane-Zinc-Titanium(IV) Chloride reagent. This chemistry addresses a shortcoming of the Wittig reagent by methylenating enolisable carbonyl groups without loss of stereochemical integrity

(Lombardo Methylenation). It can for example also be applied in a conversion of a ketene into an allene:

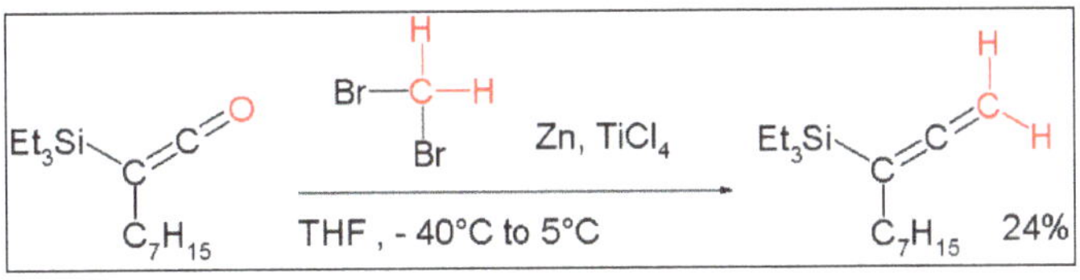

Titanocene Derivatives

A particularly rich area of organotitanium chemistry involves derivatives of titanocene dichloride. Tebbe's reagent is prepared from titanocene dichloride and trimethylaluminium. It is used as a methylenation agent for carbonyl compounds (conversion of $R_2C = O$ to $R_2C=CH_2$). It is an alternative for Wittig reagents when the carbonyl group is sterically challenged or when it easily forms the enol. Tebbe's reagent itself does not react with carbonyl compounds, but must first be treated with a mild Lewis base, such as pyridine, which generates the active Schrock carbene.

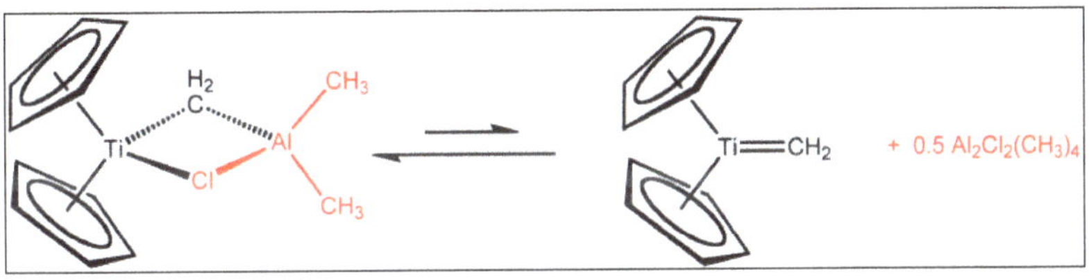

Tebbe's reagent adds simple alkenes to give titanocyclobutanes, which can be regarded as stable olefin metathesis intermediates. These compounds are reagents in itself such as 1,1-bis(cyclopentadienyl)-3,3-dimethyltitanocyclobutane, the adduct of Tebbe's reagent with isobutene catalysed with 4-dimethylaminopyridine.

The Petasis reagent or dimethyl titanocene is prepared from titanocene dichloride and methyllithium in diethyl ether. Compared to Tebbe's reagent it is easier to prepare and easier to handle. It is also a methylenation reagent.

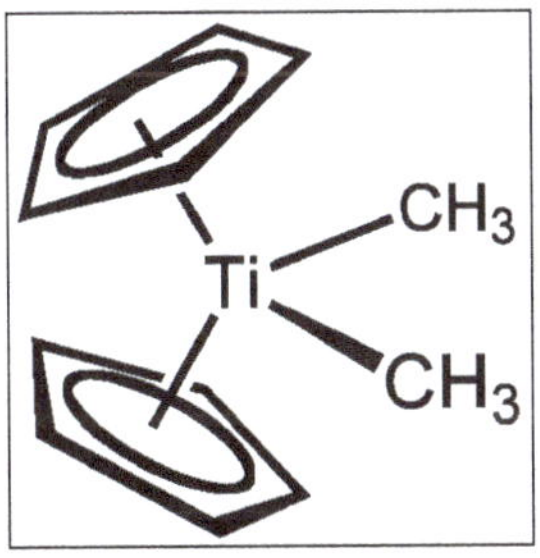

The Petasis reagent, $[(\eta^5\text{-Cp})_2Ti(CH_3)_2]$.

The Nugent-RajanBabu reagent is a one-electron reductant used in synthetic organic chemistry for the generation of alcohols via anti-Markovnikov ring-opening of epoxides, and is generated as a dimer $[(\eta^5\text{-Cp})_2Ti(\mu\text{-Cl})]_2$ and used in situ from titanocene dichloride.

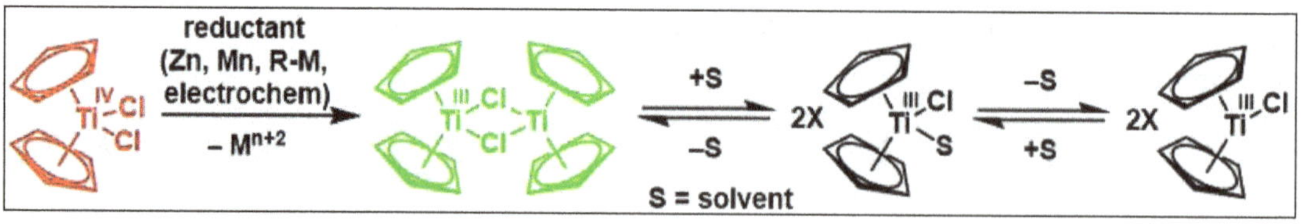

Titanocene

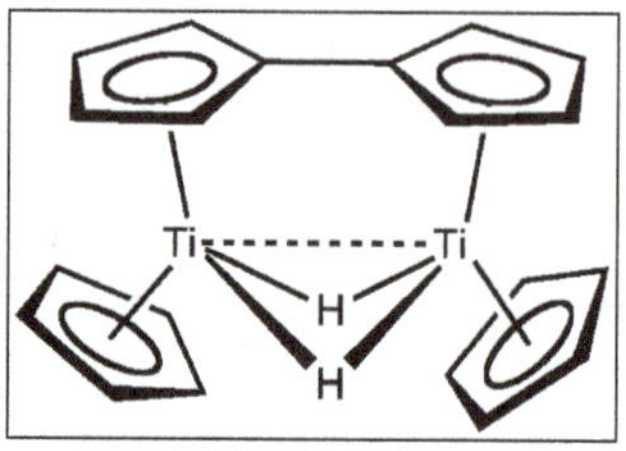

The structure of "titanocene" is not $Ti(C_5H_5)_2$, but a fulvalene complex.

Early work on "titanocene" itself eventually revealed that this species was a fulvalene complex. The titanocene dimer was recognised in the 1970s but not structurally characterised until 1992, and the investigations led to many innovations on cyclopentadienyl complexes of titanium. Only in 1998 was a true titanocene derivative identified, the paramagnetic species $(C_5Me_4SiMe_3)_2Ti$.

MonoCp Compounds

Less useful in organic chemistry but still prominent are many derivatives of (cyclopentadienyl)titanium trichloride, $(C_5H_5)TiCl_3$. This piano-stool complex is obtained by the redistribution reaction of titanocene dichloride and titanium tetrachloride. With an electron count of 12, it is far more electrophilic than the 16e titanocene dichloride.

Titanocene Dichloride

Titanocene dichloride is the organotitanium compound with the formula $(\eta^5\text{-}C_5H_5)_2TiCl_2$, commonly abbreviated as Cp_2TiCl_2. This metallocene is a common reagent in organometallic and organic synthesis. It exists as a bright red solid that slowly hydrolyzes in air. It shows antitumour activity and was the first non-platinum complex to undergo clinical trials as a chemotherapy drug.

Structure

Cp_2TiCl_2 adopts a distorted tetrahedral geometry (counting Cp as a monodentate ligand). The Ti-Cl distance is 2.37 Å and the Cl-Ti-Cl angle is 95°.

Preparation

The standard preparations of Cp_2TiCl_2 start with titanium tetrachloride. The original synthesis by Wilkinson and Birmingham uses sodium cyclopentadienide is still commonly used:

$$2\,NaC_5H_5 + TiCl_4 \rightarrow (C_5H_5)_2TiCl_2 + 2\,NaCl$$

It can also be prepared by using freshly distilled cyclopentadiene rather than its sodium derivative:

$$2\,C_5H_6 + TiCl_4 \rightarrow (C_5H_5)_2TiCl_2 + 2\,HCl$$

Reactions

Halide Replacement Reactions

Cp_2TiCl_2 is a generally useful reagent that effectively behaves as a source of Cp_2Ti^{2+}. A large range of nucleophiles will displace chloride. With NaSH and with polysulfide salts, one obtains the sulfido derivatives $Cp_2Ti(SH)_2$ and Cp_2TiS_5.

The Petasis reagent, $Cp_2Ti(CH_3)_2$, is prepared from the action of methylmagnesium chloride or methyllithium on Cp_2TiCl_2. This reagent is useful for the conversion of esters into vinyl ethers.

The Tebbe reagent $Cp_2TiCl(CH_2)Al(CH_3)_2$, arises by the action of 2 equivalents $Al(CH_3)_3$ on Cp_2TiCl_2.

Reactions affecting Cp Ligands

One Cp ligand can be removed from Cp_2TiCl_2 to give tetrahedral $CpTiCl_3$. This conversion can be effected with $TiCl_4$ or by reaction with $SOCl_2$.

Titanocene itself, $TiCp_2$, is so highly reactive that it rearranges into a Ti^{III} hydride dimer and has been the subject of much investigation. This dimer can be trapped by conducting the reduction of titanocene dichloride in the presence of ligands; in the presence of benzene, a fulvalene complex, μ(η⁵:η⁵-fulvalene)-di-(μ-hydrido)-bis(η⁵-cyclopentadienyltitanium), can be prepared and the resulting solvate structurally characterised by X-ray crystallography. The same compound had been reported earlier by a lithium aluminium hydride reduction and sodium amalgam reduction of titanocene dichloride, and studied by ^1H NMR prior to its definitive characterisation.

Reduction

Reduction with zinc gives the dimer of bis(cyclopentadienyl)titanium(III) chloride in a solvent-mediated chemical equilibrium:

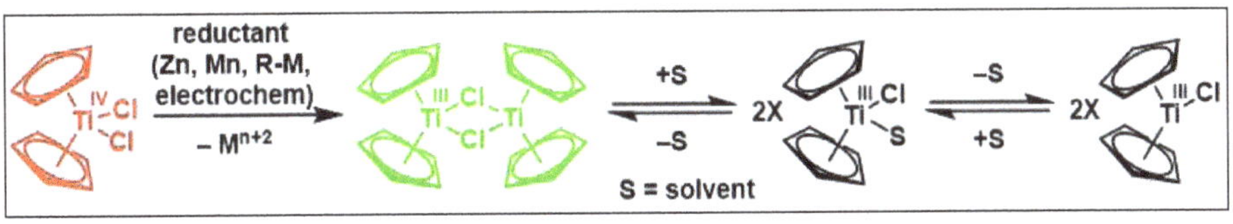

Cp_2TiCl_2 is a precursor to Ti^{II} derivatives. Reductions have been investigated using Grignard reagent and alkyl lithium compounds. More conveniently handled reductants include Mg, Al, or Zn. The following syntheses demonstrate some of the compounds that can be generated by reduction of titanocene dichloride in the presence of π acceptor ligands:

$$Cp_2TiCl_2 + 2CO + Mg \rightarrow Cp_2Ti(CO)_2 + MgCl_2$$
$$Cp_2TiCl_2 + 2PR_3 + Mg \rightarrow Cp_2Ti(PR_3)_2 + MgCl_2$$
$$Cp_2TiCl_2 + 2Me_3SiCCSiMe_3 + Mg \rightarrow Cp_2TiMe_3SiCCSiMe_3 + MgCl_2$$

Alkyne derivatives of titanocene have the formula $(C_5H_5)_2Ti(C_2R_2)$. The corresponding benzyne complexes are known. One family of derivatives are the titanocyclopentadienes.

Titanocene equivalents react with alkenyl alkynes followed by carbonylation and hydrolysis to form bicyclic cyclopentadienones, related to the Pauson–Khand reaction). A similar reaction is the reductive cyclization of enones to form the corresponding alcohol in a stereoselective manner.

Reduction of titanocene dichloride in the presence of conjugated dienes such as 1,3-butadiene gives η^3-allyltitanium complexes. Related reactions occur with diynes. Furthermore, titanocene can catalyze C-C bond metathesis to form asymmetric diynes.

Derivatives of $(C_5Me_5)_2TiCl_2$

Many analogues of Cp_2TiCl_2 are known. Prominent examples are the ring-methylated derivatives $(C_5H_4Me)_2TiCl_2$ and $(C_5Me_5)_2TiCl_2$. The ethylene complex $(C_5Me_5)_2Ti(C_2H_4)$ can be synthesised by Na reduction of $(C_5Me_5)_2TiCl_2$ in the presence of ethylene. The Cp compound has not been prepared. This pentamethylcyclopentadienyl (Cp*) species undergoes many reactions such as cycloadditions of alkynes.

TRANSITION METAL ALKENE COMPLEX

In organometallic chemistry, a transition metal alkene complex is a coordination compound containing one or more alkene ligands. Such compounds are intermediates in many catalytic reactions that convert alkenes to other organic products.

Mono- and dialkenes are often used as ligands in stable complexes.

Monoalkenes

The simplest monoalkene is ethylene. Many complexes of ethylene are known, including Zeise's salt, $Rh_2Cl_2(C_2H_4)_4$, $Cp^*_2Ti(C_2H_4)$, and the homoleptic $Ni(C_2H_4)_3$. Substituted monoalkene include the cyclic cyclooctene, as found in chlorobis(cyclooctene)rhodium dimer. Alkenes with electron-withdrawing groups commonly bind strongly to low-valent metals. Examples of such ligands are TCNE, tetrafluoroethylene, maleic anhydride, esters of fumaric acid. These acceptors form adducts with many zero-valent metals.

Dienes, Trienes, Polyenes, Keto-alkenes and other Complicated Alkene Ligands

Butadiene, cyclooctadiene, and norbornadiene are well-studied chelating agents. Trienes and even some tetraenes can bind to metals through several adjacent carbon centers. Common examples of such ligands are cycloheptatriene and cyclooctatetraene. The bonding is often denoted using the hapticity formalism. Keto-alkenes are tetrahapto ligands that stabilize highly unsaturated low valent metals as found in (benzylideneacetone)iron tricarbonyl and tris(dibenzylideneacetone)dipalladium(0).

Metal alkene complexes:

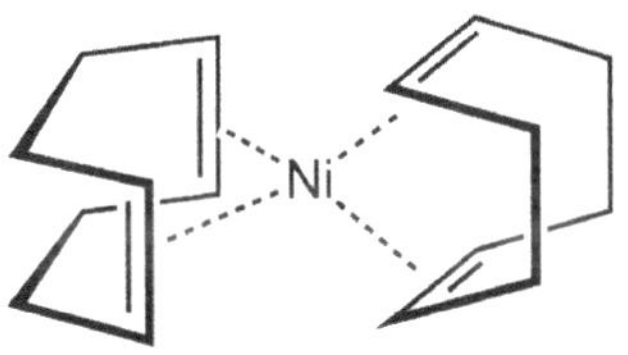	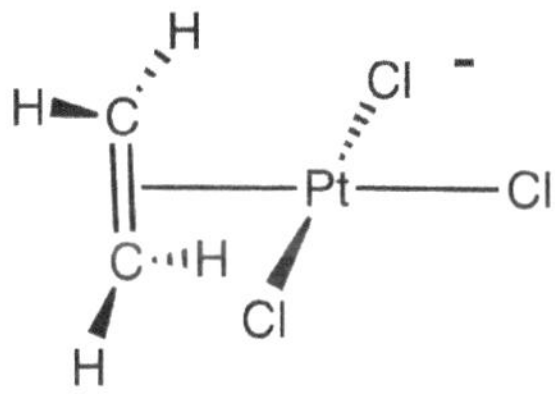	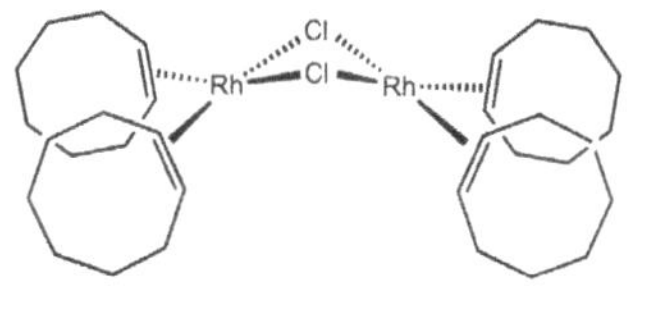
Bis(cyclooctadiene)nickel(0), a catalyst and source of "naked nickel."	The anion in Zeise's Salt, the first alkene complex.	Chlorobis(cyclooctene)rhodium dimer, source of "RhCl".
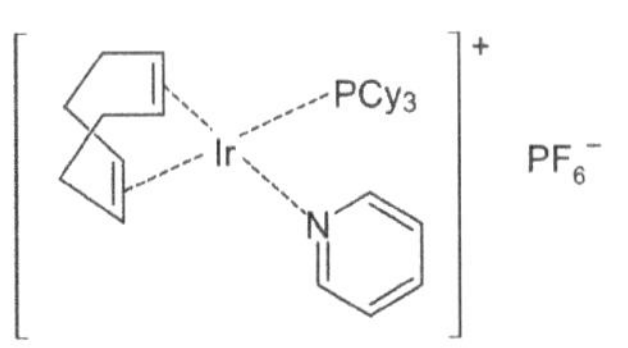	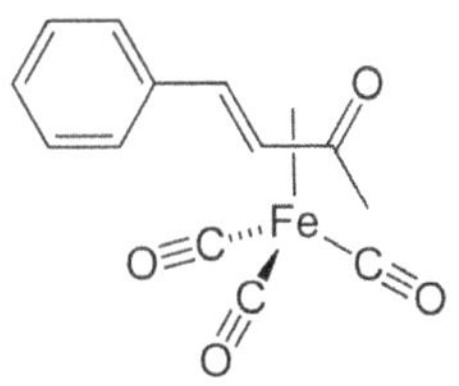	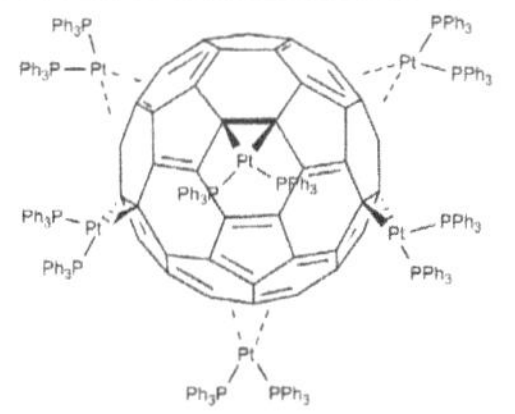
Crabtree's catalyst, a very active catalyst for hydrogenation.	(Benzylideneacetone)iron tricarbonyl, source of "$Fe(CO)_3$".	$[[Et_3P]_2Pt]_6(\eta^2{:}\eta^2{:}\eta^2{:}\eta^2{:}\eta^2{:}\eta^2\text{-}C_{60})$, a fullerene complex.
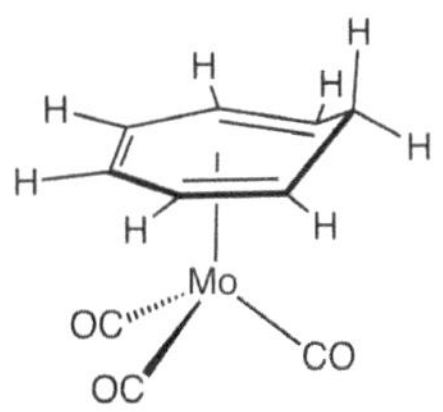	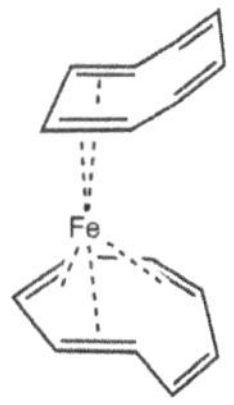	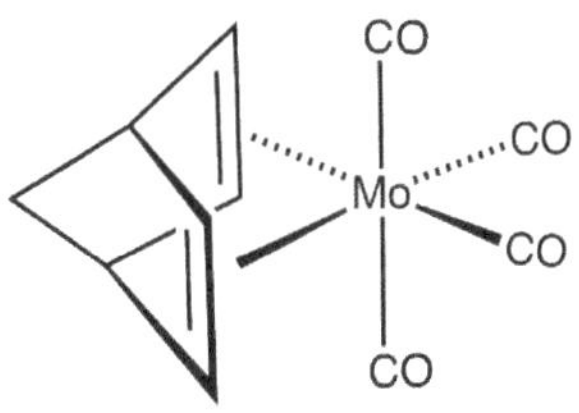
$Mo(C_7H_8)(CO)_3$, a complex of cycloheptatriene.	$Fe(C_8H_8)_2$, a complex of cyclooctatetraene	(Norbornadiene)molybdenum tetracarbonyl, a source of "$Mo(CO)_4$".

Bonding

The bonding between alkenes and transition metals is described by the Dewar-Chatt-Duncanson model, which involves donation of electrons in the pi-orbital on the alkene to empty orbitals on the metal. This interaction is reinforced by back bonding that entails sharing of electrons in other metal orbitals into the otherwise empty pi-antibonding level on the alkene. Early metals of low oxidation state (Ti(II), Zr(II), Nb(III) etc.) are strong pi donors, and their alkene complexes are often described as metallacyclopropanes. Treatment of such species with acids gives the alkanes. Late metals (Ir(I), Pt(II)), which are poorer pi-donors, tend to engage the alkene as a Lewis acid-Lewis base interaction.

Bonding images:

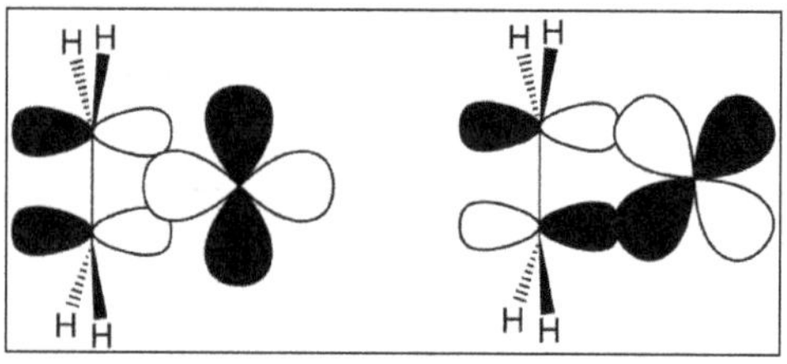

Orbital interactions in a metal-ethylene complex, as described by the Dewar-Chatt-Duncanson model.

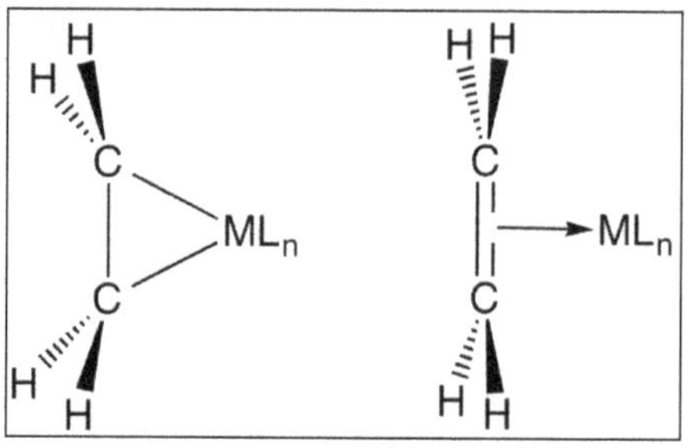

Two extremes depictions of M-C_2H_4 interactions.

Rotational Barrier

The barrier for the rotation of the alkene about the M-centroid vector is a measure of the strength of the M-alkene pi-bond. Low symmetry complexes of ethylene, e.g. $CpRh(C_2H_4)_2$, are suitable for analysis of the rotational barriers associated with the metal-ethylene bond. In Zeise's anion ($[PtCl_3(C_2H_4)]^-$) this rotational barrier cannot be assessed by NMR spectroscopy because all four protons are equivalent.

Reactions

Alkene ligands lose much of their unsaturated character upon complexation. Most famously, the alkene ligand undergoes migratory insertion, wherein it is attacked intramolecularly by alkyl and hydride ligands to form new alkyl complexes. Cationic alkene complexes are susceptible to attack by nucleophiles.

Catalysis

Metal alkene complexes are intermediates in many or most transition metal catalyzed reactions of alkenes: polymerization, hydrogenation, hydroformylation, and many other reactions.

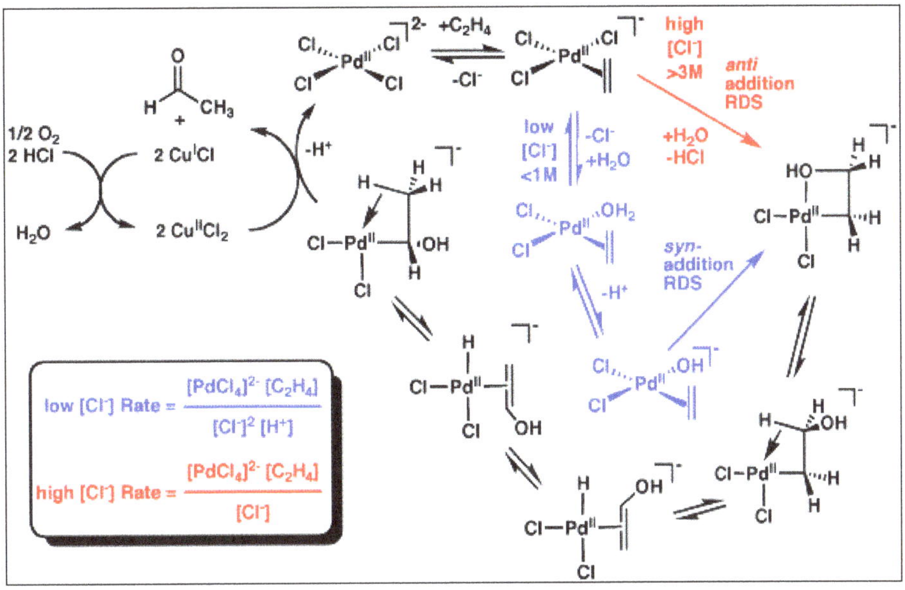

Natural Occurrence

Metal-alkene complexes are uncommon in nature, with one exception. Ethylene affects the ripening of fruit and flowers by complexation to a Cu(I) center in a transcription factor.

TRANSITION METAL ALKYL COMPLEXES

Transition metal alkyl complexes are coordination complexes that contain a bond between a transition metal and an alkyl ligand. Such complexes are not only pervasive but are of practical and theoretical interest.

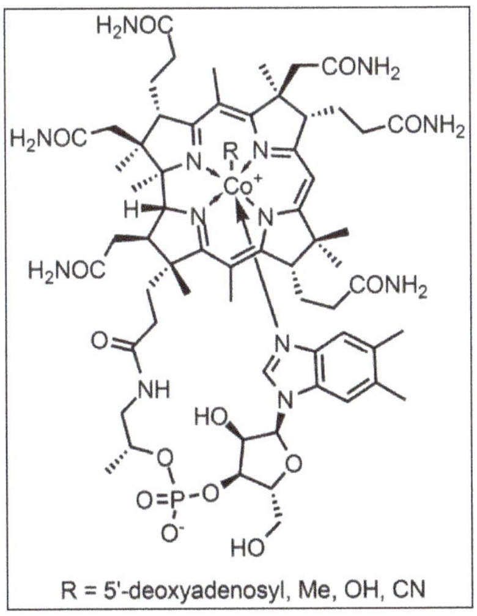

Vitamin B12 is a naturally occurring metal-alkyl complex.

Most metal alkyl complexes contain other, non-alkyl ligands. Great interest, mainly theoretical, has focused on the homoleptic complexes. Indeed, the first reported example of a complex containing a metal-sp^3 carbon bond was the homoleptic complex diethylzinc. Other examples include hexamethyltungsten, tetramethyltitanium, and tetranorbornylcobalt.

Mixed ligand, or heteroleptic, complexes containing alkyls are numerous. In nature, vitamin B12 and its many derivatives contain reactive Co-alkyl bonds.

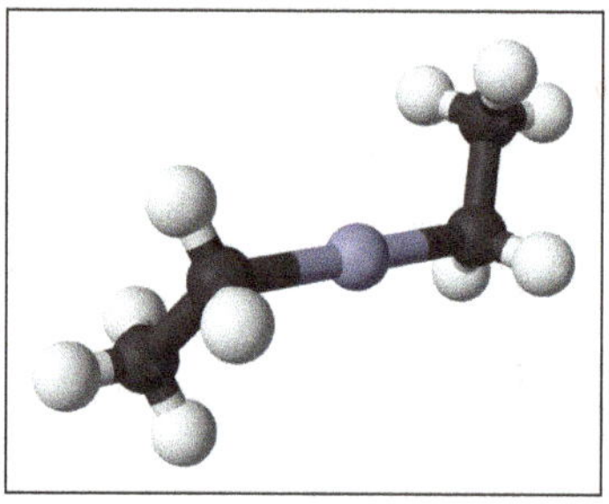

Structure of diethylzinc. The Zn-C bonds measure 194.8(5) pm, while the C-Zn-C angle is slightly bent with 176.2(4)°.

Preparation

Metal alkyl complexes are prepared generally by two pathways, use of alkyl nuncleophiles and use of alkyl electrophiles. Nucleophilic sources of alkyl ligands include Grignard reagents and organolithium compounds. Since many strong nucleophiles are also potent reductants, mildly nucleophilic alkylating agents are sometimes employed to avoid redox reactions. Organozinc compounds and organoaluminium compounds are such milder reagents.

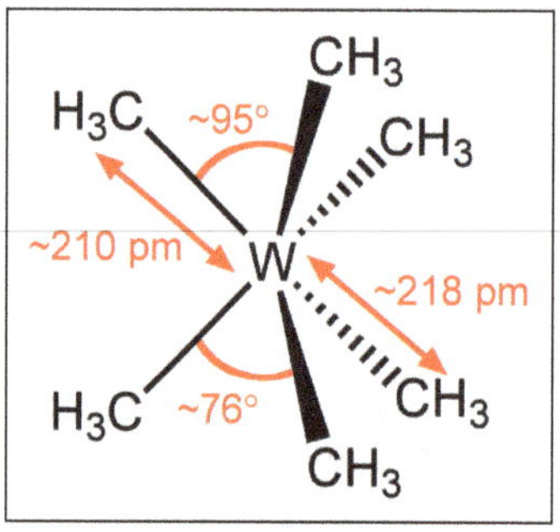

Hexamethyltungsten is an example of a "homoleptic" (all ligands being the same) metal alkyl complex.

Electrophilic alkylation commonly starts with low valence metal complexes. Typical electrophilic reagents are alkyl halides. Illustrative is the preparation of the methyl derivative of cyclopentadienyliron dicarbonyl anion:

$$CpFe(CO)_2Na + CH_3I \rightarrow CpFe(CO)_2CH_3 + NaI$$

Many metal alkyls are prepared by oxidative addition:

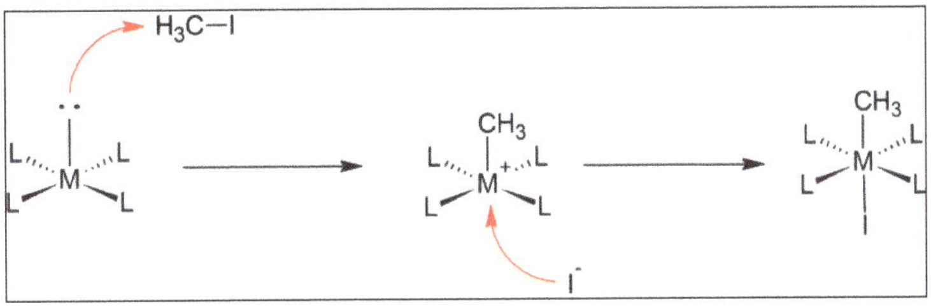

Agostic Interactions and Beta-hydride Elimination

Some metal alkyls feature agostic interactions between a C-H bond on the alkyl group and the metal. Such interactions are especially common for complexes of early transition metals in their highest oxidation states.

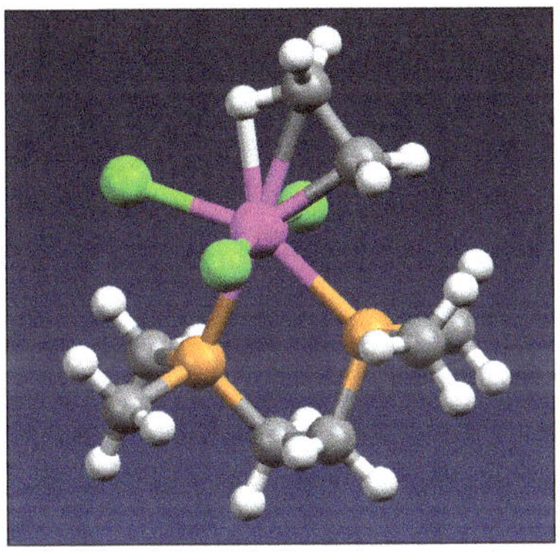

Structure of the alkyl complex $(C_2H_5)TiCl_3(dmpe)$,
highlighting an agostic interaction between the methyl group and the Ti(IV) center.

One determinant of the kinetic stability of metal-alkyl complexes is the presence of hydrogen at the position beta to the metal. If such hydrogens are present and if the metal center is coordinatively unsaturated, then the complex can undergo beta-hydride elimination to form a metal-alkene complex:

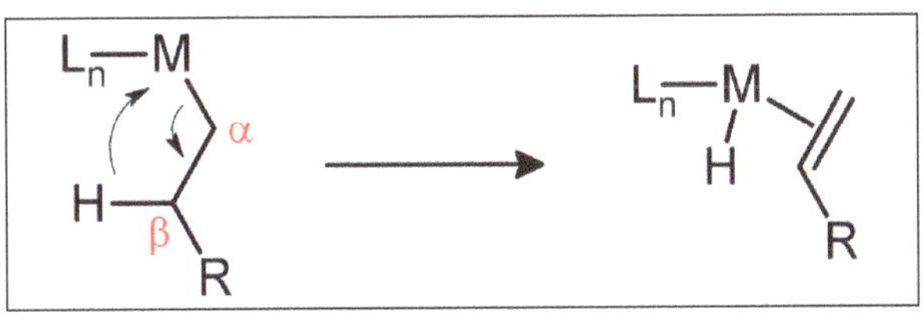

These conversions are assumed to proceed via the intermediacy of agostic interactions.

Catalysis

Many homogeneous catalysts operate via the intermediacy of metal alkyls. These reactions include hydrogenation, hydroformylation, alkene isomerization, and olefin polymerization. It is assumed that the corresponding heterogeneous reactions also involve metal-alkyl bonds.

TRANSITION METAL HYDRIDE

Transition metal hydrides are chemical compounds containing a transition metal bonded to hydrogen. Most transition metals form hydride complexes and some are significant in various catalytic and synthetic reactions. The term "hydride" is used loosely: some so-called hydrides are acidic (e.g., $H_2Fe(CO)_4$), whereas some others are hydridic, having H^--like character (e.g., ZnH_2).

Classes of Metal Hydrides

Binary Metal Hydrides

Many transition metals form compounds with hydrogen, called binary hydrides: binary, because these compounds contain only contain two elements, and hydride, because the hydrogenic ligand is assumed to have hydridic (H–like) character. These compounds are invariably insoluble in all solvents, reflecting their polymeric structures. They often exhibit metal-like electrical conductivity. Many are nonstoichiometric compounds. Electropositive metals (Ti, Zr, Hf, Zn) and some other metals form hydrides with the stoichiometry MH or sometimes MH_2 (M = Ti, Zr, Hf, V, Zn). The best studied are the binary hydrides of palladium, which readily forms a limiting monohydride. In fact, hydrogen gas diffuses through Pd windows via the intermediacy of PdH.

Ternary Metal Hydrides

Ternary metal hydrides have the formula AxMHn, where A+ is an alkali or alkaline earth metal cation, e.g. K^+ and Mg^{2+}. A celebrated example is K_2ReH_9, a salt containing two K^+ ions and the ReH_9^{2-} anion. Other homoleptic metal hydrides include the anions in Mg_2FeH_6 and Mg_2NiH_4. Some of these anionic polyhydrides satisfy the 18-electron rule, many do not. Because of their high lattice energy, these salts are typically not soluble in any solvents, a well-known exception being K_2ReH_9.

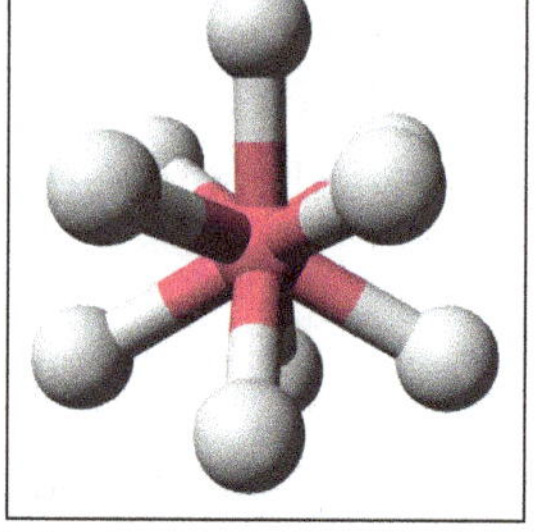

Structure of the ReH^{2-}_9 anion in the salt K_2ReH_9.

Coordination Complexes

The most prevalent hydrides of the transition metals are metal complexes that contain a mix of other ligands in addition to hydride. The range of coligands is large. Virtually all of the metals form such derivatives. The main exceptions include the late metals silver, gold, cadmium, and mercury, which form few or unstable complexes with direct M-H bonds. Examples of an industrially useful hydrides are $HCo(CO)_4$ and $HRh(CO)(PPh_3)_3$, which are catalysts for hydroformylation.

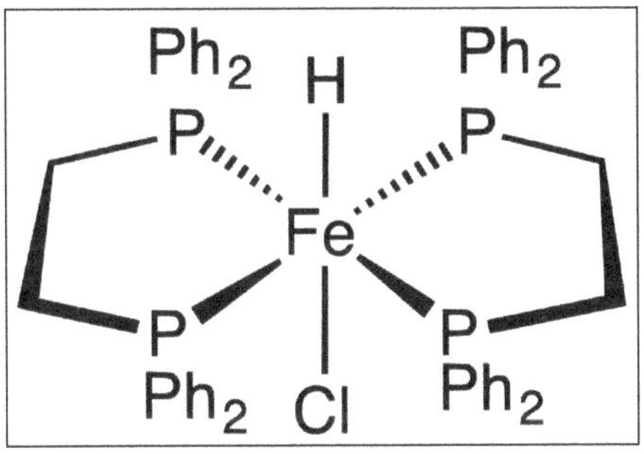

$HFeCl(dppe)_2$ is one of the most accessible transition metal hydrides.

The first molecular hydrides of the transition metals were first reported in the 1930s by Walter Hieber and coworkers. They described $H_2Fe(CO)_4$ and $HCo(CO)_4$. After a hiatus of several years, and following the release of German war documents on the postulated role of $HCo(CO)_4$ in hydroformylation, several new hydrides were reported in the mid-1950s by three prominent groups in organometallic chemistry: $HRe(C_5H_5)_2$ by Wilkinson, $HMo(C_5H_5)(CO)_3$ by Fischer, and $HPtCl(PEt_3)_2$ by Chatt. Thousands of such compounds are now known.

Cluster Hydrides

Like hydrido coordination complexes, many clusters feature terminal (bound by one M–H bond) hydride ligands. Hydride ligands can also bridge pairs of metals, as illustrated by $[HW_2(CO)_{10}]^-$. The cluster $H_2Os_3(CO)_{10}$ features both terminal and doubly bridging hydride ligands. Hydrides can also span the triangular face of a cluster as in $[Ag_3\{(PPh_2)_2CH_2\}_3(\mu_3\text{-H})(\mu_3\text{-Cl})]BF_4$. In the cluster $[Co_6H(CO)_{15}]^-$, the hydride is "interstitial", occupying a position at the center of the Co_6 octahedron. The assignment for cluster hydrides can be challenging as illustrated by studies on Stryker's reagent $[Cu_6(PPh_3)_6H_6]$.

Synthesis

Hydride Transfer

Nucleophilic main group hydrides convert many transition metal halides and cations into the corresponding hydrides:

$$ML_nX + LiBHEt_3 \rightarrow HML_n + BEt_3 + LiX$$

These conversions are metathesis reactions, and the hydricity of the product is generally less than of the hydride donor. Classical (and relatively cheap) hydride donor reagents include sodium borohydride and lithium aluminium hydride. In the laboratory, more control is often offered by "mixed hydrides" such as lithium triethylborohydride and Red-Al. Alkali metal hydrides, e.g. sodium hydride, are not typically useful reagents.

Elimination Reactions

Beta-hydride elimination and alpha-hydride elimination are processes that afford hydrides. The former a common termination pathway in homogeneous polymerization. It also allows some transition metal hydride complexes to be synthesized from organolithium and Grignard reagents:

$$ML_nX + LiC_4H_9 \rightarrow C_4H_9ML_n + LiX$$

$$C_4H_9ML_n \rightarrow HML_n + H_2C{=}CHC_2H_5$$

Oxidative Additions

Oxidative addition of dihydrogen to a low valent transition metal center is common aspect of hydrogenation. A classic example involves Vaska's complex:

$$Ir^{I}Cl(CO)(PPh_3)_2 + H_2 \rightleftharpoons H_2Ir^{III}Cl(CO)(PPh_3)_2$$

Oxidative addition also can occur to dimetallic complexes, e.g.:

$$Co_2(CO)_8 + H_2 \rightleftharpoons 2\ HCo(CO)_4$$

Many acids participate in oxidative additions, as illustrated by the addition of HCl to Vaska's complex:

$$Ir^{I}Cl(CO)(PPh_3)_2 + HCl \rightarrow HIr^{III}Cl_2(CO)(PPh_3)_2$$

Heterolytic Cleavage of Dihydrogen

Some metal hydrides form when a metal complex is treated with hydrogen in the presence of a base. The reaction involves no changes in the oxidation state of the metal and can be viewed as splitting H_2 into hydride (which binds to the metal) and proton (which binds to the base).

$$ML_n^{x+} + Base + H_2 \rightleftharpoons HML_n^{(x-1)+} + HBase^+$$

Such reaction is assumed to involve the intermediacy of dihydrogen complexes. Bifunctional catalysts activate H_2 in this way.

Thermodynamic Considerations

The M-H bond can in principle cleave to produce a proton, hydrogen radical, or hydride.

$$HML_n \rightleftharpoons ML_n^- + H^+$$
$$HML_n \rightleftharpoons ML_n + H$$
$$HML_n \rightleftharpoons ML_n^+ + H^-$$

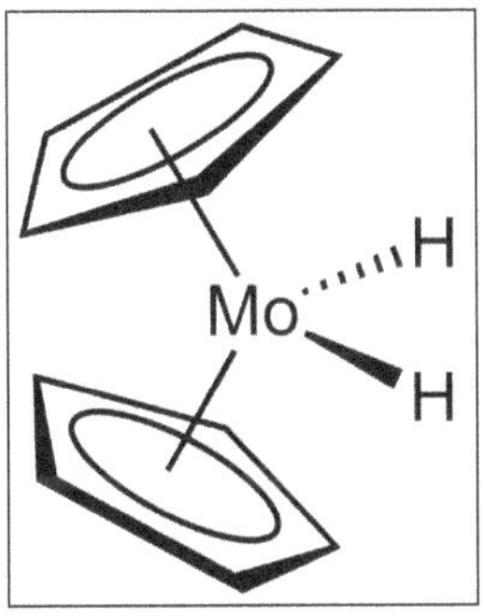

Molybdocene dihydride is produced using $NaBH_4$ as the hydride source.

Although these properties are interrelated, they are not interdependent. A metal hydride can be a thermodynamically a weak acid and a weak H^- donor; it could also be strong in one category but not the other or strong in both. The H^- strength of a hydride also known as its hydride donor ability or hydricity corresponds to the hydride's Lewis base strength. Not all hydrides are powerful Lewis bases. The base strength of hydrides varies as much as the pKa of protons. This hydricity can be measured by heterolytic cleaving hydrogen between a metal complex and base with a known pK_a then measuring the resulting equilibrium. This presupposes that the hydride doesn't heterolytically or homolytically react with itself to reform hydrogen. A complex would homolytically react with itself if the homolytic M-H bond is worth less than half of the homolytic H-H bond. Even if the homolytic bond strength is above that threshold the complex is still susceptible to radical reaction pathways.

$$2\ HML_n^z \rightleftharpoons 2\ ML_n^z + H_2$$

A complex will heterolytically react with itself when it's simultaneously a strong acid and a strong hydride. This conversion results in disproportionation producing a pair of complexes with oxidation states that differ by two electrons. Further electrochemical reactions are possible.

$$2HML_n^z \rightleftharpoons ML_n^{z+1} + ML_n^{z-1} + H_2$$

As noted some complexes heterolytically cleave dihydrogen in the presence of a base. A portion of these complexes result in hydride complexes acidic enough to be deprotonated a second time by the base. In this situation the starting complex can be reduced by two electrons with hydrogen and base. Even if the hydride is not acidic enough to be deprotonated it can homolytically react with it for an overall one electron reduction.

Two deprotonations: $ML_n^{z} + H_2 + 2Base \rightleftharpoons ML_n^{z-2} + 2H^+ Base$

Deprotonation followed by homolysis: $2ML_n^{z} + H_2 + 2Base \rightleftharpoons 2ML_n^{z-1} + 2H^+Base$

Hydricity

For a Lewis acid, its affinity for a hydride ligand is called its hydricity:

$$ML_nH^{n-} \rightleftharpoons ML_n^{(n+1)-} + H^-$$

Since hydride does not exist as a stable anion in solution, this equilibrium constant (and its associated free energy) is calculated from measurable equilibria. The reference point is the hydricity of a proton, which in acetonitrile solution is calculated at −76 kcal/mol:

$$H^+ + H^- \rightleftharpoons H_2$$

Relative to a proton, most cations exhibit a lower affinity for H-. Some examples include:

$$\left[Ni(dppe)_2\right]^{2+} + H^- \rightleftharpoons \left[HNi(dppe)_2\right]^+ \ \Delta G(298) = -63 \text{ kcal/mol}$$

$$\left[Ni(dmpe)_2\right]^{2+} + H^- \rightleftharpoons \left[HNi(dmpe)_2\right]^+ \ \Delta G(298) = -50.7$$

$$\left[Pt(dppe)_2\right]^{2+} + H^- \rightleftharpoons \left[HPt(dppe)_2\right]^+ \ \Delta G(298) = -53$$

$$\left[Pt(dmpe)_2\right]^{2+} + H^- \rightleftharpoons \left[HPt(dmpe)_2\right]^+ \ \Delta G(298) = -42.6$$

These data suggest that $[HPt(dmpe)_2]^+$ would be a strong hydride donor, reflecting the relatively high stability of $[Pt(dmpe)_2]^{2+}$. The ionic bond character in transition metal hydrides can be measured for D ligands via deuterium quadrupole coupling constants.

Kinetics and Mechanism

The rates of proton-transfer to and between metal complexes are often slow. Many hydrides are inaccessible to study through Bordwell thermodynamic cycles. As a result, kinetic studies are employed to elucidate both the relevant thermodynamic parameters. Generally hydrides derived from first row transition metals display the most rapid kinetics followed by the second and third row metal complexes.

Structure and Bonding

The determination of structures of metal hydrides can be challenging since hydride ligands do not scatter X-ray well, especially in comparison to the attached metal. Consequently M-H distances are often underestimated, especially in early studies. Often the presence of a hydride ligand was deduced by the absence of a ligand at an apparent coordination site. Classically, the structure of metal hydrides was addressed by neutron diffraction since hydrogen strongly scatters neutrons.

Metal complexes containing terminal hydrides are common. In bi- and polynuclear compounds, hydrides usually are bridging ligands. Of these bridging hydrides many are oligomeric, such as Stryker's reagent. $[(Ph_3P)CuH]_6$ and clusters such as $[Rh_6(PR_3)_6H_{12}]^{2+}$. The final bonding motif is the non-classical dihydride also known as sigma bond dihydrogen adducts or simply dihydrogen complexes. The $[W(PR_3)_2(CO)_3(H_2)]$ complex was the first well characterized example of both a non-classical dihydride and sigma-bond complex in general. X-ray diffraction is generally insufficient to locate hydrides in crystal structures and thus their location must be assumed. It requires Neutron diffraction to unambiguously locate a hydride near a heavy atom crystallographically. Non-classical hydrides have also been studied with a variety of variable temperature NMR techniques and HD Couplings.

Classical Terminal: M—H

Classical Bridging: M—H—M

Non-Classical: M—H_2

Spectroscopy

Late transition metal hydrides characteristically show up-field shifts in their proton NMR spectra. It is common for the M-H signal to appear between δ-5 and -25 with many examples outside this range but generally all appear below 0 ppm. The large shifts arise from the influence of the excited states and due to strong spin-orbit coupling (in contrast, ^{1}H NMR shifts for organic compounds typically occur in the range δ12-1). At one extreme is the 16e complex $IrHCl_2(PMe(t\text{-}Bu)_2)_2$ with a shift of -50.5. The signals often exhibit spin-spin coupling to other ligands, e.g. phosphines.

Metal hydrides exhibit IR bands near 2000 cm^{-1} for $\nu_{M\text{-}H}$, although the intensities are variable. These signals can be identified by deuterium labeling.

CARBONYL COMPLEXES

Carbonyl Complexes are compounds that contain carbon monoxide as a coordinated ligand. Carbon monoxide is a common ligand in transition metal chemistry, in part due to the synergistic nature of its bonding to transition metals. We can describe the bonding of CO to a metal as consisting of two components. The first component is a two electron donation of the lone pair on carbon (coordination exclusively through the oxygen is extremely rare) into a vacant metal d-orbital. This electron donation makes the metal more electron rich, and in order to compensate for this increased electron density, a filled metal d-orbital may interact with the empty pi* orbital on the carbonyl ligand to relieve itself of the added electron density. This second component is called pi-backbonding or pi-backdonation. This is shown diagramatically as well as through a simple MO picture below. The MO has been color coded so you can identify each component of the bonding interaction:

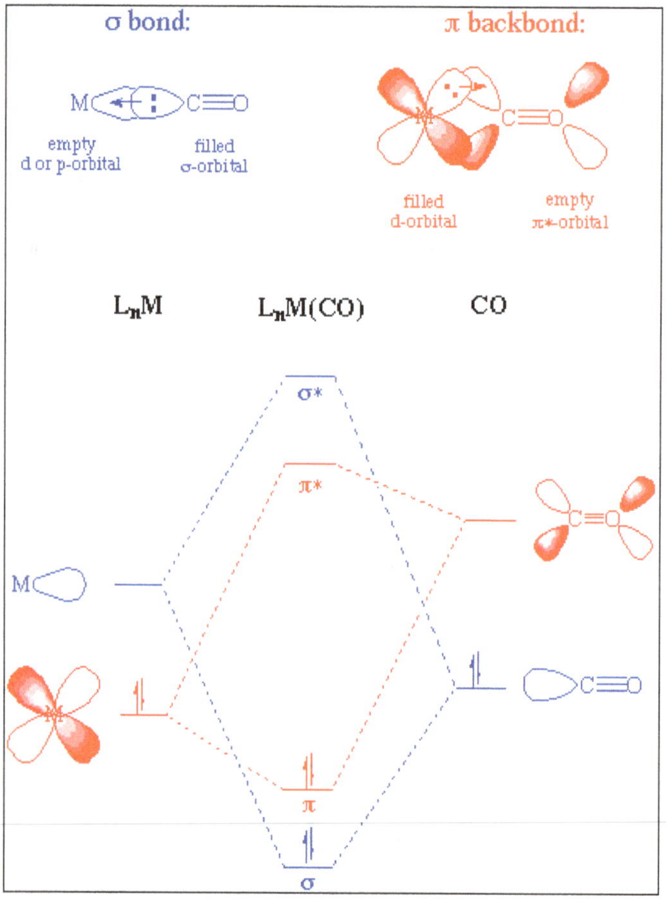

The two components of this bonding are synergistic. The more sigma donation by the carbonyl (or other sigma-donors on the metal center), the stronger the pi-backbonding interaction. Notice that although this involves the occupation of a pi* orbital on the CO, it is still a bonding interaction as far as the metal center is concerned. There is a fundamental similarity between the nature of carbonyl-metal bonding and that of alkenes, acetylenes, phosphines, and dihydrogen.

This occupation of the pi* on CO does lead to a decreased bond order in the carbon monoxide molecule itself. As we might expect, as the pi-backdonation becomes stronger, the CO bond order should decrease from that of the free ligand. Two consequences that we might expect if the CO bond order was reduced would be a lengthening of the C-O bond and a decrease in the carbonyl stretching frequency in the IR. Both of these hold true.

Bonding Modes of CO

CO typically bonds in an end-on fashion through carbon as shown above. However, bridging carbonyls are not uncommon and often undergo exchange with terminal carbonyls. A variant of a bridging carbonyl is the "semi-bridging" carbonyl in which the M-CO-M bond is asymmetric rather than symmetric. Some less common bonding modes are shown below; notice that the metrical data provided in each case are consistent with a decrease in the CO bond order:

Free C≡O

d_{CO} = 1.128Å
ν_{CO} = 2143 cm^{-1}

Terminal carbonyls d_{CO} = 1.12-1.18Å

$Ni(CO)_4$ ν_{CO} = 2057 cm^{-1}
$(PR_3)_2ClIr(CO)$ ν_{CO} = 1960 cm^{-1}
$Co(CO)_4^-$ ν_{CO} = 1886 cm^{-1}

μ_2-Bridging carbonyls d_{CO} = 1.17-1.22Å

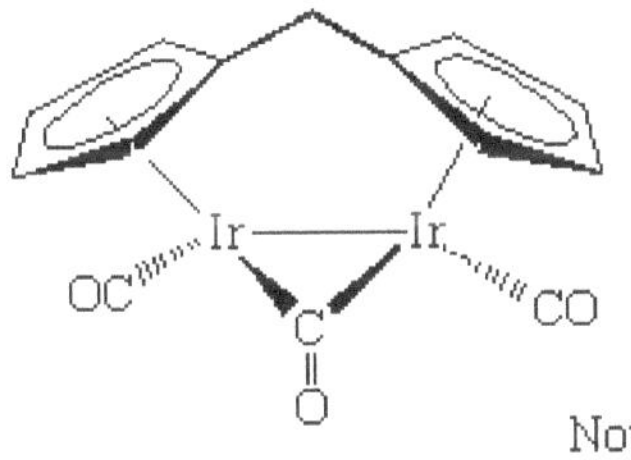

ν_{CO} = 2002(s), 1959(w), and 1776(m) cm^{-1}
d_{CO} (bridge) = 1.207(58) Å
d_{CO} (terminal) = 1.177(59) Å
^{13}C: δ 203.4 (terminal), 171.40 (bridge)
Notice that the C=O bond is ⊥ to the M-M bond

μ_3-Bridging carbonyls d_{CO} = 1.17-1.22Å

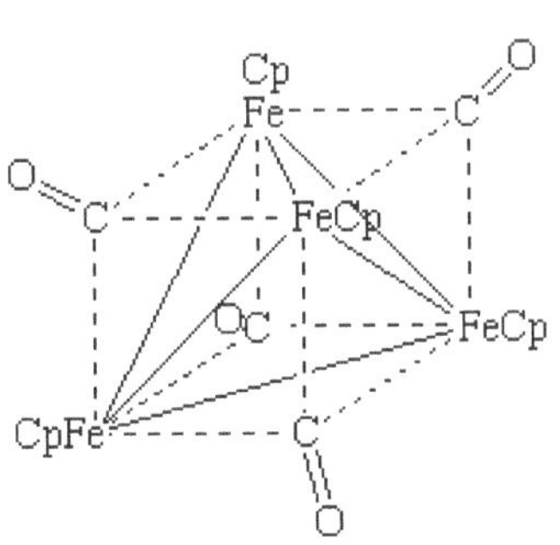

ν_{CO} = 1620 cm^{-1}
d_{CO} = 1.197(7) - 1.209(7) Å

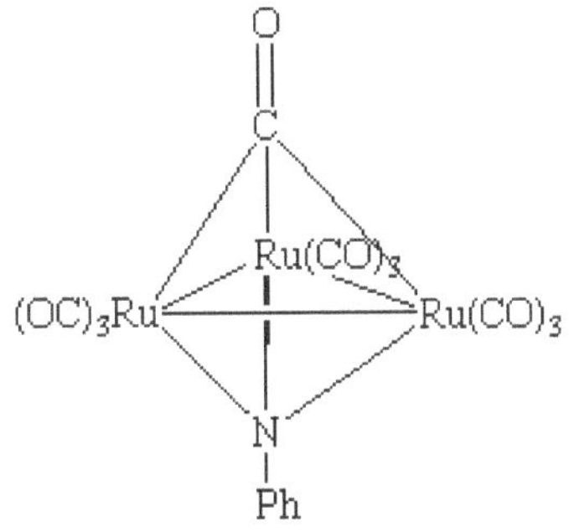

ν_{CO} = 2104(m), 2070(s), 2032(vs) 2011(s) 1978(w), 1743 cm^{-1}

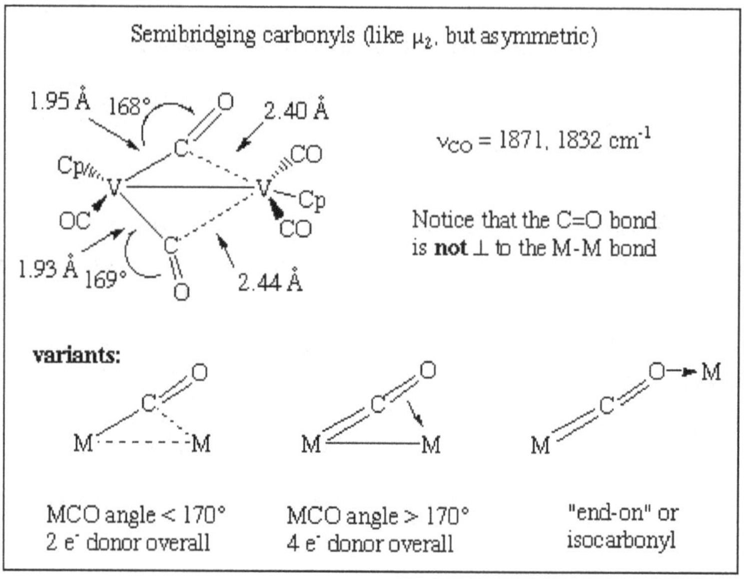

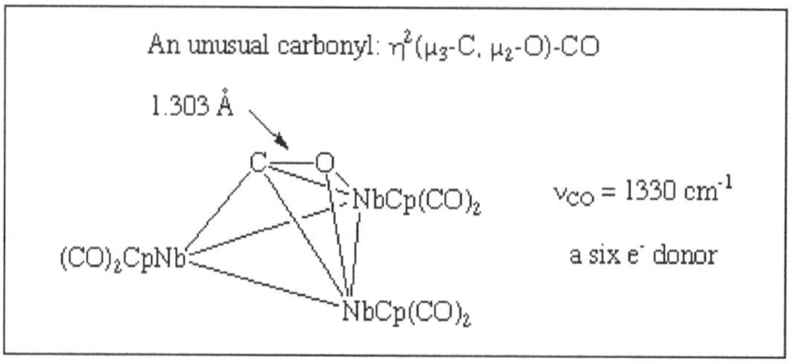

Spectroscopic Features of Carbonyl Complexes

In the IR, typical stretching frequencies are:

- Uncoordinated or "free" CO: 2143 cm^{-1}.
- Terminal M-CO: 2125 to 1850 cm^{-1}.
- Doubly bridging (mu^{-2}): 1850 to 1750 cm^{-1}.
- Triply bridging (mu^{-3}): 1675 to 1600 cm^{-1}.
- Semibridging: somewhere between terminal and mu^{-2}.

Two convenient trends are observed in the IR spectra of carbonyl complexes that are both consistent with the concept of pi-backbonding:

- With each charge added to the metal center, the CO stretching frequency decreases by approximately 100 cm^{-1}.

- The better the sigma-donating capability (or worse the pi-acceptor ability) of the other ligands on the metal, the lower the CO stretching frequency.

For simple carbonyl complexes, counting the number of IR and Raman CO stretching frequencies will often permit one to make a structural assignment. The number of CO stretches expected for possible geometries/isomers can be predicted using group theory and the calculated results compared to the experimental data.

In the ^{13}C NMR spectrum, coordinated carbonyl ligands typically appear in the range of 180 to 250 ppm. Isotopically enriched carbonyl complexes are often prepared to simplify mechanistic investigations or to facilitate the collection of the spectrum. In addition, the coupling of a ^{13}C-enriched complex to other spin active nuclei such as ^{103}Rh or ^{31}P, can aid in structural assignments.

Synthesis of Metal Carbonyls

Metal carbonyls can be made in a variety of ways. Here are just a few examples:

- For Ni and Fe, the homoleptic or binary metal carbonyls can be made by the direct interaction with the metal:

$$Ni^{o} + xs\ CO \xrightarrow{25\ ^{\circ}C} Ni(CO)_4$$

- In other cases, a reduction of a metal precursor in the presence of CO (or using CO as the reductant) is used:

$$WMe_6\ xs\,CO \rightarrow W(CO)_6 + 3Me_2CO$$

$$Re_2O_7 + xs\,CO \rightarrow (OC)_5Re - Re(CO)_5 + 7CO_2$$

- Carbon monoxide also reacts with various metal complexes, most typically filling a vacant coordination site:

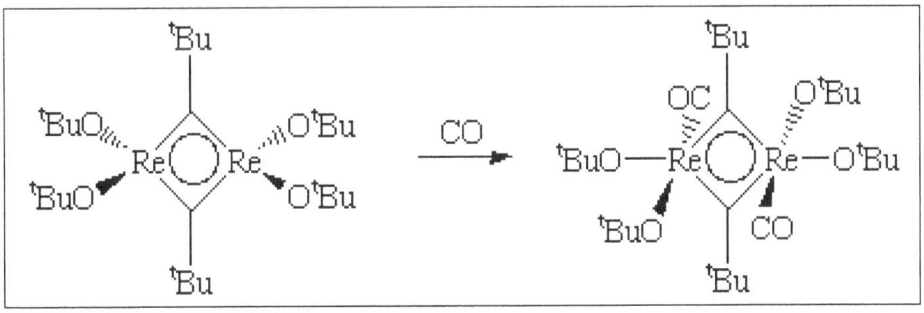

Or performing ligand substitution reactions:

$$[Ru(NH_3)_5H_2O]^{2+} \xrightarrow{CO} [Ru(NH_3)_5(CO)]^{2+}$$

- Occasionally, CO ligands are derived from the reaction of a coordinated ligand through a deinsertion reaction.

$$RhCl(PPh_3)_3 + RCHO \rightarrow\rightarrow\rightarrow RhCl(CO)(PPh_3)_2 + RH$$

References

- Organotin-compound, medicine-and-dentistry: sciencedirect.com, Retrieved 02 August, 2020
- Organo-tin-compounds: npi.gov.au, Retrieved 25 May, 2020
- Iron-Catalyzed Reactions in Organic SynthesisCarsten Bolm, Julien Legros, Jacques Le Paih, and Lorenzo Zani Chem. Rev. 2004, 104, 6217-6254 doi:10.1021/cr040664h
- Mohammadi, Narges; Ganesan, Aravindhan; Chantler, Christopher T.; Wang, Feng (2012). "Differentiation of ferrocene D5d and D5h conformers using IR spectroscopy". Journal of Organometallic Chemistry. 713: 51–59. doi:10.1016/j.jorganchem.2012.04.009
- Carbonyl, organomet: ilpi.com, Retrieved 02 August, 2020
- Staeb, T.H.; Chavez, J.; Gleiter, R.; Nuber, Bernhard (2005). "The Role of [η2-Bis(tert-butylsulfonyl)acetylene](carbonyl)(η5-cyclopentadienyl) cobalt(I) as an Intermediate in the Alkyne Dimerisation". Eur. J. Inorg. Chem. 2005 (20): 4090. doi:10.1002/ejic.200500394
- Delbaere, L. T. J.; McBride, D. W.; Ferguson, R. B. (1970). "Crystal structure of π-cyclopentadienyl(triethylphosphine)copper(I), π-C5H5CuP(C2H5)3". Acta Crystallographica B. 26 (5): 515–21. doi:10.1107/S056774087000273X

5

Organometallic Reactions

Organometallic reactions form a new bond through reduction in a formal oxidation state of metals. They show the basic and nucleophilic character of the carbon atoms in organometallic compounds. Some of the reactions are hydrogenation reaction, carbometalation, transmetalation, oxidative addition reaction, etc. This chapter delves into the aspects of organometallic reactions for a thorough understanding of the subject.

HYDROGENATION REACTION

Hydrogenation refers to the treatment of substances with molecular hydrogen (H_2), adding pairs of hydrogen atoms to compounds (generally unsaturated compounds). These usually require a catalyst for the reaction to occur under normal conditions of temperature and pressure. Most hydrogenation reactions use gaseous hydrogen as the hydrogen source, but alternative sources have been developed. The reverse of hydrogenation, where hydrogen is removed from the compounds, is known as dehydrogenation. Hydrogenation differs from protonation or hydride addition because in hydrogenation the products have the same charge as the reactants.

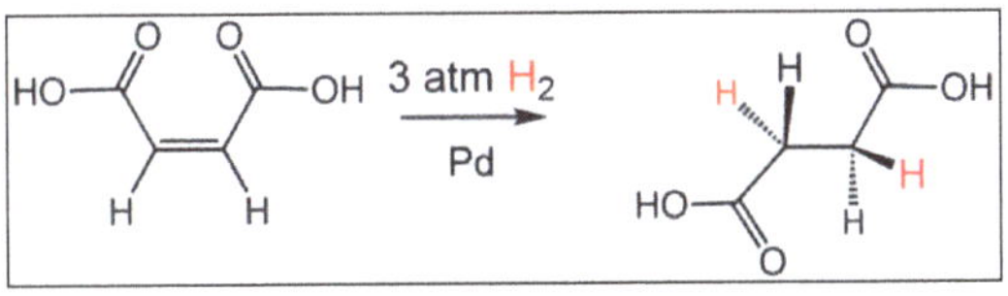

Hydrogenation: Hydrogen can be added across a double bond—such as the olefin in maleic acid shown—by utilizing a catalyst, such as palladium.

Hydrogenation reactions generally require three components: the substrate, the hydrogen source, and a catalyst. The reaction is carried out at varying temperatures and pressures depending on the catalyst and substrate used. The hydrogenation of an alkene produces an alkane. The addition of hydrogen to compounds happens in a syn addition fashion, adding to the same face of the compound and entering from the least hindered side. Generally, alkenes will convert to alkanes, alkynes to alkenes, aldehydes and ketones to alcohols, esters to secondary alcohols, and amides to amines via hydrogenation reactions.

Catalysts of Hydrogenation

Generally, hydrogenation reactions will not occur between hydrogen and organic compounds below 480 degrees Celsius without metal catalysts. Catalysts are responsible for binding the H_2molecule and facilitating the reaction between the hydrogen and the substrate. Platinum, palladium, rhodium, and ruthenium are known to be active catalysts which can operate at lower temperatures and pressures. Research is ongoing to procure non-precious metal catalysts which can produce similar activity at lower temperatures and pressures. Nickel-based catalysts, such as Raney nickel, have been developed, but still require high temperatures and pressures.

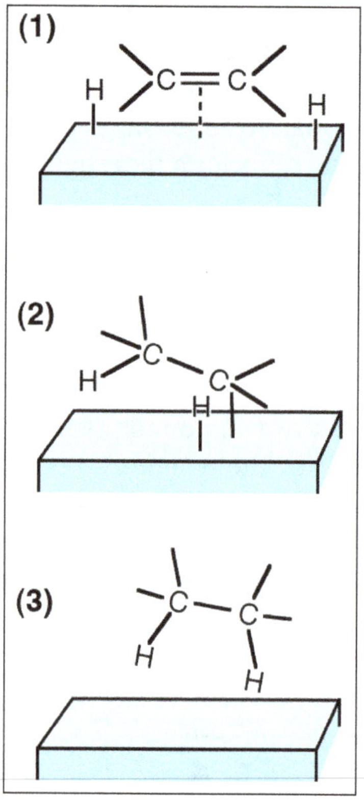

Heterogeneous Catalysis: The hydrogenation of ethylene (C2H4) on a solid support is an example of heterogeneous catalysis.

Catalysts can be divided into two categories: homogeneous or heterogeneous catalysts. Homogeneous catalysts are soluble in the solvent that contains the unsaturated substrate. Heterogeneous catalysts are found more commonly in industry, and are not soluble in the solvent containing the substrate. Often, heterogeneous catalysts are metal-based and are attached to supports based on carbon or oxide. The choice of support for these materials is important, as the supports can affect the activity of the catalysts. Hydrogen gas is the most common source of hydrogen used and is commercially available.

Hydrogenation is an exothermic reaction, releasing about 25 kcal/mol in the hydrogenation of vegetable oils and fatty acids. For heterogenous catalysts, the Horiuti-Polanyi mechanism explains how hydrogenation occurs. First, the unsaturated bond binds to the catalyst, followed by H2dissociation into atomic hydrogen onto the catalyst. Then one atom of hydrogen attaches to the substrate in a reversible step, followed by the addition of a second atom, rendering the hydrogenation process irreversible. For homogeneous catalysis, the metal binds to hydrogen to give a dihydride complex via oxidative addition. The metal binds the substrate and then transfers one of the hydrogen atoms from the metal to the substrate via migratory insertion. The second hydrogen atom from the metal is transferred to the substrate with simultaneous dissociation of the newly formed alkane via reductive elimination.

Industrial uses of Hydrogenation Reactions

Heterogeneous catalytic hydrogenation is very important in industrial processes. In petrochemical processes, hydrogenation is used to saturate alkenes and aromatics, making them less toxic and reactive. Hydrogenation is also important in processing vegetable oils because most vegetable oils are derived from polyunsaturated fatty acids. Partial hydrogenation reduces most, but not all, of the carbon-carbon double bonds, making them better for sale and consumption. The degree of saturation of fats changes important physical properties such as the melting range of the oils; an example of this is how liquid vegetable oils become semi-solid at various temperatures.

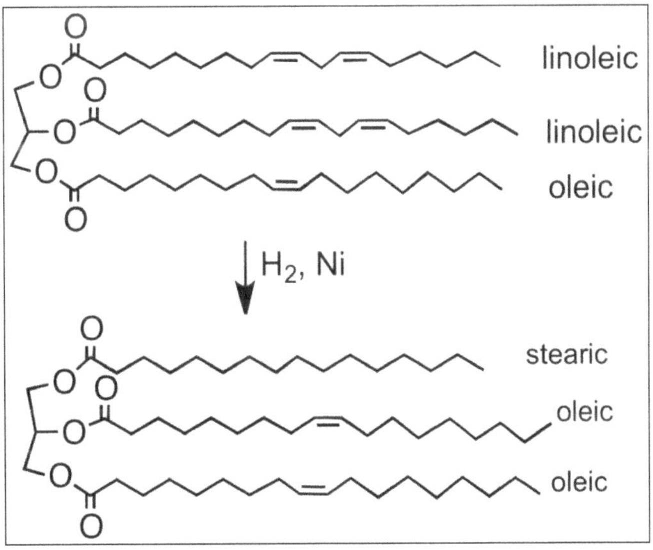

Partial hydrogenation in margarine: Margarine is a semi-solid butter substitute created from vegetable oil, which is typically unsaturated and therefore liquid at room temperature. The process of partial hydrogenation adds hydrogen atoms and reduces the double bonds in the fatty acids, creating a semi-solid vegetable oil at room temperature.

Incomplete hydrogenation of the double bonds has health implications; some double bonds can isomerize from the cis to the trans state. This isomerization occurs because the trans configuration has lower energy than the cis configuration. The trans isomers have been implicated in contributing to pathological blood circulatory system conditions (i.e., atherosclerosis and heart disease).

Catalytic Hydrogenation of Alkenes

The double bond of an alkene consists of a sigma (σ) bond and a pi (π) bond. Because the carbon-carbon π bond is relatively weak, it is quite reactive and can be easily broken and reagents can be added to carbon. Reagents are added through the formation of single bonds to carbon in an addition reaction.

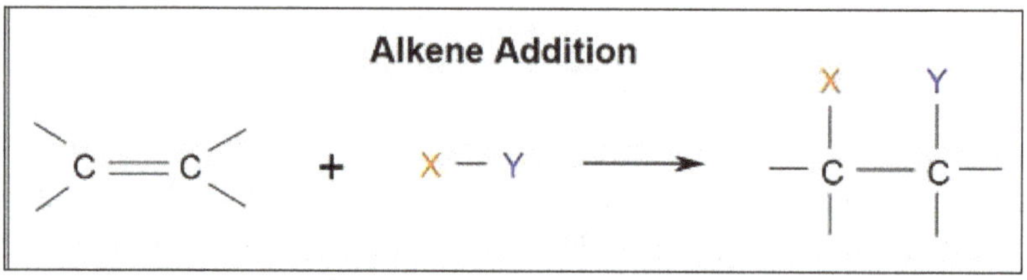

An example of an alkene addition reaction is a process called hydrogenation. In a hydrogenation reaction, two hydrogen atoms are added across the double bond of an alkene, resulting in a saturated alkane. Hydrogenation of a double bond is a thermodynamically favorable reaction because it forms a more stable (lower energy) product. In other words, the energy of the product is lower than the energy of the reactant; thus it is exothermic (heat is released). The heat released is called the heat of hydrogenation, which is an indicator of a molecule's stability.

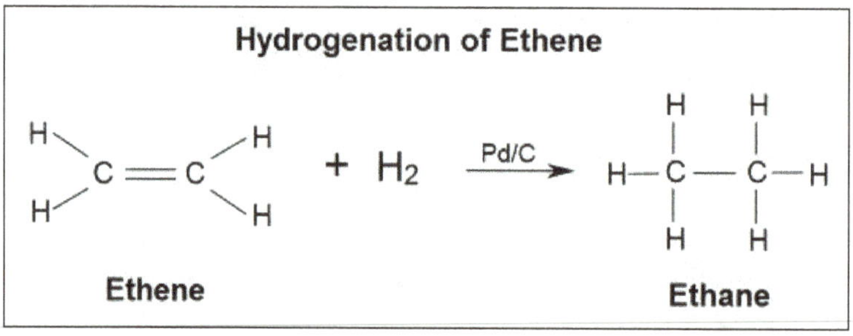

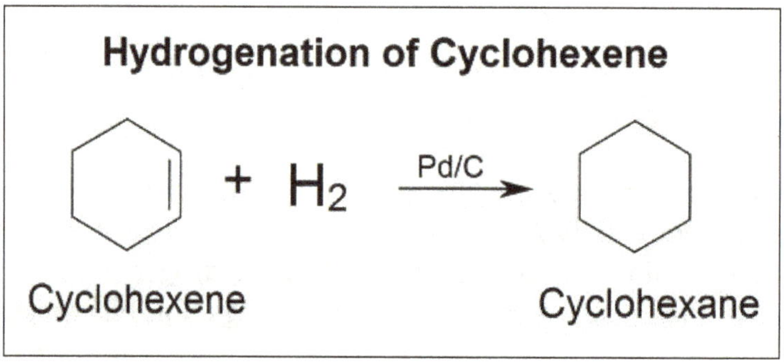

Although the hydrogenation of an alkene is a thermodynamically favorable reaction, it will not proceed without the addition of a catalyst.

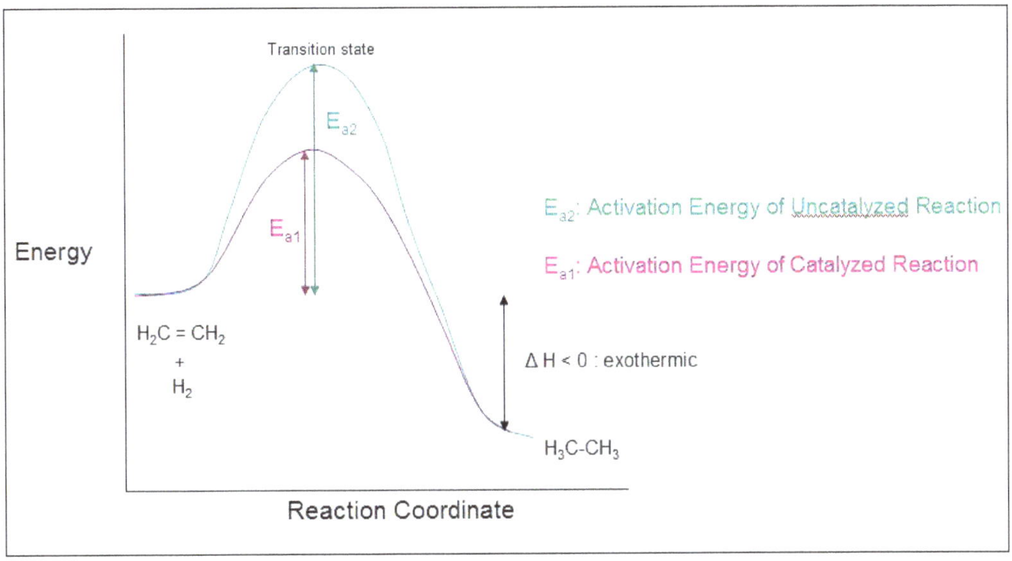

Hydrogeneration Reaction Energy Diagram.

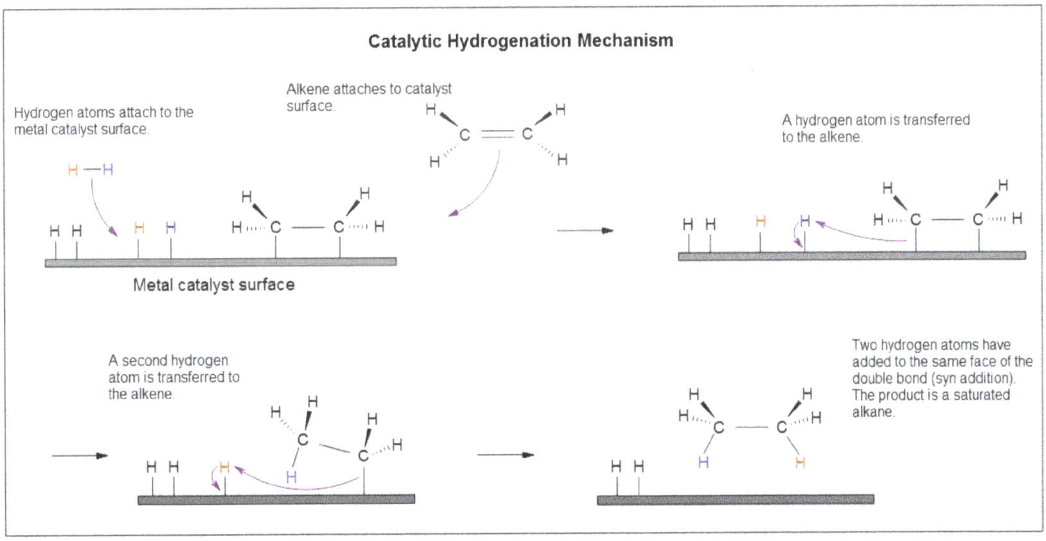

Common catalysts used are insoluble metals such as palladium in the form Pd-C, platinum in the form PtO_2, and nickel in the form Ra-Ni. With the presence of a metal catalyst, the H-H bond in H_2 cleaves, and each hydrogen attaches to the metal catalyst surface, forming metal-hydrogen bonds. The metal catalyst also absorbs the alkene onto its surface. A hydrogen atom is then transferred to the alkene, forming a new C-H bond. A second hydrogen atom is transferred forming another C-H bond. At this point, two hydrogens have added to the carbons across the double bond. Because of the physical arrangement of the alkene and the hydrogens on a flat metal catalyst surface, the two hydrogens must add to the same face of the double bond, displaying syn addition.

Common Applications

Hydrogenation reactions are extensively used to create commercial goods. Hydrogenation is used in the food industry to make a large variety of manufactured goods, like spreads and shortenings, from liquid oils. This process also increases the chemical stability of products and yields semi-solid products like margarine. Hydrogenation is also used in coal processing. Solid coal is converted to a liquid through the addition of hydrogen. Liquefying coal makes it available to be used as fuel.

Catalytic Hydrogenation of Alkynes

Reactions between alkynes and catalysts are a common source of alkene formation. Because alkynes differ from alkenes on account of their two procurable π bonds, alkynes are more susceptible to additions. Aside from turning them into alkenes, these catalysts affect the arrangement of substituents on the newly formed alkene molecule. Depending on which catalyst is used, the catalysts cause anti- or syn-addition of hydrogens. Alkynes can readily undergo additions because of their availability of two π bonds.

Hydrogenation of an Alkyne

Alkynes can be fully hydrogenated into alkanes with the help of a platinum catalyst. However, the use of two other catalysts can be used to hydrogenate alkynes to alkanes. These catalysts are: Palladium dispersed on carbon (Pd/C) and finely dispersed nickel (Raney-Ni).

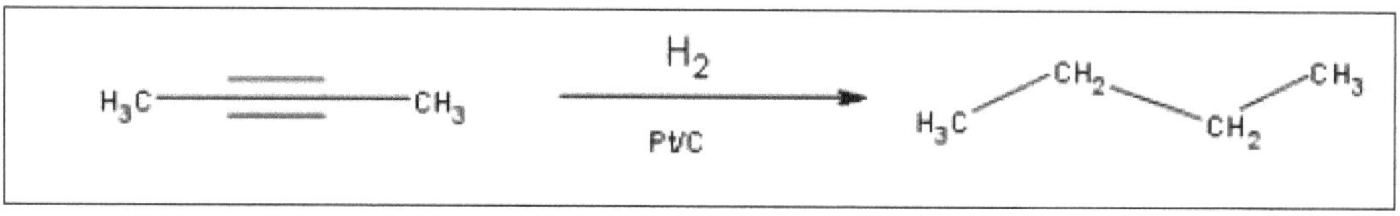

Hydrogenation of an Alkyne to a Cis-alkene

Because hydrogenation is an interruptible process involving a series of steps, hydrogenation can be stopped, using modified catalysts (e.g., Lindlar's Catalyst) at the transitional alkene stage. Lindar's catalyst has three components: Palladium-Calcium Carbonate, lead acetate and quinoline. The quinoline serves to prevent complete hydrogenation of the alkyne to an alkane. Lindlar's Catalyst transforms an alkyne to a cis-alkene.

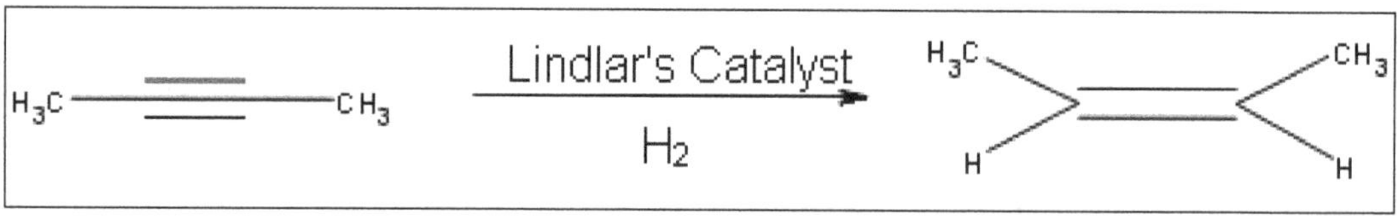

Lindlar's Catalyst:

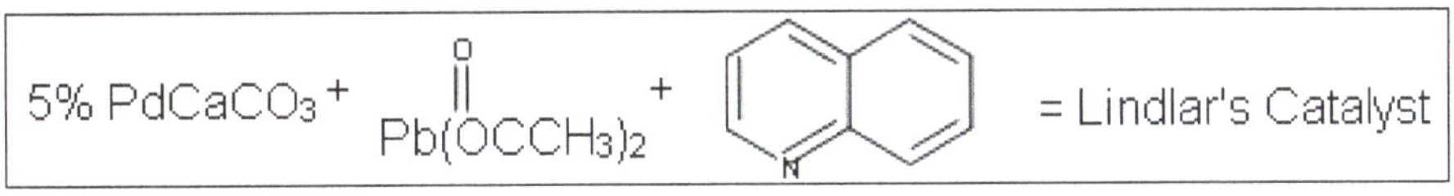

Hydrogenation of an Alkyne to a Trans-alkene

Alkynes can be reduced to trans-alkenes with the use of sodium dissolved in an ammonia solvent. A Na radical donates an electron to one of the P bonds in a carbon-carbon triple bond. This forms an anion, which can be protonated by hydrogen in an ammonia solvent. This prompts another Na radical to donate an electron to the second P orbital. Soon after this anion is also protonated by hydrogen from the ammonia solvent, resulting in a trans-alkene.

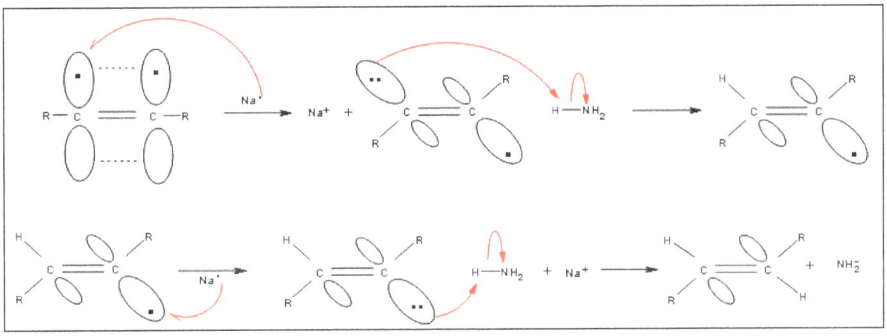

HYDROFORMYLATION REACTION

Hydroformylation, also known as oxo synthesis or oxo process, is an industrial process for the production of aldehydes from alkenes. This chemical reaction entails the net addition of a formyl group (CHO) and a hydrogen atom to a carbon-carbon double bond. This process has undergone continuous growth since its invention: Production capacity reached 6.6×10^6 tons in 1995. It is important because aldehydes are easily converted into many secondary products. For example, the resulting aldehydes are hydrogenated to alcohols that are converted to detergents. Hydroformylation is also used in speciality chemicals, relevant to the organic synthesis of fragrances and drugs. The development of hydroformylation is one of the premier achievements of 20th-century industrial chemistry.

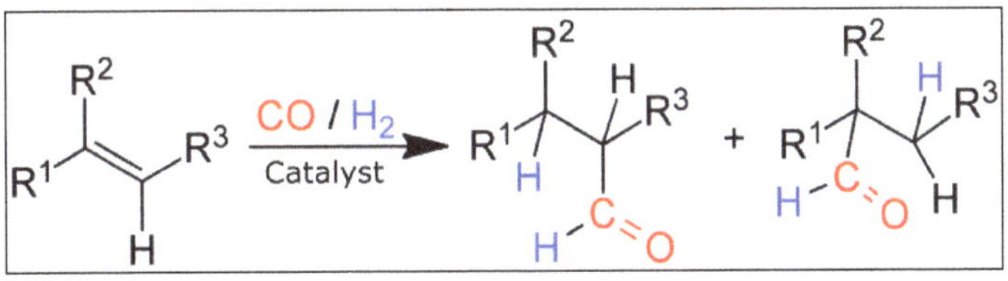

Hydroformylation of an Alkene (R_1 to R_3 organyl groups (i. e. Alkyl- or Aryl group) or hydrogen).

The process entails treatment of an alkene typically with high pressures (between 10 and 100 atmospheres) of carbon monoxide and hydrogen at temperatures between 40 and 200 °C. Transition metal catalysts are required. Invariably, the catalyst dissolves in the reaction medium, i.e. hydroformylation is an example of homogeneous catalysis.

Mechanism

Selectivity

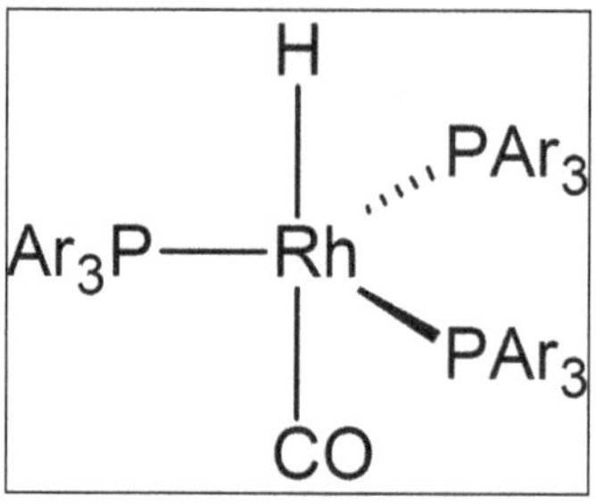

A generic rhodium catalyst, where PAr_3 = triphenylphosphine or its sulfonated analogue Tppts.

A key consideration of hydroformylation is the "normal" vs. "iso" selectivity. For example, the hydroformylation of propylene can afford two isomeric products, butyraldehyde or isobutyraldehyde:

$$H_2 + CO + CH_3CH{=}CH_2 \rightarrow CH_3CH_2CH_2CHO \ (\text{"normal"})$$

vs.

$$H_2 + CO + CH_3CH{=}CH_2 \rightarrow (CH_3)_2\,CHCHO \ (\text{"iso"})$$

These isomers reflect the regiochemistry of the insertion of the alkene into the M–H bond. Since both products are not equally desirable (normal is more stable than iso), much research was dedicated to the quest for catalyst that favored the normal isomer.

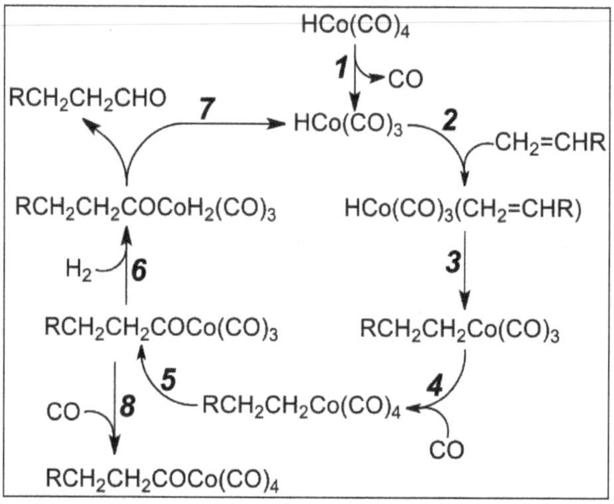

Figure shows mechanism of cobalt-catalyzed hydroformylation. The process begins with dissociation of CO from cobalt tetracarbonyl hydride to give the 16-electron species (step 1). Subsequent binding of alkene gives an 18e species (step 2). In step 3, the olefin inserts to give the 16e alkyl tricarbonyl. Coordination of another equivalent of CO gives alkyl tetracarbonyl (step 4). Migratory insertion of CO gives the 16e acyl in step 5. In step 6, oxidative addition of hydrogen gives a dihydrido complex, which in step 7 releases aldehyde by reductive elimination. Step 8 is unproductive and reversible.

Steric Effects

Markovnikov's rule addition of the cobalt hydride to primary alkenes is disfavored by steric hindrance between the cobalt centre and the secondary alkyl ligand. Bulky ligands exacerbate this steric hindrance. Hence, the mixed carbonyl/phosphine complexes offer a greater selectivity for anti-Markovnikov addition, thus favoring straight chain products (n-) aldehydes. Modern catalysts rely increasingly on chelating ligands, especially diphosphites.

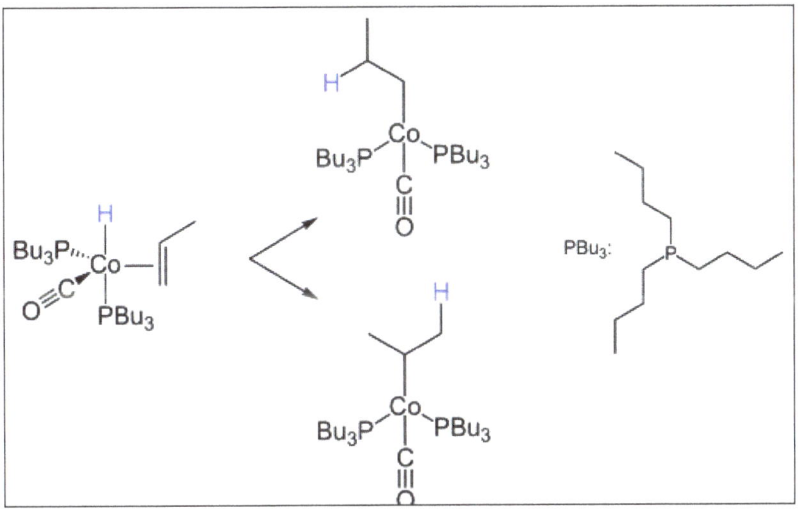

n (top) vs. iso (bottom)-selectivity.

Electronic Effects

Additionally, electron-rich the hydride complexes are less proton-like. Thus, as a result, the electronic effects that normally favour the Markovnikov addition to an alkene are less applicable. Thus, electron-rich hydrides are more selective.

Acyl Formation

To suppress competing isomerization of the alkene, the rate of migatory insertion of the carbonyl into the carbon-metal bond of the alkyl be fast. The rate of insertion of the carbonyl carbon into the C-M bond is likely to be greater than the rate of beta-hydride elimination.

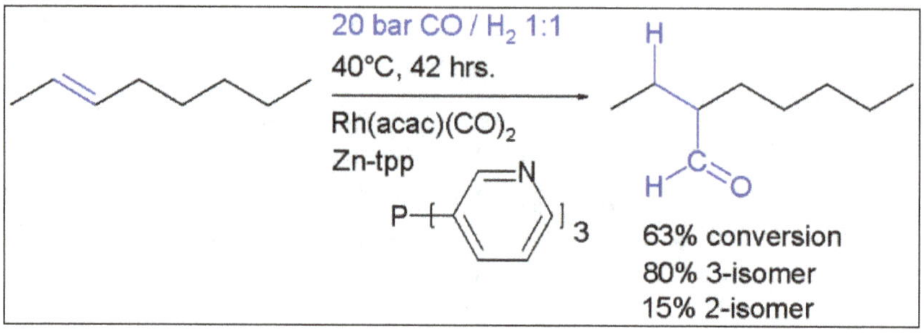

Asymmetric Hydroformylation

Hydroformylation of prochiral alkenes creates new stereocenters. Using chiral phosphine ligands, the hydroformylation can be tailored to favor one enantiomer. Thus, for example, dexibuprofen, the (+)–(s)-enantiomer of ibuprofen, can be produced by enantioselective hydroformylation followed by oxidation.

Processes

The industrial processes vary depending on the chain length of the olefin to be hydroformylated, the catalyst metal and ligands, and the recovery of the catalyst. The original Ruhrchemie process produced propanal from ethene and syngas using cobalt tetracarbonyl hydride. Today, industrial processes based on cobalt catalysts are mainly used for the production of medium- to long-chain olefins, whereas the rhodium-based catalysts are usually used for the hydroformylation of [propene]. The rhodium catalysts are significantly more expensive than cobalt catalysts. In the hydroformylation of higher molecular weight olefins the separation of the catalyst from the produced aldehydes is difficult.

BASF-oxo Process

The BASF-oxo process starts mostly with higher olefins and relies on cobalt carbonyl-based catalyst. By conducting the reaction at low temperatures, one observes increased selectivity favoring the linear product. The process is carried out at a pressure of about 30 MPa and in a temperature range of 150 to 170 °C. The cobalt is recovered from the liquid product by oxidation to water-soluble Co^{2} +, followed by the addition of aqueous formic or acetic acids. This process gives an aqueous phase of cobalt, which can then recycled. Losses are compensated by the addition of cobalt salts.

Exxon Process

The Exxon process, also Kuhlmann- or PCUK– oxo process, is used for the hydroformylation of C6–C12 olefins. The process relies on cobalt catalysts. In order to recover the catalyst, an aqueous sodium hydroxide solution or sodium carbonate is added to the organic phase. By extraction with olefin and neutralization by addition of sulfuric acid solution under carbon monoxide pressure the metal carbonyl hydride can recover. This is stripped out with syngas, absorbed by the olefin, and returned to the reactor. Similar to the BASF process, the Exxon process is carried out at a pressure of about 30 MPa and at a temperature of about 160 to 180 °C.

Shell Process

The Shell process uses cobalt complexes modified with phosphine ligands for the hydroformylation of C7–C14 olefins. The resulting aldehydes are directly hydrogenated to the fatty alcohols, which are separated by distillation, which allows the catalyst to be recycled. The process has good selectivity to linear products, which find use as feedstock for detergents. The process is carried out at a pressure of about 4 to 8 MPa and at a temperature range of about 150–190 °C.

Union Carbide Process

The Union Carbide (UCC) process, also known as low-pressure oxo process (LPO), relies on a rhodium catalyst dissolved in high-boiling thick oil, a higher molecular weight condensation product of the primary aldehydes, for the hydroformylation of propene. The reaction mixture is separated in a falling film evaporator from volatile components. The liquid phase is distilled and butyraldehyde is removed as head product while the catalyst containing bottom product is recycled to the process. The process is carried out at about 1.8 MPa and 95–100 °C.

Ruhrchemie/Rhone–Poulenc Process

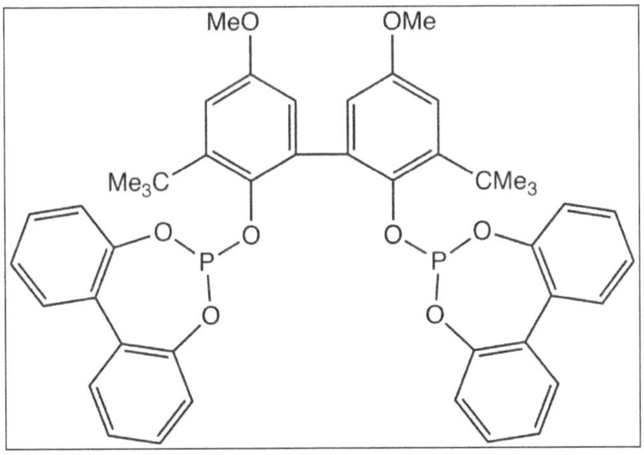

BiPhePhos is representative diphosphite ligand popularized by workers at Union Carbide.

The Ruhrchemie/Rhone–Poulenc process (RCRPP) relies on a rhodium catalyst with water-soluble TPPTS as ligand (Kuntz Cornils catalyst) for the hydroformylation of propene. The tri-sulfonation of triphenylphosphane ligand provides hydrophilic properties to the organometallic complex. The catalyst complex carries nine sulfonate-groups and is highly soluble in water (about 1 kg L^{-1}), but not in the emerging product phase.

The water-soluble TPPTS is used in about 50-fold excess, whereby the leaching of the catalyst is effectively suppressed. Reactants are propene and syngas consisting of hydrogen and carbon monoxide in a ratio of 1.1:1. A mixture of butyraldehyde and isobutyraldehyde in the ratio 96:4 is generated with few by-products such as alcohols, esters and higher boiling fractions. The Ruhrchemie/Rhone-Poulenc-process is the first commercially available two-phase system in which the catalyst is present in the aqueous phase. In the progress of the reaction an organic product phase is formed which is separated continuously by means of phase separation, wherein the aqueous catalyst phase remains in the reactor.

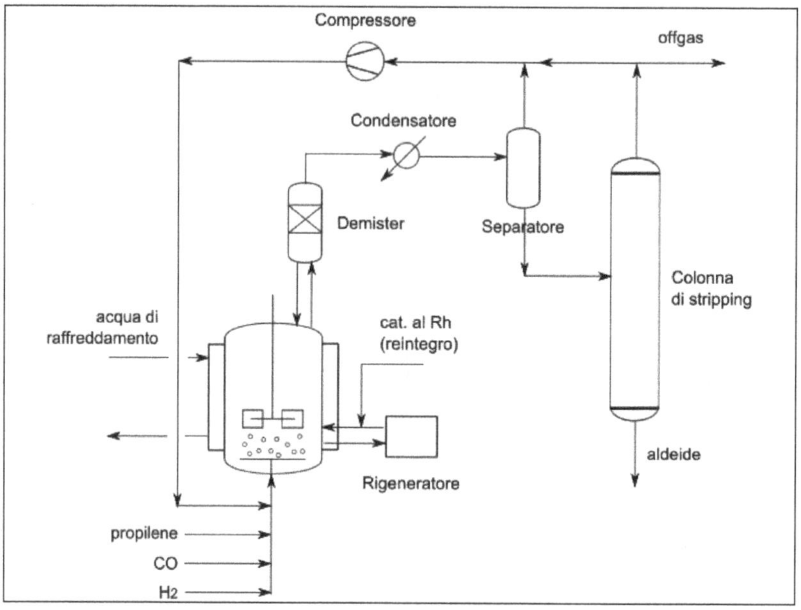

Process scheme of the Ruhrchemie/Rhône–Poulenc process.

The process is carried out in a stirred tank reactor where the olefin and the syngas are entained from the bottom of the reactor through the catalyst phase under intensive stirring. The resulting crude aldehyde phase is separated at the top from the aqueous phase. The aqueous catalyst-containing solution is re-heated via a heat exchanger and pumped back into the reactor. The excess olefin and syngas is separated from the aldehyde phase in a stripper and fed back to the reactor. The generated heat is used for the generation of process steam, which is used for subsequent distillation of the organic phase to separate into butyraldehyde and isobutyraldehyde. Potential catalyst poisons coming from the synthesis gas migrate into the organic phase and removed from the reaction with the aldehyde. Thus there is no accumulation of catalyst poisons, and the elaborate fine purification of the syngas can be omitted.

A plant was built in Oberhausen in 1984, which was debottlenecked in 1988 and again in 1998 up to a production capacity of 500,000 t/a butanal. The conversion rate of propene is 98% and the selectivity to n-butanal is high. During the life time of a catalyst batch in the process less than 1 ppb rhodium is lost.

Laboratory Process

Recipes have been developed for the hydroformylation on a laboratory scale, e.g. of cyclohexene.

Substrates other than Alkenes

Cobalt carbonyl and rhodium complexes catalyse the hydroformylation of formaldehyde and ethylene oxide to give 2-hydroxyacetaldehyde and 3-hydroxypropanaldehyde, which can then be hydrogenated to ethylene glycol and 1,3-propanediol, respectively. The reactions work best when the solvent is basic (such as pyridine).

In the case of dicobalt octacarbonyl or $Co_2(CO)_8$ as a catalyst, 3-pentanone can arise from ethylene and CO, in the absence of hydrogen. A proposed intermediate is the ethylene-propionyl species $[CH_3C(O)Co(CO)_3(ethylene)]$ which undergoes a migratory insertion to form $[CH_3COCH_2CH_2Co(CO)_3]$. The required hydrogen arises from the water shift reaction.

If the water shift reaction is not operative, the reaction affords a polymer containing alternating carbon monoxide and ethylene units. Such aliphatic polyketones are more conventionally prepared using palladium catalysts.

Functionalized olefins such as allyl alcohol can be hydroformylated. The target product 1,4-butanediol and its isomer is obtained with isomerization free catalysts such as rhodium-triphenylphosphine complexes. The use of the cobalt complex leads by isomerization of the double bond to n-propanal. The hydroformylation of alkenyl ethers and alkenyl esters occurs usually in the α-position to the ether or ester function.

The hydroformylation of acrylic acid and methacrylic acid leads in the rhodium-catalyzed process in the first step to the Markovnikov product. By variation of the reaction conditions the reaction can be directed to different products. A high reaction temperature and low carbon monoxide pressure favors the isomerization of the Markovnikov product to the thermodynamically more stable β-isomer, which leads to the n-aldehyde. Low temperatures and high carbon monoxide pressure and an excess of phosphine, which blocks free coordination sites, can lead to faster hydroformylation in the α-position to the ester group and suppress the isomerization.

Side and Consecutive Reactions

Olefins

Side reactions of the olefins are the isomerization and hydrogenation of the olefinic double bond. While the alkanes resulting from hydrogenation of the double bond do not participate further in the reaction, the isomerization of the double bond with subsequent formation of the n-alkyl complexes is a desired reaction. The hydrogenation is usually of minor importance; However, Cobalt-phosphine-modified catalysts can have an increased hydrogenation activity, where up to 15% of the olefin is hydrogenated.

Aldehydes

A usually desired consecutive reaction is the hydrogenation of the aldehydes to alcohols. Higher temperatures and hydrogen partial pressures favor the hydrogenation of the resulting aldehyde to the alcohol. For the reaction mechanism it is believed that the aldehyde initially forms a CO-π-complex with the catalyst. This is rearranged to the alkoxide complex and by subsequent oxidative addition of hydrogen the alcohol is eliminated and the starting complex is formed.

The aldehydic carbon-oxygen double bond can also be subject to hydroformylation, which leads to formic acid and its esters. The reaction requires the carbon monoxide insertion into the oxygen-metal bond of the alkoxide complex. The resulting Formyl complex can converted into the formic acid esters and the starting complex by oxidative addition of hydrogen. The initially produced aldehydes can react further by aldol condensation to either target product precursors like 2-ethylhexene-al or higher molecular weight condensation products, so-called thick oil.

Catalyst and Ligands

Conditions for hydroformylation catalysis can induce degradation of supporting organophosphorus ligands. Triphenylphosphineis subject to hydrogenolysis, releasing benzene and diphenylphosphine. The insertion of carbon monoxide in an intermediate metal-phenyl bond can lead to the formation of benzaldehyde or by subsequent hydrogenation to benzyl alcohol. One of the ligands phenyl-groups can be replaced by propene, and the resulting diphenylpropylphosphine ligand can inhibit the hydroformylation reaction due to its increased basicity.

ASSOCIATIVE SUBSTITUTION

Associative substitution describes a pathway by which compounds interchange ligands. The terminology is typically applied to coordination and organometallic complexes, but resembles the Sn2 mechanism in organic chemistry. The opposite pathway is dissociative substitution, being analogous to the Sn1 pathway. Intermediate pathways exist between the pure associative and pure dissociative pathways, these are called interchange mechanisms.

Associative pathways are characterized by binding of the attacking nucleophile to give a discrete, detectable intermediate followed by loss of another ligand. Complexes that undergo associative substitution are either coordinatively unsaturated or contain a ligand that can change its bonding to the metal, e.g. change in hapticity or bending of a nitrogen oxide ligand (NO). In homogeneous catalysis, the associative pathway is desirable because the binding event, and hence the selectivity of the reaction, depends not only on the nature of the metal catalyst but also on the substrate.

Examples of associative mechanisms are commonly found in the chemistry of 16e square planar metal complexes, e.g. Vaska's complex and tetrachloroplatinate. These compounds (MX_4) bind the incoming (substituting) ligand Y to form pentacoordinate intermediates MX_4Y that in a subsequent step dissociates one of their ligands. Dissociation of Y results in no reaction, but dissociation of X results in net substitution, giving the 16e complex MX_3Y. The first step is typically rate determining. Thus, the entropy of activation is negative, which indicates an increase in order in the system. These reactions follow second order kinetics: the rate of the appearance of product depends on the concentration of MX_4 and Y. The rate law is governed by the Eigen–Wilkins Mechanism.

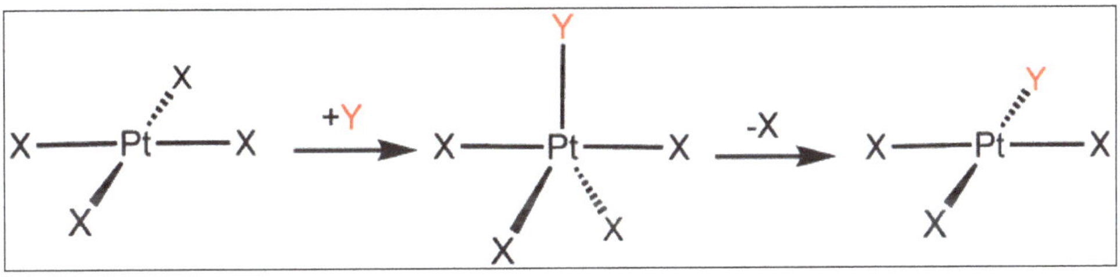

Associative Interchange Pathway

In many substitution reactions, well-defined intermediates are not observed, when the rate of such processes are influenced by the nature of the entering ligand, the pathway is called associative interchange, abbreviated I_a. Representative is the interchange of bulk and coordinated water in $[V(H_2O)_6]^{2+}$. In contrast, the slightly more compact ion $[Ni(H_2O)_6]^{2+}$ exchanges water via the I_d.

Effects of Ion Pairing

Polycationic complexes tend to form ion pairs with anions and these ion pairs often undergo reactions via the I_a pathway. The electrostatically held nucleophile can exchange positions with a ligand in the first coordination sphere, resulting in net substitution. An illustrative process comes from the "anation" (reaction with an anion) of chromium(III) hexaaquo complex:

$$\left[Cr(H_2O)_6\right]^{3+} + SCN^- \rightleftharpoons \left\{\left[Cr(H_2O)_6\right], NCS\right\}^{2+}$$

$$\left\{\left[Cr(H_2O)_6\right], NCS\right\}^{2+} \rightleftharpoons \left[Cr(H_2O)_5 NCS\right]^{2+} + H_2O$$

Special Ligand Effects

In special situations, some ligands participate in substitution reactions leading to associative pathways. These ligands can adopt multiple motifs for binding to the metal, each of which involves a different number of electrons "donated." A classic case is the indenyl effect in which an indenyl ligand reversibly "slips' from pentahapto (η^5) coordination to trihapto (η^3). Other pi-ligands behave in this way, e.g. allyl (η^3 to η^1) and naphthalene (η^6 to η^4). Nitric oxide typically binds to metals to make a linear MNO arrangement, wherein the nitrogen oxide is said to donate $3e^-$ to the metal. In the course of substitution reactions, the MNO unit can bend, converting the $3e^-$ linear NO ligand to a $1e^-$ bent NO ligand.

S_N1cB Mechanism

The rate for the hydrolysis of cobalt(III) ammine halide complexes are deceptive, appearing to be associative but proceeding by an alternative pathway. The hydrolysis of $[Co(NH_3)_5Cl]^{2+}$ follows second order kinetics: the rate increases linearly with concentration of hydroxide as well as the starting complex. Based on this information, the reactions would appear to proceed via nucleophilic attack of hydroxide at cobalt. Studies show, however, that the hydroxide deprotonates one NH_3 ligand to give the conjugate base of the starting complex, i.e., $[Co(NH_3)_4(NH_2)Cl]^+$. In this monocation, the chloride spontaneously dissociates. This pathway is called the SN1cB mechanism.

Eigen-Wilkins Mechanism

The Eigen-Wilkins mechanism, named after chemists Manfred Eigen and R. G. Wilkins, is a mechanism and rate law in coordination chemistry governing associative substitution reactions of octahedral complexes. It was discovered for substitution by ammonia of a chromium-(III) hexaaqua complex. The key feature of the mechanism is an initial rate-determining pre-equilibrium to form an encounter complex ML_6-Y from reactant ML_6 and incoming ligand Y. This equilibrium is represented by the constant KE:

$$ML_6 + Y \rightleftharpoons ML_6 - Y$$

The subsequent dissociation to form product is governed by a rate constant k:

$$ML_6 - Y \rightarrow ML_5Y + L$$

A simple derivation of the Eigen-Wilkins rate law follows:

$$\begin{aligned} [ML_6 - Y] &= K_E[ML_6][Y] \\ [ML_6 - Y] &= [M]_{tot} - [ML_6] \\ \text{rate} &= k[ML_6 - Y] \\ \text{rate} &= kK_E[Y][ML_6] \end{aligned}$$

Leading to the final form of the rate law, using the steady-state approximation ($d[ML_6\text{-}Y]/dt = 0$):

$$\text{rate} = kK_E[Y][M]_{tot} / (1 + K_E[Y])$$

Eigen-Fuoss Equation

A further insight into the pre-equilibrium step and its equilibrium constant K_E comes from the Fuoss-Eigen equation proposed independently by Eigen and R. M. Fuoss:

$$K_E = (4\pi a^3 / 3000) \times N_A \exp(-V / RT)$$

Where a represents the minimum distance of approach between complex and ligand in solution (in cm), N_A is the Avogadro constant, R is the gas constant and T is the reaction temperature. V is the electrostatic potential energy of the ions at that distance:

$$V = z_1 z_2 e^2 / 4\pi a \varepsilon$$

Where z is the charge number of each species and ε is the vacuum permittivity.

A typical value for KE is 0.0202 $dm^3 mol^{-1}$ for neutral particles at a distance of 200 pm. The result of the rate law is that at high concentrations of Y, the rate approximates k[M] tot while at low concentrations the result is $kK_E[M]_{tot}[Y]$. The Eigen-Fuoss equation shows that higher values of K_E (and thus a faster pre-equilibrium) are obtained for large, oppositely-charged ions in solution.

NUCLEOPHILIC ABSTRACTION

Nucleophilic abstraction is a type of an organometallic reaction which can be defined as a nucleophilic attack on a ligand which causes part or all of the original ligand to be removed from the metal along with the nucleophile. The nucleophile essentially attaches itself to the metal, where many holes are present, and soon this begins to fall apart, meaning the metal is full, and no longer requires electrical bonding with any of the associated structures involved with the central metal ion, which after becoming filled decides to leave the actual structure. If strong enough the ligands thus begin to fall apart, and this means they can be dismantled from the original structure.

Alkyl Abstraction

While nucleophilic abstraction of an alkyl group is relatively uncommon, there are examples of this type of reaction. In order for this reaction to be favorable, the metal must first be oxidized because reduced metals are often poor leaving groups. The oxidation of the metal causes the M-C bond to weaken, which allows for the nucleophilic abstraction to occur. G.M. Whitesides and D.J. Boschetto use the halogens Br_2 and I_2 as M-C cleaving agents in the following example of nucleophilic abstraction.

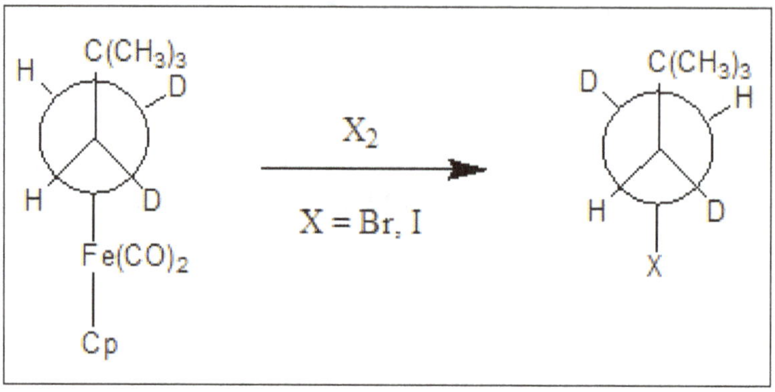

It is important to note that the product of this reaction is inverted with respect to the stereochemical center attached to the metal. There are several possibilities for the mechanism of this reaction which are shown in the following schematic.

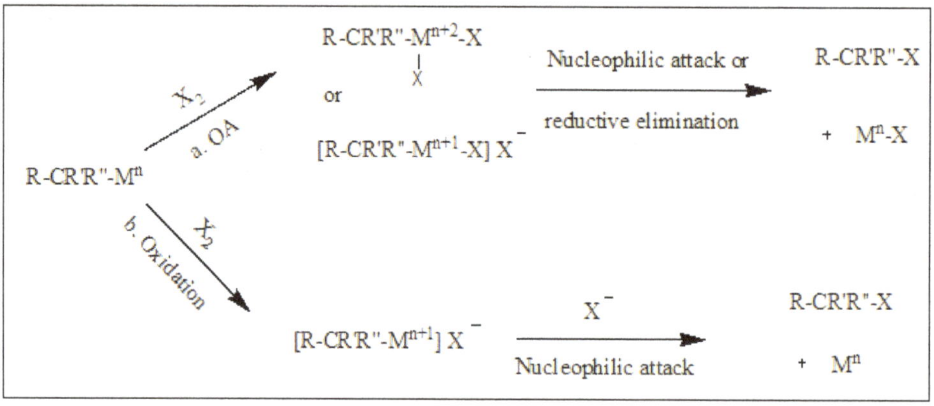

In path a, the first step proceeds with the oxidative addition of the halogen to the metal complex. This step results in the oxidized metal center that is needed to weaken the M-C bond. The second step can proceed with either the nucleophilic attack of the halide ion on the α-carbon of the alkyl group or reductive elimination, both of which result in the inversion of stereochemistry. In path b, the metal is first oxidized without the addition of the halide. The second step occurs with a nucleophilic attack of the α-carbon which again results in the inversion of stereochemistry.

Carbonyl Abstraction

Trimethylamine N-oxide (Me_3NO) can be used in the nucleophilic abstraction of carbonyl. There is a nucleophilic attack of Me_3NO on the carbon of the carbonyl group which pushes electrons on the metal. The reaction then proceeds to kick out CO_2 and NMe_3.

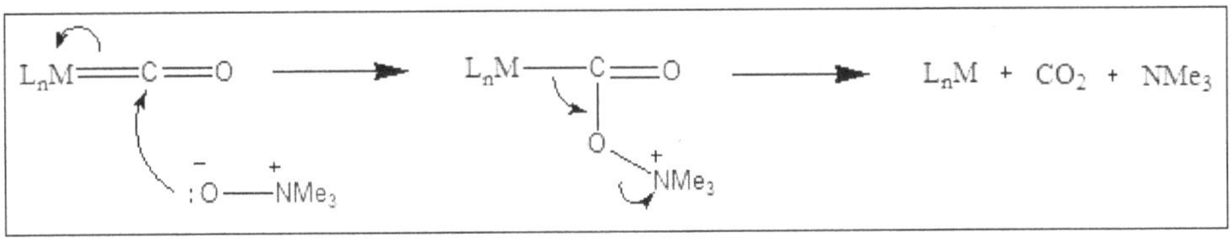

One iridium complex undergoes carbonyl abstraction while a very similar iridium complex undergoes hydride extraction.

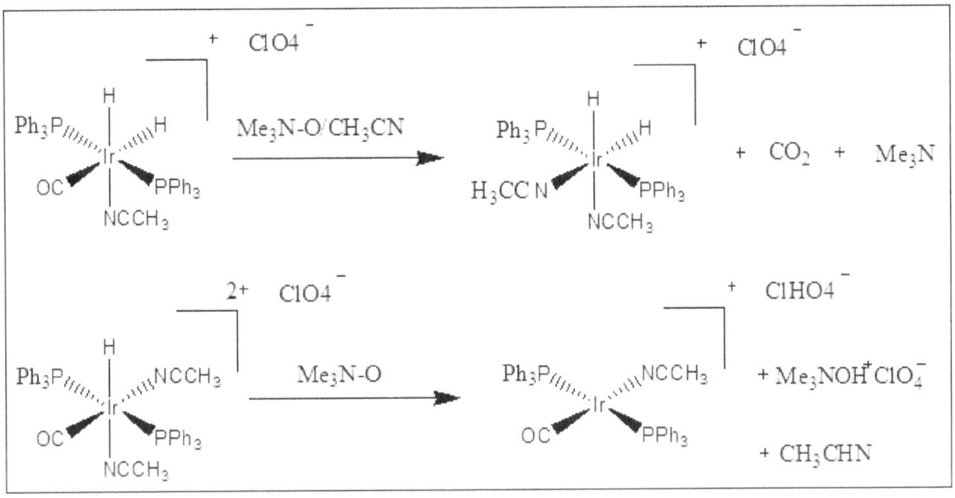

Hydrogen Abstraction

Nucleophilic abstraction can occur on a ligand of a metal if the conditions are right. For instance the following example shows the nucleophilic abstraction of H^+ from an arene ligand attached to chromium. The electron withdrawing nature of the chromium allows for the reaction to occur as a facile reaction.

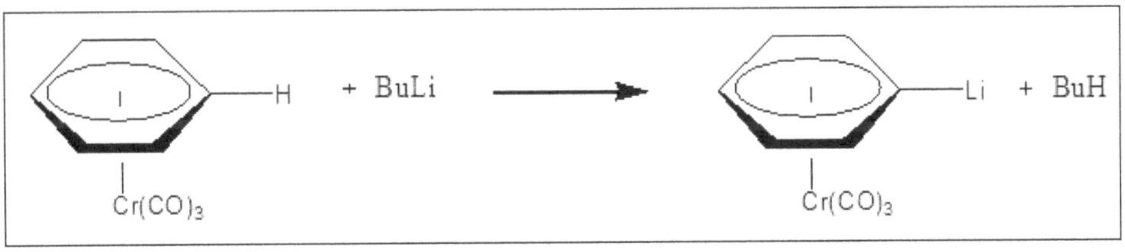

Methyl Abstraction

A Fischer carbene can undergo nucleophilic abstraction where a methyl group is removed. With the addition of a small abstracting agent, the abstracting agent would normally add to the carbene carbon. In this case however, the steric bulk of the abstracting agent that is added causes the abstraction of the methyl group. If the methyl group is replaced with ethyl, the reaction proceeds 70 times slower which is to be expected with a S_N2 displacement mechanism.

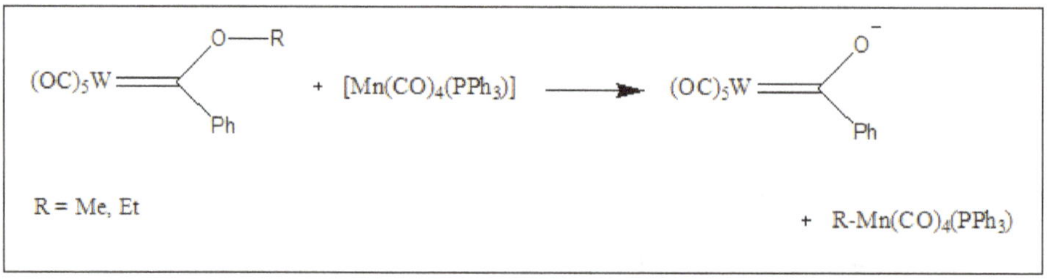

Silylium Abstraction

A silylium ion is a silicon cation with only three bonds and a positive charge. The abstraction of the silylium ion is seen from the ruthenium complex.

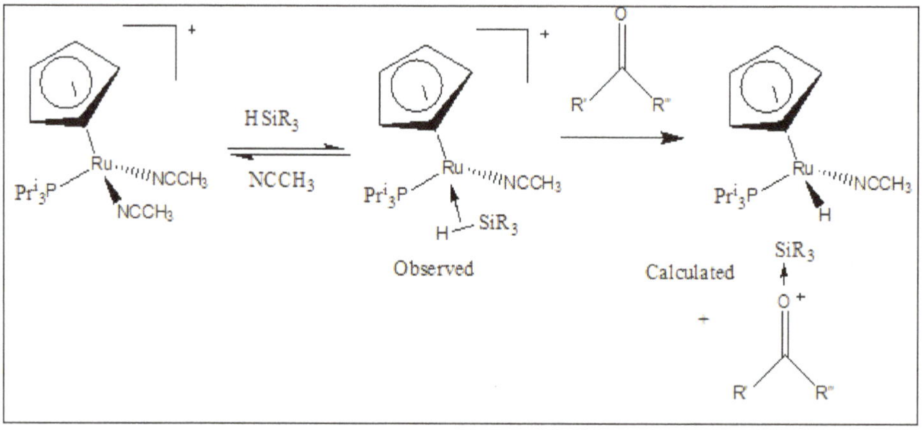

In the first step of this mechanism one of the acetonitrile groups is replaced by a silicon molecule where the bond between the silicon and the hydrogen is coordinating to the ruthenium. In the second step a ketone is added for the nucleophilic abstraction of the silylium ion and the hydrogen is left on the metal.

α-Acyl Abstraction

One example of nucleophilic abstraction of an α-acyl group is seen when MeOH is added to the following palladium complex. The mechanism follows a tetrahedral intermediate which results in the methyl ester and the reduced palladium complex shown.

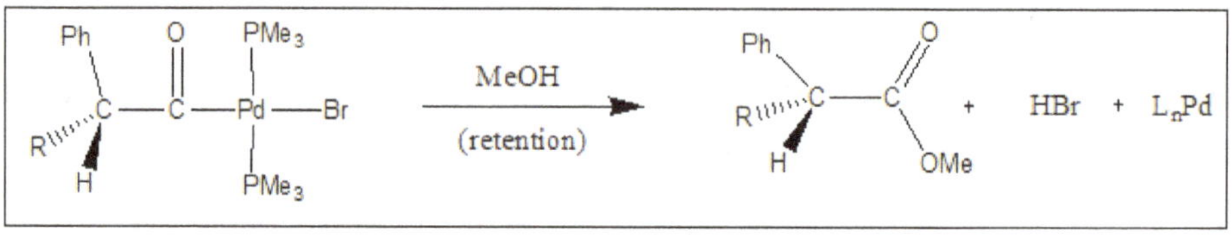

The following year a similar mechanism was proposed where oxidative addition of an aryl halide followed by migratory CO insertion and is followed by nucleophilic abstraction of the α-acyl by MeOH. One of the advantages of this intermolecular nucleophilic abstraction is the production of linear acyl derivatives. The intramolecular attack of these linear acyl derivatives gives rise to cyclic compounds such as lactones or lactams.

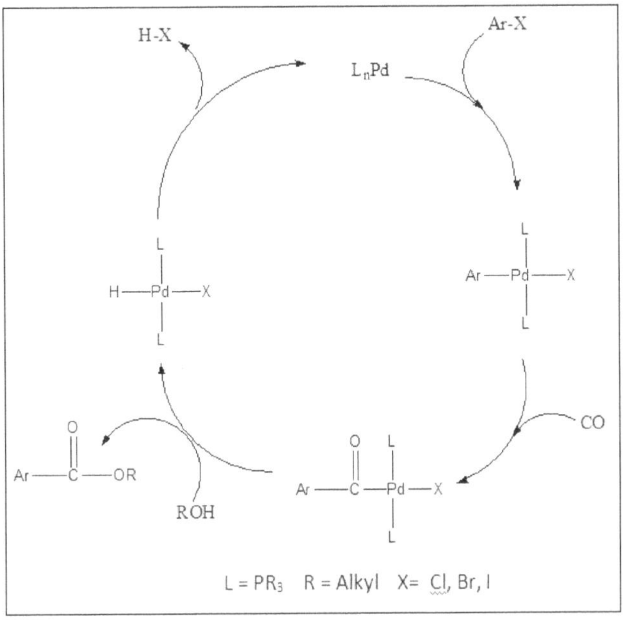

MIGRATORY INSERTION

A migratory insertion is a type of reaction in organometallic chemistry wherein two ligands on a metal complex combine. It is a subset of reactions that very closely resembles the insertion reactions, and both are differentiated by the mechanism that leads to the resulting stereochemistry of the products. However, often the two are used interchangeably because the mechanism is sometimes unknown. Therefore, migratory insertion reactions or insertion reactions, for short, are defined not by the mechanism but by the overall regiochemistry wherein one chemical entity interposes itself into an existing bond of typically a second chemical entity e.g.:

$$A+B-C \rightarrow B-A-C$$

In the migratory insertion, a ligand that is viewed as an anion (X) ligand in and a ligand that is viewed as neutral couple, generating a new anionic ligand. The anion and neutral ligands that react are adjacent. If the precursor complex is coordinatively saturated, migratory insertion often result in a coordinatively unsaturated product. A new (neutral) ligand can then react with the metal leading to a further insertion. The process can occur many times on a single metal, as in olefin polymerization.

The anionic ligand can be: H^- (hydride), R^- (alkyl), acyl, Ar^- (aryl), or OR^- (alkoxide).

The ability of these groups to migrate is called their migratory aptitude. The neutral ligand can be CO, alkene, alkyne, or in some cases, even carbene.

Diverse reactions apply to the migratory insertion. One mechanism involves the attack of the anionic ligand on the electrophilic part of the neutral ligand (the anionic ligand migrates to the neutral ligand). The other mechanism involves the neutral ligand inserts itself between the metal and the anionic ligand.

CO Insertion

The insertion of carbon monoxide into a metal-carbon bond to form an acyl group is the basis of carbonylation reactions, which provides many commercially useful products.

Mechanism

CO inserts into a metal-alkyl bond via migratory insertion. The key concept is that both the CO and the alkyl groups are ligands on the same metal. For example, the reaction of ^{13}CO with $Mn(CO)_5CH_3$ exclusively form $Mn(CO)_4(^{13}CO)COCH_3$. The alkyl group migrates intramolecularly to an adjacent CO ligand within the coordination sphere of the Mn(I) centre. Subsequent to the migration, the metal binds free CO.

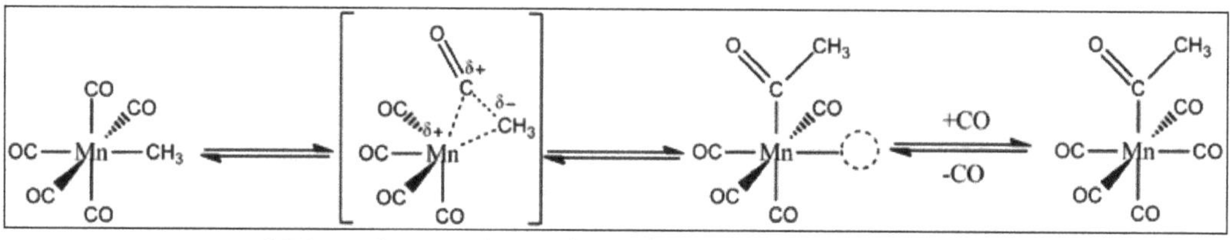

CO Insertion reaction pathway for an octahedral complex.

CO insertion does not always involve migration. Treatment of $CpFe(L)(CO)CH_3$ with ^{13}CO yields a mix of both alkyl migration products and products formed by true insertion of bound carbonyls into the methyl group. Product distribution is influenced by the choice of solvent.

Alkyl derivatives of square planar complexes undergo CO insertions particularly readily. Insertion reactions on square planar complexes are of particular interest because of their industrial applications. Since square planar complexes are often coordinatively unsaturated, they are susceptible to formation of 5-coordinate adducts, which undergo migratory insertion readily. In most cases the in-plane migration pathway is preferred, but, unlike the nucleophilic pathway, it is inhibited by an excess of CO.

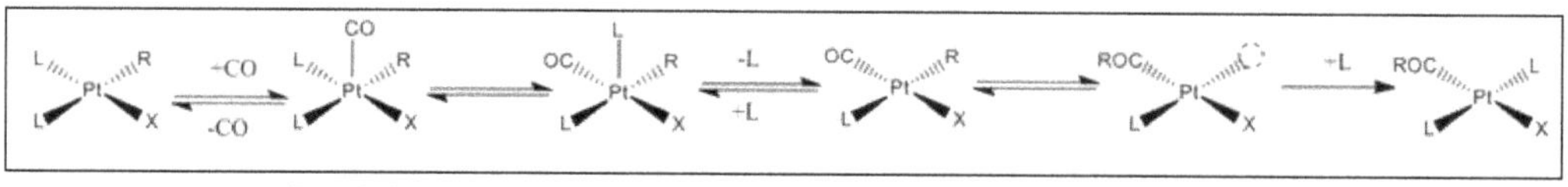

Nucleophilic insertion and rearrangement of a square planar complex.

Effects on Reaction Rates

- Steric effects strain: Increasing the steric strain of the chelate backbone in square planar complexes pushes the carbonyl and methyl groups closer together, increasing the reactivity of insertion reactions.
- Oxidation state: Oxidation of the metal tends to increase insertion reaction rates. The main rate-limiting step in the mechanism is the migration of the methyl group onto a carbonyl ligand, oxidizing the metal by imparting a greater partial positive charge on the acetyl carbon, and thus increasing the rate of reaction.
- Lewis acids: Lewis acids also increase the reaction rates, for reasons similar to metal oxidation increasing the positive charge on the carbon. Lewis acids bind to the CO oxygen and remove charge, increasing the electrophilicity of the carbon. This can increase the reaction rate by a factor of up to 10^8, and the complex formed is stable enough that the reaction proceeds even without additional CO to bind to the metal.
- Electronegativity of the leaving group: Increasing the electronegativity of the leaving alkyl group stabilizes the metal-carbon bond interaction and thus increases the activation energy required for migration, decreasing the reaction rate.
- Trans-effect: Ligands in an octahedral or square planar complex are known to influence the reactivity of the group to which they are trans. This ligand influence is often referred to as the trans-influence, and it varies in intensity between ligands. A partial list of trans-influencing ligands is as follows, from highest trans-effect to lowest: aryl, alkyl > NR_3 > PR_3 > AsR_3 > CO > Cl. Ligands with a greater trans-influence impart greater electrophilicity to the active site. Increasing the electrophilicity of the CO group has been shown experimentally to greatly increase the reaction rate, while decreasing the electrophilicity of the methyl group slightly increases the reaction rate. This can be demonstrated by reacting a square planar $[(PN)M(CO)(CH_3)]$ complex with CO, where PN is a bidentate phosphorus- or nitrogen-bound ligand. This reaction proceeds in much greater yield when the methyl group is trans-P and the CO trans-N, owing to the higher trans-influence of the more electronegative nitrogen.

Reverse Reaction

Decarbonylation of aldehydes, the reverse of CO insertion, is a well-recognized reaction:

$$RCHO \rightarrow RH + CO$$

The reaction is not widely practiced in part because the alkanes are less useful materials than are the aldehyde precursors. Furthermore, the reaction is not often conducted catalytically because the extruded CO can be slow to dissociate. Extrusion of CO from an organic aldehyde is most famously demonstrated using Wilkinson's catalyst:

$$RhCl(PPh_3)_3 + RCHO \rightarrow RhCl(CO)(PPh_3)_2 + RH + PPh_3$$

Insertion of other Oxides

Many electrophilic oxides insert into metal carbon bonds; these include sulfur dioxide, carbon dioxide, and nitric oxide. These reactions have limited or no practical significance, but are of historic interest. With transition metal alkyls, these oxides behave as electrophiles and insert into the bond between metals and their relatively nucleophilic alkyl ligands. The insertion of SO_2 has been examined in particular detail. SO_2 inserts to give both O-sulphinates and S-sulphinates, depending on the metal centre. With square planar alkyl complexes, a pre-equilibrium is assumed involving formation of an adduct.

Insertion of Alkenes into Metal-carbon Bonds

The insertion of alkenes into both metal-carbons is important. The insertion of ethylene and propylene into titanium alkyls is the cornerstone of Ziegler–Natta catalysis, the main source of polyethylene and polypropylene. The majority of this technology involves heterogeneous catalysts, but it is widely assumed that the principles and observations on homogeneous systems are applicable to the solid-state versions. Related technologies include the Shell Higher Olefin Process which produces detergent precursors.

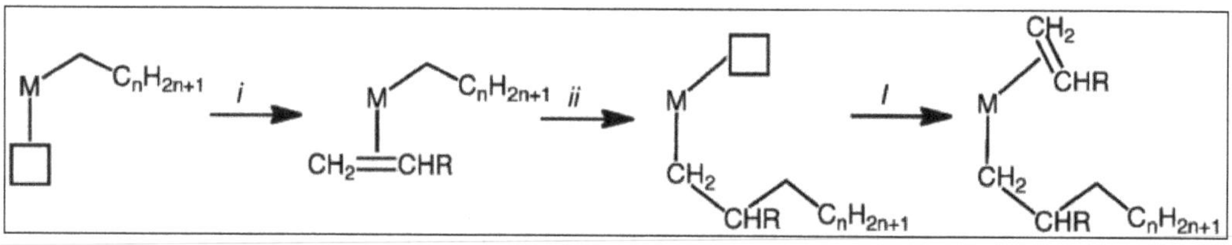

Steps in alkene polymerization. Step (i) involves binding of the monomer to the metal and step (ii) involves the migratory insertion step. These steps, which alternate from one side of the metal center to the other, are repeated many times for each polymer chain. The box represents a vacant (or extremely labile) coordination site.

Mechanism

Factors affecting the rate of olefin insertions include the formation of the cyclic, planar, four-center transition state involving incipient formation of a bond between the metal and an olefin carbon. From this transition state, it can be seen that a partial positive charge forms on the β-carbon with a partial negative charge formed on the carbon initially bonded to the metal. This polarization explains the subsequently observed formation of the bond between the negatively charged carbon/hydrogen and the positively charged β-carbon as well as the simultaneous formation of the metal-α-carbon bond. This transition state also highlights the two factors that most strongly contribute to the rate of olefin insertion reactions: (i) orbital overlap of the alkyl group initially attached to the metal and (ii) the strength of the metal-alkyl bond. With greater orbital overlap between the partially positive β-carbon and the partially negative hydrogen/alkyl group carbon, the formation of the new C-C bond is facilitated. With increasing strength of the metal-alkyl bond, the breaking of the bond between the metal and the hydrogen/alkyl carbon bond to form the two new bonds with the α-carbon and β-carbon (respectively) is slower, thus decreasing the rate of the insertion reaction.

Insertion of Alkenes into M–H Bonds

The insertion of alkenes into metal-hydrogen bonds is a key step in hydrogenation and hydroformylation reactions. The reaction involves the alkene and the hydride ligands combining within the coordination sphere of a catalyst. In hydrogenation, the resulting alkyl ligand combines with a second hydride to give the alkane. Analogous reactions apply to the hydrogenation of alkynes: an alkenyl ligand combines with a hydride to eliminate an alkene.

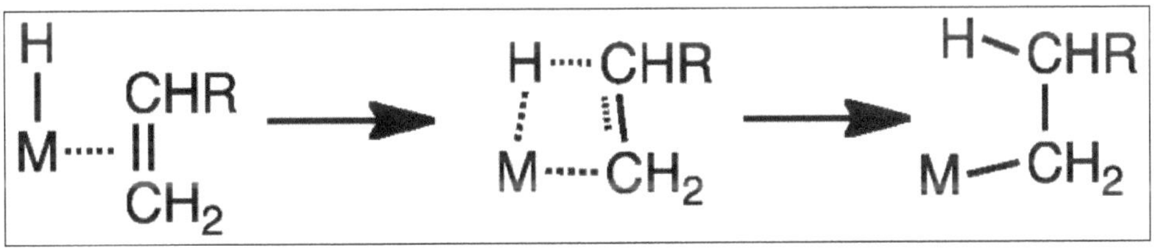

Migratory insertion of an alkene into a metal hydride. The middle image depicts the transition state. Notice that the less substituted carbon becomes bonded to the metal, at least usually. The reverse reaction is called beta-hydride elimination.

Mechanism

In terms of mechanism, the insertion of alkenes into M–H bond and into M–C bonds are described similarly. Both involve four-membered transition states that place the less substituted carbon on the metal.

The reverse of olefin insertion into a metal-hydrogen bond is β-hydride elimination. The Principle of Microscopic Reversibility requires that the mechanism of β-hydride elimination follow the same pathway as the insertion of alkenes into metal hydride bonds. The first requirement for β-hydride elimination is the presence of a hydrogen at a position that is β with respect to the metal. β-elimination requires a vacant coordination position on the metal that will accommodate the hydrogen that is abstracted.

Industrial Applications

Carbonylation

Two widely employed applications of migratory insertion of carbonyl groups are hydroformylation and the production of acetic acid by carbonylation of methanol. The former converts alkenes, hydrogen, and carbon monoxide into aldehydes. The production of acetic acid by carbonylation proceeds via two similar industrial processes. More traditional is the Monsanto acetic acid process, which relies on a rhodium-iodine catalyst to transform methanol into acetic acid. This process has been superseded by the Cativa process which uses a related iridium catalyst, $[Ir(CO)_2I_2]^-$ (1). By 2002, worldwide annual production of acetic acid stood at 6 million tons, of which approximately 60% is produced by the Cativa process.

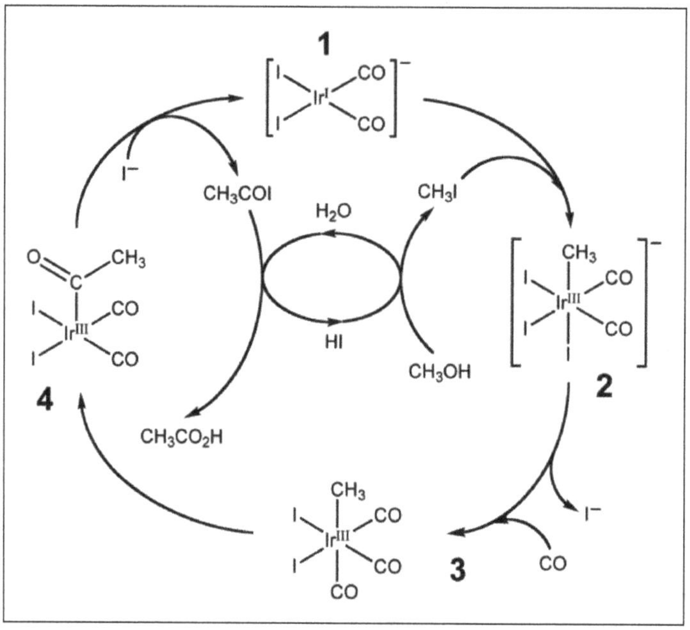

The Cativa process catalytic cycle, shown above, includes both insertion and de-insertion steps. The oxidative addition reaction of methyl iodide with (1) involves the formal insertion of the iridium(I) centre into the carbon-iodine bond, whilst step (3) to (4) is an example of migratory insertion of carbon monoxide into the iridium-carbon bond. The active catalyst species is regenerated by the reductive elimination of acetyl iodide from (4), a de-insertion reaction.

Alkene Polymerization

Industrial applications of alkene insertions include metal-catalyzed routes to polyethylene and polypropylene. Typically these conversions are catalyzed heterogeneously by titanium trichlorides which are activated by aluminium alkyls. This technology is known as Ziegler–Natta catalysts. In these reactions, ethylene coordinates to titanium metal followed by its insertion. These steps can be repeated multiple times, potentially leading to high molecular weight polymers.

LIGAND SUBSTITUTION

Ligand substitution is an organometallic reaction that involves the exchange of one ligand for another, with no change in oxidation state at the metal center. The incoming and outgoing ligands may be L- or X-type, but the charge of the complex changes if the ligand type changes. Keep charge conservation in mind when writing out ligand substitutions.

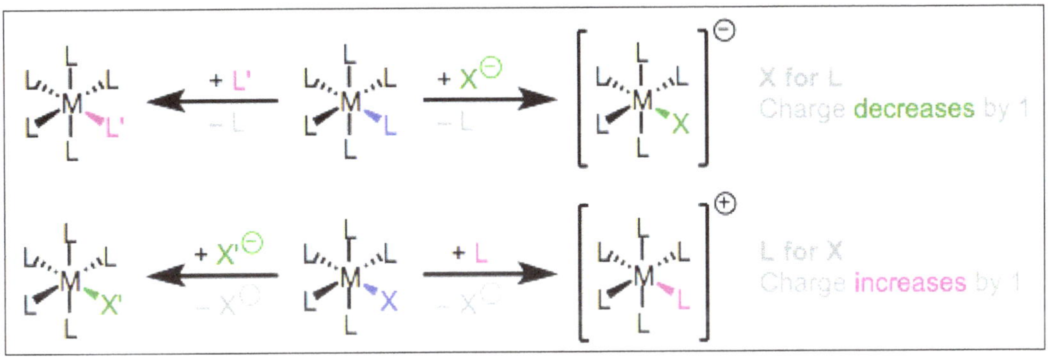

Charge is conserved in ligand substitution reactions. Four general types of substitution are shown here—note the X-for-L and L-for-X types.

How do we know when a ligand substitution reaction is favorable? The thermodynamics of the reaction depend on the relative strength of the two metal-ligand bonds, and the stability of the departing and incoming ligands (or salt sof the ligand, if they're X type). It's often useful to think of X-for-X substitutions like acid-base reactions, with the metal and spectator ligands serving as a "glorified proton." Like acid-based equilibria in organic chemistry, we look to the relative stability of the two charged species (the free ligands) to draw conclusions. Of course, we don't necessarily need to rely just on primal thermodynamics to drive ligand substitution reactions. Photochemistry, neighboring-group participation, and other tools can facilitate otherwise difficult substitutions.

Ligand substitution is characterized by a continuum of mechanisms bound by associative (A) and dissociative (D) extremes. At the associative extreme, the incoming ligand first forms a bond to the metal, then the departing ligand takes its lone pair and leaves. At the dissociative extreme, the order of events is opposite—the departing ligand leaves, then the incoming ligand comes in. Associative substitution is common for 16-electron complexes (like d8 complexes of Ni, Pd, and Pt), while dissociative substitution is the norm for 18-electron complexes. Then again, reality is often more complicated than these extremes. In some cases, evidence is available for simultaneous dissociation and association, and this mechanism has been given the name interchange (IA or ID).

Photochemical Substitution

Substitution reactions of dative ligands—most famously, CO—may be facilitated by photochemical excitation. Two examples are shown below. The first reaction yields only monosubstituted product without ultraviolet light, even in the presence of a strongly donating phosphine.

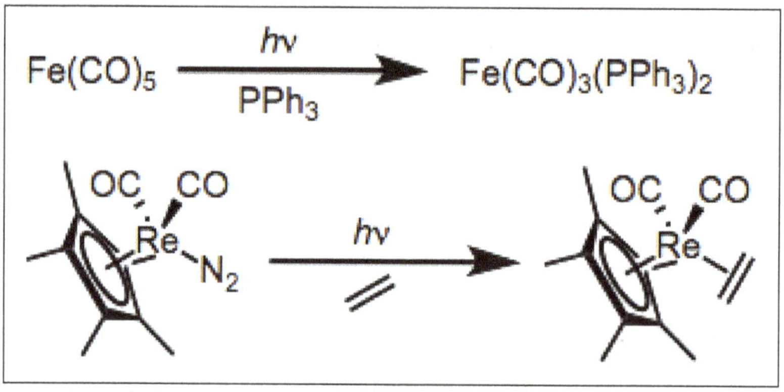

Dissociative photochemical substitutions of CO and dinitrogen.

All signs point to dissociative mechanisms for these reactions (the starting complexes have 18 total electrons each). Excitation, then, must increase the M–L antibonding character of the complex's electrons; exactly how this increase in antibonding character happens has been a matter of some debate. Originally, the prevailing explanation was that the LUMO bears M–L antibonding character, and excitation kicks an electron up from the HOMO to the LUMO, encouraging cleavage of the M–L bond. A more recent, more subtle explanation backed by calculations supports the involvement of a metal-to-ligand charge-transfer state along with the "classical" ligand-field excited state.

Oxidation/Reduction

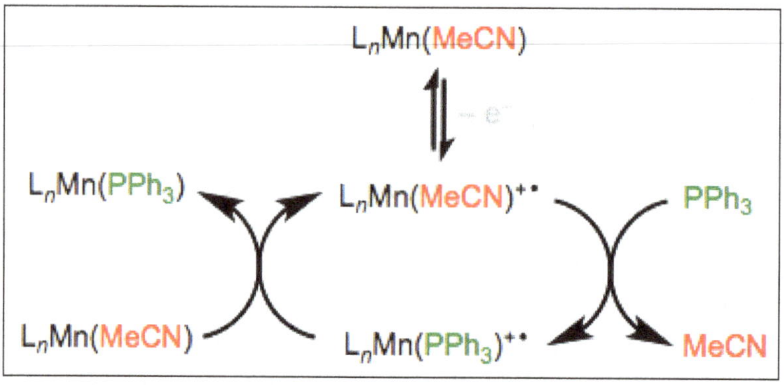

Oxidation accelerates substitution in electron-rich complexes through a chain process.

Complexes bearing 17 and 19 total electrons are much more reactive toward substitution than their even-electron counterparts. Single-electron oxidation and reduction can thus be used to efficiently turn on substitution. As you might expect, oxidation and reduction work best on electron-rich and electron-poor complexes, respectively. The Mn complex in the oxidative example below, for instance, includes a strongly donating MeCp group.

Reduction works well for electron-poor metal carbonyl complexes, which are happy to accept an additional electron.

There is a two-electron oxidation method that's also worth knowing: the oxidation of CO with amine oxides. This nifty little method releases carbon dioxide, amine, and an unsaturated complex that may be quenched by a ligand hanging around. The trick is addition to the CO ligand followed by elimination of the unsaturated complex. As the oxidized CO_2 and reduced amine float away, the metal complex finds another ligand.

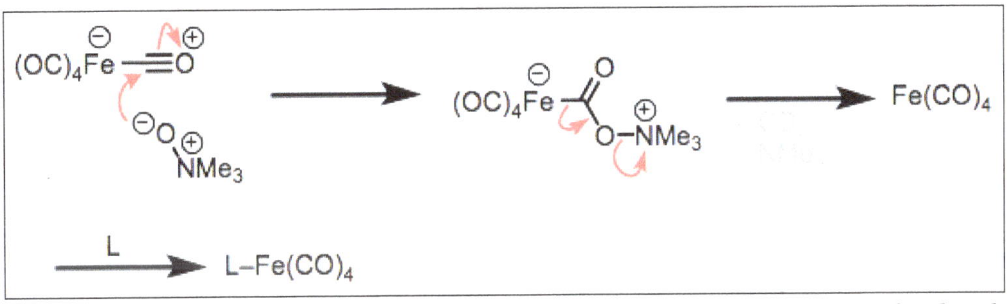

Oxidation of CO with amine oxides. A fun method for dissociative substitution of metal carbonyls.

Radical Chain Processes

Atom abstraction from 18-electron complexes produces neutral 17-electron intermediates, which are susceptible to ligand substitution via radical chain mechanisms. The fact that the intermediates are neutral distinguishes these methods from oxidation-based methods. First-row metal hydrides are great for these reactions, owing to their relatively weak M–H bonds.

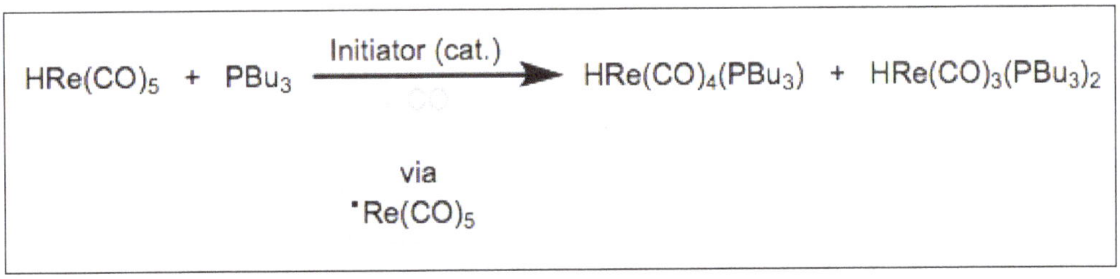

Radical-chain substitution involving atom abstraction.

After abstraction of the hydrogen atom by initiator, substitution is rapid and may occur multiple times. Propagation begins anew when the substituted radical abstracts hydrogen from the starting material to regenerate the propagating radical and form the product.

OXIDATIVE ADDITION

Oxidative addition and reductive elimination are two important and related classes of reactions in organometallic chemistry. Oxidative addition is a process that increases both the oxidation state and coordination number of a metal centre. Oxidative addition is often a step in catalytic cycles, in conjunction with its reverse reaction, reductive elimination.

Role in Transition Metal Chemistry

For transition metals, oxidative reaction results in the decrease in the d^n to a configuration with fewer electrons, often 2e fewer. Oxidative addition is favored for metals that are (i) basic and/or (ii) easily oxidized. Metals with a relatively low oxidation state often satisfy one of these requirements, but even high oxidation state metals undergo oxidative addition, as illustrated by the oxidation of Pt(II) with chlorine:

$$[PtCl_4]^{2-} + Cl_2 \rightarrow [PtCl_6]^{2-}$$

In classical organometallic chemistry, the formal oxidation state of the metal and the electron count of the complex both increase by two. One-electron changes are also possible and in fact some oxidative addition reactions proceed via series of 1e changes. Although oxidative additions can occur with the insertion of a metal into many different substrates, oxidative additions are most commonly seen with H–H, H–X, and C–X bonds because these substrates are most relevant to commercial applications.

Oxidative addition requires that the metal complex have a vacant coordination site. For this reason, oxidative additions are common for four- and five-coordinate complexes.

Reductive elimination is the reverse of oxidative addition. Reductive elimination is favored when the newly formed X–Y bond is strong. For reductive elimination to occur the two groups (X and Y) should be mutually adjacent on the metal's coordination sphere. Reductive elimination is the key product-releasing step of several reactions that form C–H and C–C bonds.

Mechanisms of Oxidative Addition

Oxidative additions proceed via many pathways that depend on the metal center and the substrates.

Concerted Pathway

Oxidative additions of nonpolar substrates such as hydrogen and hydrocarbons appear to proceed via concerted pathways. Such substrates lack π-bonds, consequently a three-centered σ complex is invoked, followed by intramolecular ligand bond cleavage of the ligand (probably by donation of electron pair into the sigma* orbital of the inter ligand bond) to form the oxidized complex. The resulting ligands will be mutually cis, although subsequent isomerization may occur.

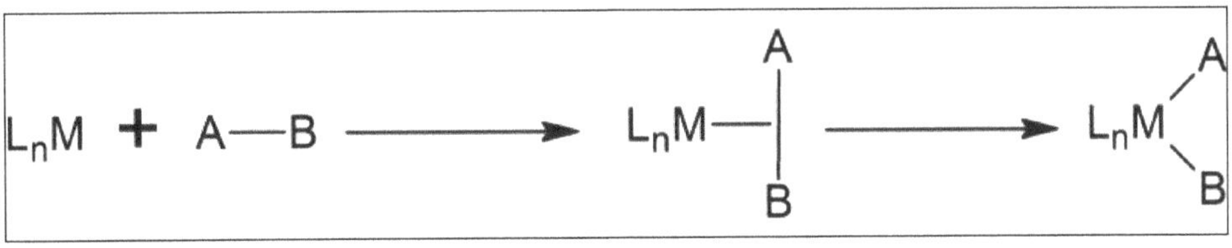

This mechanism applies to the addition of homonuclear diatomic molecules such as H_2. Many C–H activation reactions also follow a concerted mechanism through the formation of an M–(C–H) agostic complex.

A representative example is the reaction of hydrogen with Vaska's complex, trans-IrCl(CO)[P(C_6H_5)$_3$]$_2$. In this transformation, iridium changes its formal oxidation state from +1 to +3. The product is formally bound to three anions: one chloride and two hydride ligands. As shown below, the initial metal complex has 16 valence electrons and a coordination number of four whereas the product is a six-coordinate 18 electron complex.

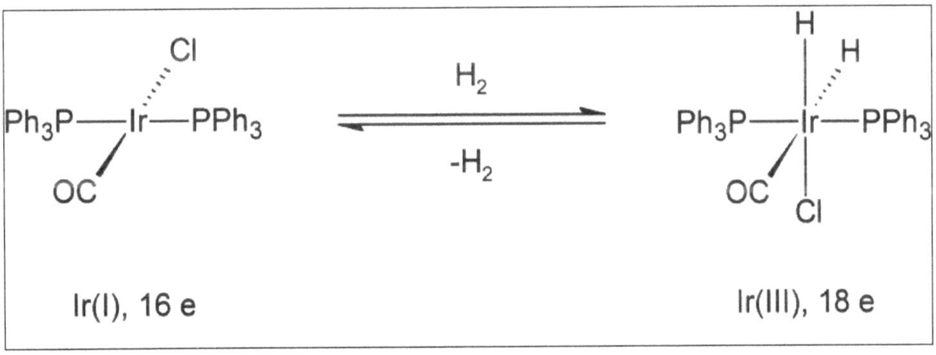

Formation of a trigonal bipyramidal dihydrogen intermediate is followed by cleavage of the H–H bond, due to electron back donation into the H–H σ*-orbital. This system is also in chemical equilibrium, with the reverse reaction proceeding by the elimination of hydrogen gas with simultaneous reduction of the metal center.

The electron back donation into the H–H σ*-orbital to cleave the H–H bond causes electron-rich metals to favor this reaction. The concerted mechanism produces a cis dihydride, while the stereochemistry of the other oxidative addition pathways do not usually produce cis adducts.

S_N2-type

Some oxidative additions proceed analogously to the well-known bimolecular nucleophilic substitution reactions in organic chemistry. Nucleophillic attack by the metal center at the less electronegative atom in the substrate leads to cleavage of the R–X bond, to form an [M–R]$^+$ species. This step is followed by rapid coordination of the anion to the cationic metal center. For example, reaction of a square planar complex with methyl iodide:

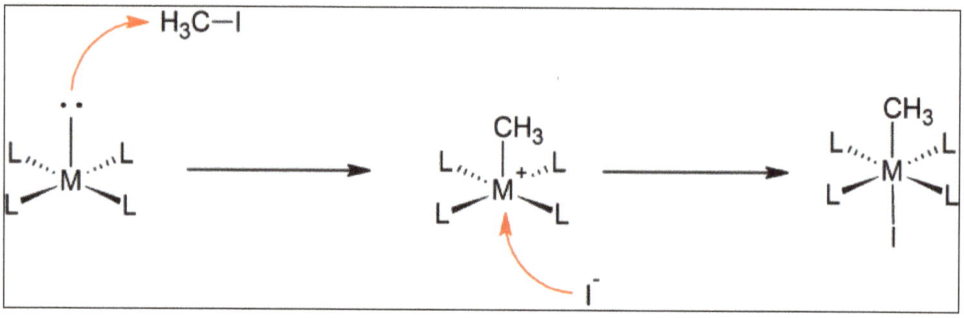

This mechanism is often assumed in the addition of polar and electrophilic substrates, such as alkyl halides and halogens.

Ionic

The ionic mechanism of oxidative addition is similar to the S_N2 type in that it involves the stepwise addition of two distinct ligand fragments. The key difference being that ionic mechanisms involve substrates which are dissociated in solution prior to any interactions with the metal center. An example of ionic oxidative addition is the addition of hydrochloric acid.

Radical

In addition to undergoing S_N2-type reactions, alkyl halides and similar substrates can add to a metal center via a radical mechanism, although some details remain controversial. Reactions which are generally accepted to proceed by a radical mechanism are known however. One example was proposed by Lednor and co-workers.

Initiation:

$$\left[(CH_3)_2C(CN)N\right]_2 \rightarrow 2(CH_3)_2(CN)C^{\bullet} + N_2$$

$$(CH_3)_2(CN)C^{\bullet} + PhBr \rightarrow (CH_3)_2(CN)CBr + Ph^{\bullet}$$

Propagation:

$$Ph^{\bullet} + \left[Pt(PPh_3)_2\right] \rightarrow \left[Pt(PPh_3)_2Ph\right]^{\bullet}$$

$$\left[Pt(PPh_3)_2Ph\right]^{\bullet} + PhBr \rightarrow \left[Pt(PPh_3)_2PhBr\right] + Ph^{\bullet}$$

Applications

Oxidative addition and reductive elimination are invoked in many catalytic processes both in homogeneous catalysis (i.e., in solution) such as the Monsanto process and alkene hydrogenation using Wilkinson's catalyst. It is often suggested that oxidative addition-like reactions are also involved in mechanisms of heterogeneous catalysis, e.g. hydrogenation catalyzed by platinum metal. Metals are however characterised by band structures, so oxidation states are not meaningful. Oxidative addition is also needed in order for nucleophilic addition of an alkyl group to occur. Oxidative insertion is also a crucial step in many cross-coupling reactions like the Suzuki coupling, Negishi coupling, and the Sonogashira coupling.

REDUCTIVE ELIMINATION

Reductive elimination is an elementary step in organometallic chemistry in which the oxidation state of the metal center decreases while forming a new covalent bond between two ligands. It is the microscopic reverse of oxidative addition, and is often the product-forming step in many catalytic processes. Since oxidative addition and reductive elimination are reverse reactions, the same mechanisms apply for both processes, and the product equilibrium depends on the thermodynamics of both directions.

Reductive elimination is often seen in higher oxidation states, and can involve a two-electron change at a single metal center (mononuclear) or a one-electron change at each of two metal centers (binuclear, dinuclear, or bimetallic).

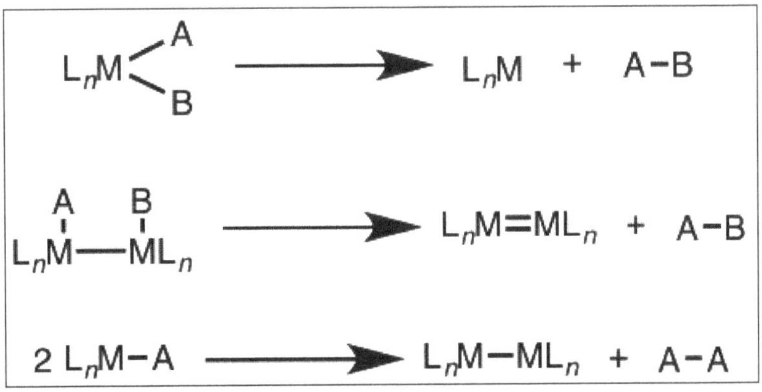

For mononuclear reductive elimination, the oxidation state of the metal decreases by two, while the d-electron count of the metal increases by two. This pathway is common for d^8 metals Ni(II), Pd(II), and Au(III) and d^6 metals Pt(IV), Pd(IV), Ir(III), and Rh(III). Additionally, mononuclear reductive elimination requires that the groups being eliminated must be cis to one another on the metal center.

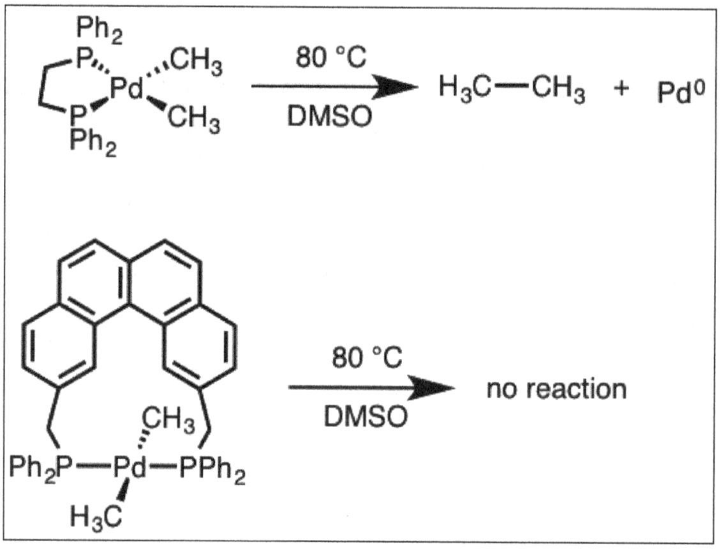

For binuclear reductive elimination, the oxidation state of each metal decreases by one, while the d-electron count of each metal increases by one. This type of reactivity is generally seen with first row metals, which prefer a one-unit change in oxidation state, but has been observed in both second and third row metals.

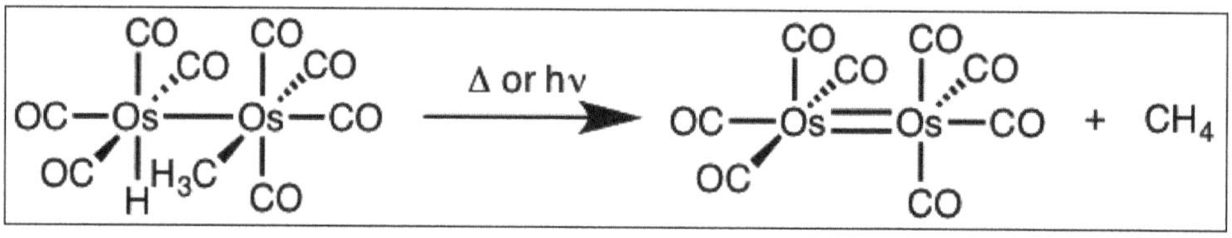

Mechanisms

As with oxidative addition, several mechanisms are possible with reductive elimination. The prominent mechanism is a concerted pathway, meaning that it is a nonpolar, three-centered transition state with retention of stereochemistry. In addition, an S_N2 mechanism, which proceeds with inversion of stereochemistry, or a radical mechanism, which proceeds with obliteration of stereochemistry, are other possible pathways for reductive elimination.

Octahedral Complexes

The rate of reductive elimination is greatly influenced by the geometry of the metal complex. In octahedral complexes, reductive elimination can be very slow from the coordinatively saturated center, and often, reductive elimination only proceeds via a dissociative mechanism, where a ligand must initially dissociate to make a five-coordinate complex. This complex adopts a Y-type distorted trigonal bipyramidal structure where a π-donor ligand is at the basal position and the two groups to be eliminated are brought very close together. After elimination, a T-shaped three-coordinate complex is formed, which will associate with a ligand to form the square planar four-coordinate complex.

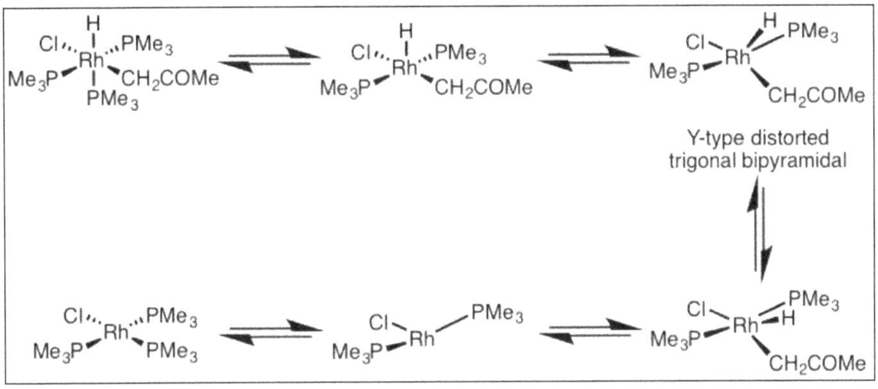

Square Planar Complexes

Reductive elimination of square planar complexes can progress through a variety of mechanisms: dissociative, nondissociative, and associative. Similar to octahedral complexes, a dissociative mechanism for square planar complexes initiates with loss of a ligand, generating a three-coordinate intermediate that undergoes reductive elimination to produce a one-coordinate metal complex. For a nondissociative pathway, reductive elimination occurs from the four-coordinate system to afford a two-coordinate complex. If the eliminating ligands are trans to each other, the complex must first undergo a trans to cis isomerization before eliminating. In an associative mechanism, a ligand must initially associate with the four-coordinate metal complex to generate a five-coordinate complex that undergoes reductive elimination synonymous to the dissociation mechanism for octahedral complexes.

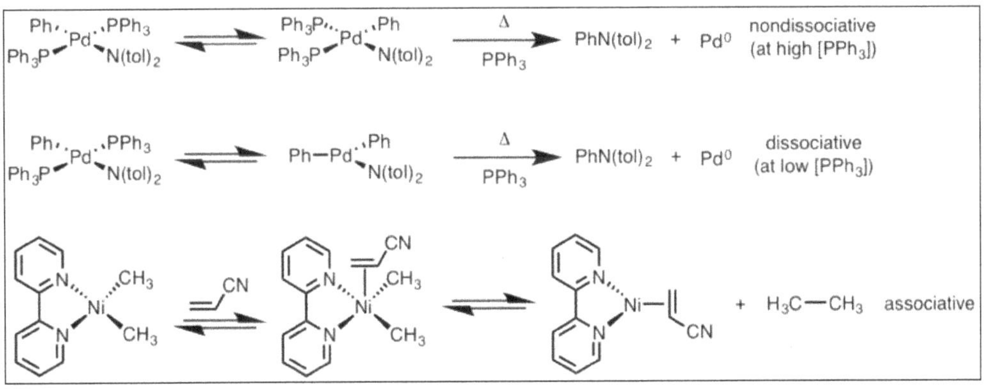

Factors that affect Reductive Elimination

Reductive elimination is sensitive to a variety of factors including: 1) metal identity and electron density; 2) sterics; 3) participating ligands; 4) coordination number; 5) geometry; and 6) photolysis/oxidation. Additionally, because reductive elimination and oxidative addition are reverse reactions, any sterics or electronics that enhance the rate of reductive elimination must thermodynamically hinder the rate of oxidative addition.

Metal Identity and Electron Density

First-row metal complexes tend to undergo reductive elimination faster than second-row metal complexes, which tend to be faster than third-row metal complexes. This is due to bond strength, with metal-ligand bonds in first-row complexes being weaker than metal-ligand bonds in third-row complexes. Additionally, electron-poor metal centers undergo reductive elimination faster than electron-rich metal centers since the resulting metal would gain electron density upon reductive elimination.

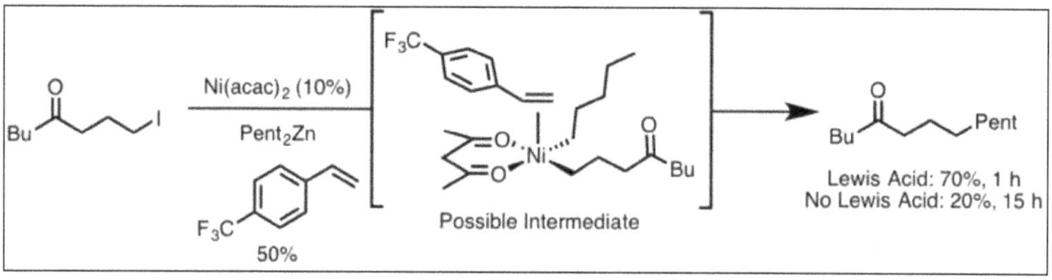

Sterics

Reductive elimination generally occurs more rapidly from a more sterically hindered metal center because the steric encumbrance is alleviated upon reductive elimination. Additionally, wide ligand bite angles generally accelerate reductive elimination because the sterics force the eliminating groups closer together, which allows for more orbital overlap.

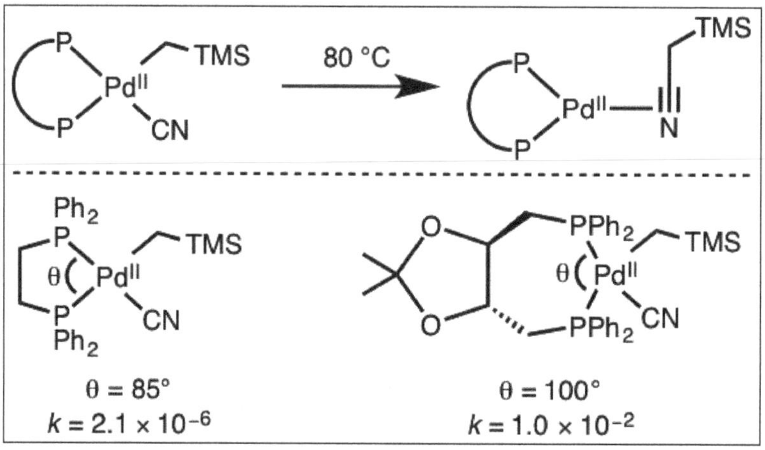

Participating Ligands

Kinetics for reductive elimination is hard to predict, but reactions that involve hydrides are particularly fast due to effects of orbital overlap in the transition state.

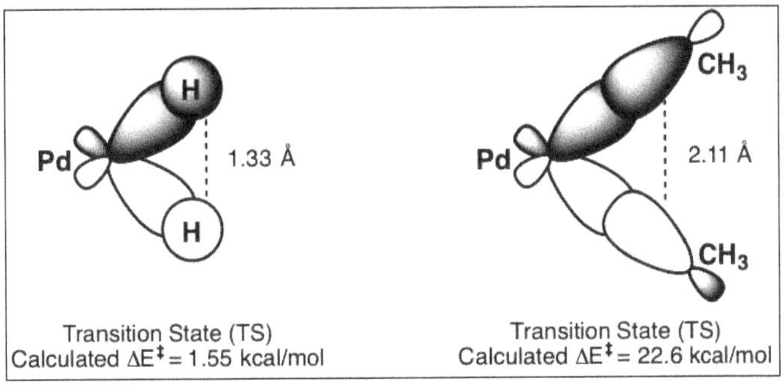

Coordination Number

Reductive elimination occurs more rapidly for complexes of three- or five-coordinate metal centers than for four- or six-coordinate metal centers. For even coordination number complexes, reductive elimination leads to an intermediate with a strongly metal-ligand antibonding orbital. When reductive elimination occurs from odd coordination number complexes, the resulting intermediate occupies a nonbonding molecular orbital.

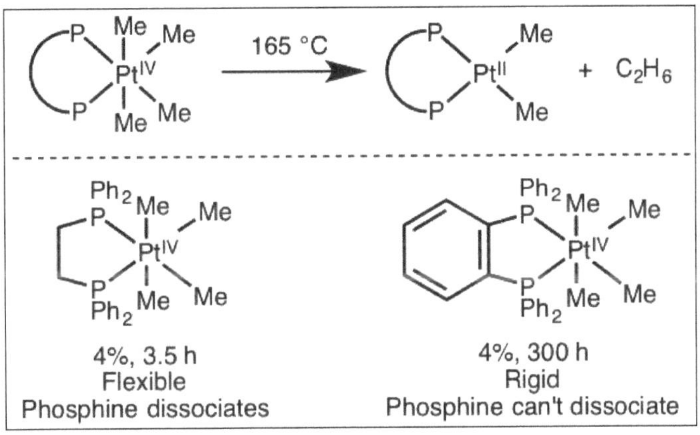

Geometry

Reductive elimination generally occurs faster for complexes whose structures resemble the product.

Photolysis/Oxidation

Reductive elimination can be induced by oxidizing the metal center to a higher oxidation state via light or an oxidant.

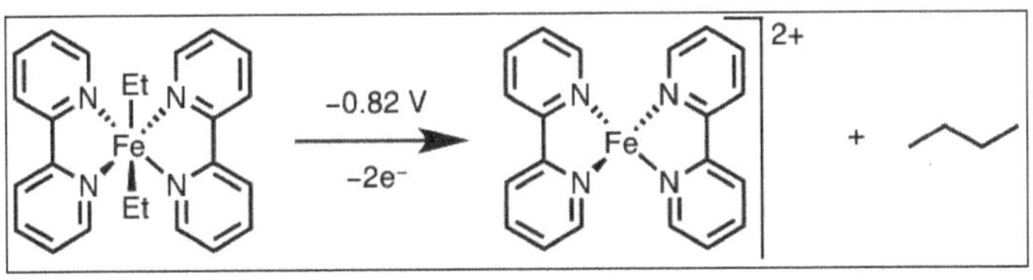

Applications

Reductive elimination has found widespread application in academia and industry, most notable being hydrogenation, the Monsanto acetic acid process, hydroformylation, and cross-coupling reactions. In many of these catalytic cycles, reductive elimination is the product forming step and regenerates the catalyst; however, in the Heck reaction and Wacker process, reductive elimination is involved only in catalyst regeneration, as the products in these reactions are formed via β–hydride elimination.

CARBOMETALATION

A carbometalation is any reaction where a carbon-metal bond reacts with a carbon-carbon π-bond to produce a new carbon-carbon σ-bond and a carbon-metal σ-bond. The resulting carbon-metal bond can undergo further carbometallation reactions or it can be reacted with a variety of electrophiles including halogenating reagents, carbonyls, oxygen, and inorganic salts to produce different organometallic reagents. Carbometalations can be performed on alkynes and alkenes to form products with high geometric purity or enantioselectivity, respectively. Some metals prefer to give the anti-addition product with high selectivity and some yield the syn-addition product. The outcome of syn and anti- addition products is determined by the mechanism of the carbometalation.

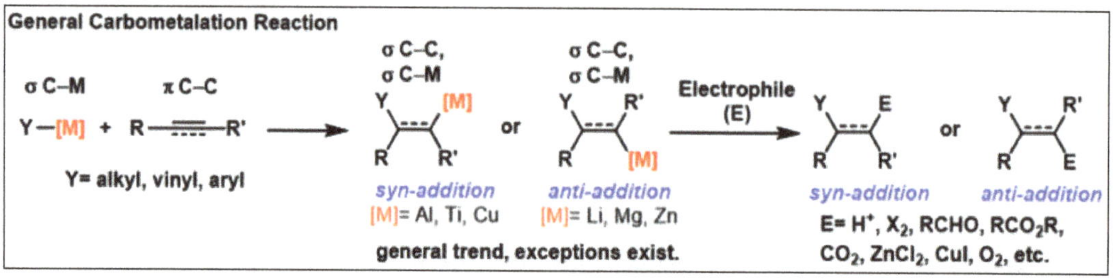

Carboalumination

The Carboalumination reaction is most commonly catalyzed by Zirconocene dichloride (or related catalyst). Some carboaluminations are performed with Titanocene complexes. This reaction is sometimes referred to as the Zr- catalyzed asymmetric carboalumination of alkenes (ZACA) or the Zr-catalyzed methylalumination of alkynes (ZMA).

The most common trialkyl aluminum reagents for this transformation are trimethyl aluminum, triethyl aluminum, and sometimes triisobutyl aluminum. When using trialkylaluminum reagents that have beta-hydrides, eliminations and hydroaluminum reactions become competing processes. The general mechanism of the ZMA reaction can be described as first the formation of the active catalytic species from the pre-catalyst zirconocene dichloride through its reaction with trimethyl aluminum. First transmetalation of a methyl from the aluminum to the zirconium occurs. Next, chloride abstraction by aluminum creates a cationic zirconium species that is closely associated with an anionic aluminum complex. This zirconium cation can coordinate an alkene or alkyne where migratory insertion of a methyl then takes place. The resultant vinyl or alkyl zirconium species can undergo a reversible, but stereoretentive transmetalation with an organoaluminum to provide the carboalumination product and regeneration of the zirconcene dichloride catalyst. This process generally provides the syn-addition product; however, conditions exist to provide the anti-addition product though a modified mechanism.

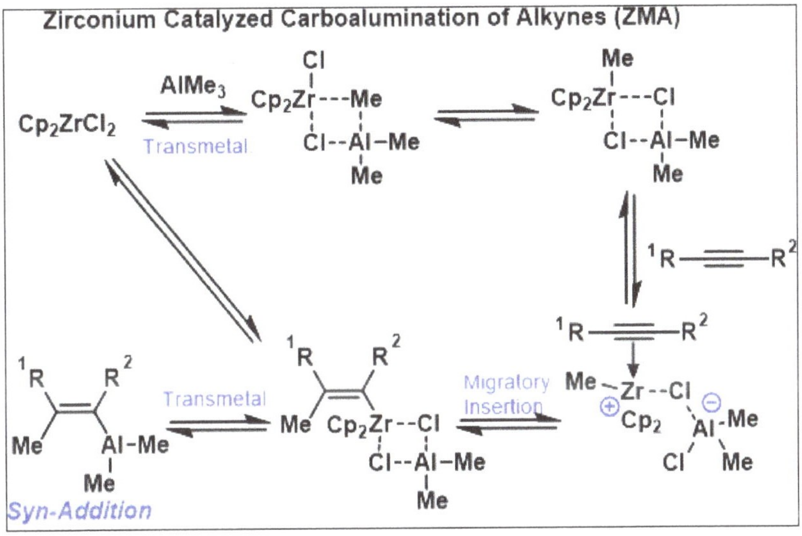

Trimethyl Silyl (TMS) protected alkynes, trimethyl germanium alkynes, and terminal alkynes can produce anti-carboalumination products at room temperature or elevated temperatures if a coordinating group is nearby on the substrate. In these reactions, first syn-carboalumination takes place under the previously outlined mechanism. Then, another equivalent of aluminum that is coordinated to the directing group can displace the vinyl aluminum, inverting the geometry at the carbon where displacement takes place.

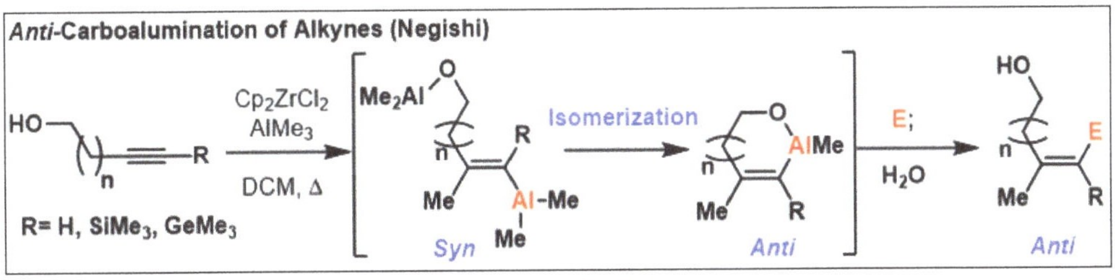

This forms a thermodynamically favorable metallacycle to prevent subsequent inversions. Formally, this process provides anti-carboalumination products that can be quenched with electrophiles. A limitation of this methodology is that the directing group must be sufficiently close to the carbon-carbon π-bond to form a thermodynamically favorable ring or else mixtures of geometric isomers will form.

Enantioselective Carboalumination of 1-Alkenes (ZACA Reaction) (Negishi)

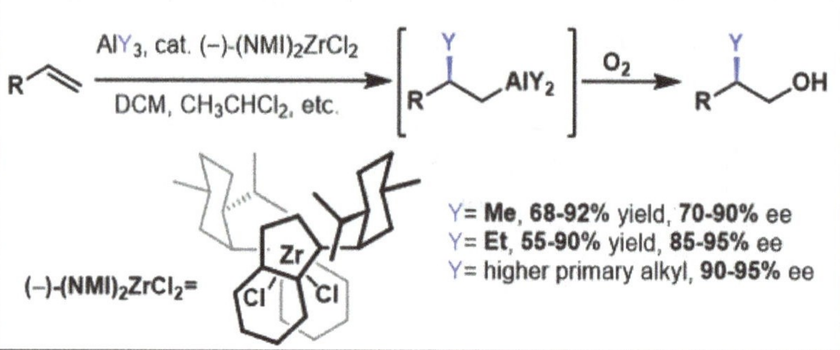

The carboalumination of alkenes to form substituted alkanes can be rendered enantioselective if prochiral alkenes are used. In these reactions, a chiral indenyl zirconium catalyst is used to induce enantioselectivity. In these reactions, high enantioselectivities were obtained for a several trialkyl aluminum reagents, however, the yield decreases dramatically with each additional carbon of the alkyl chain on the trialkyl aluminum reagent.

Carbolithiation

Enantioselective, Directed Carbolithiation Using n-BuLi

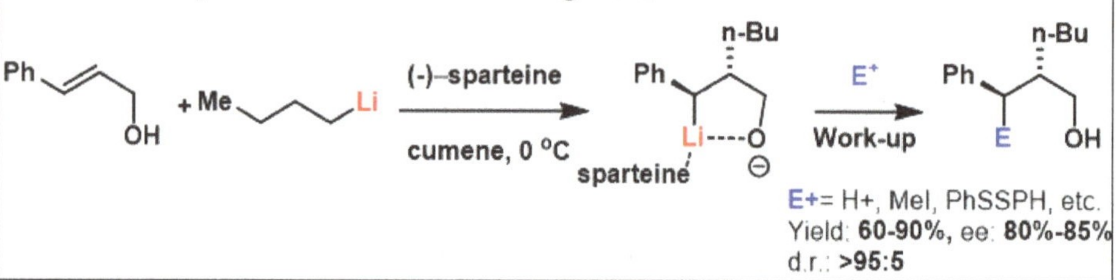

Carbolithiation is the addition of an organolithium reagent across a carbon-carbon pi-bond. The organolithium reagents used in this transformation can be commercial (such as n-butyllithium) or can be generated through deprotonation or lithium halogen exchange. Both inter- and intramolecular examples of carbolithiation exist and can be used in synthesis to generate complexity. Organolithiums are highly reactive chemicals and often the resulting organolithium reagent generated from the carbolithiation can continue to react with electrophiles or remaining starting material (resulting in polymerization). This reaction has been rendered enantioselective through the use of sparteine, which can chelate the lithium ion and induce chirality. Today, this is not a common strategy due to a shortage of natural sparteine.

Another demonstration of this reaction type is an alternative route to tamoxifen starting from diphenylacetylene and ethyllithium: The capturing electrophile here is triisopropyl borate forming the boronic acid R–B$(OH)_2$. The second step completing tamoxifen is a Suzuki reaction.

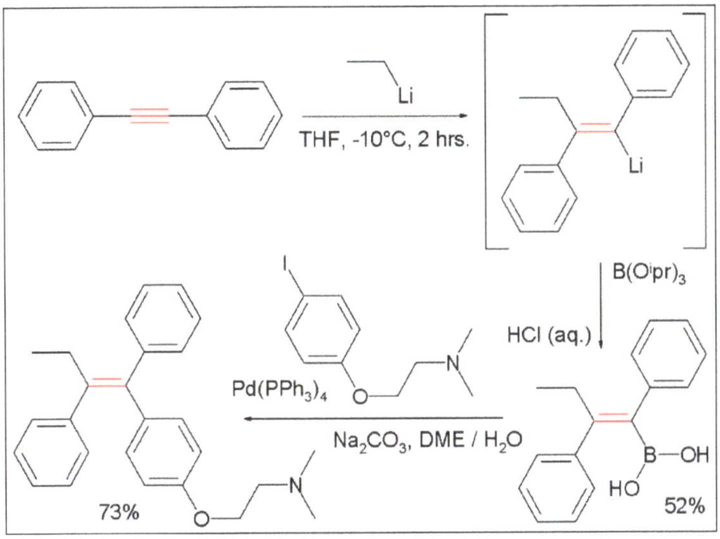

As a consequence of the high reactivity of organolithiums as strong bases and strong nucleophiles, the substrate scope of the carbolithiation is generally limited to chemicals that do not contain acidic or electrophilic functional groups.

Carbomagnesiation and Carbozincation

Due to the decreased nucleophilicity of Grignard reagents (organomagnesium reagents) and organozinc reagents, non-catalyzed carbomagnesiation and carbozincation reactions are typically only observed on activated or strained alkenes and alkynes. For example, electronwithdrawing groups like esters, nitriles or sulfones must be in conjugation with the carbon-carbon π-system or a directing group like an alcohol or amine must be nearby to direct the reaction. These reactions can be catalyzed by a variety of transition metals such as iron, copper, zirconium, nickel, cobalt and others.

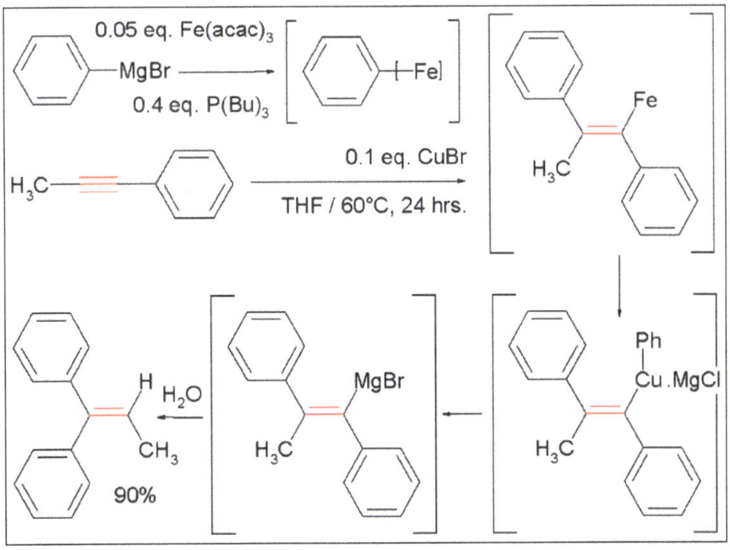

Illustrative is the Fe-catalyzed reaction of methylphenylacetylene with phenylmagnesium bromide, which generates vinyl magnesium intermediate. Hydrolysis affords the diphenylalkene as shown in figure.

Carbopalladation

Carbopalladations can be a description of the elementary step of a reaction catalyzed by a palladium catalyst (Mizoroki-Heck reaction) and can also refer to a carbometalation reaction with a palladium catalyst (alkene difunctionalization, hydrofunctionalization, or reductive heck).

CARBON–HYDROGEN BOND ACTIVATION

Carbon–hydrogen bond functionalization (C–H functionalization) is a type of reaction in which a carbon–hydrogen bond is cleaved and replaced with a carbon–X bond (where X is usually carbon, oxygen, or nitrogen). The term usually implies that a transition metal is involved in the C-H cleavage process. Reactions classified by the term typically involve the hydrocarbon first to react with a metal catalyst to create an organometallic complex in which the hydrocarbon is coordinated to the inner-sphere of a metal, either via an intermediate "alkane or arene complex" or as a transition state leading to a "M–C" intermediate. The intermediate of this first step (known as C–H activation and sometimes used interchangeably with C-H functionalization) can then undergo subsequent reactions to produce the functionalized product. Important to this definition is the requirement that during the C–H cleavage event, the hydrocarbyl species remains associated in the inner-sphere and under the influence of "M".

As an alternative definition, other authors use the term C–H functionalization to mean any organic transformation in which the net result in the transformation of a relatively inert C–H bond into a C–X bond (i.e., a functional group), irrespective of the mechanism (or with an agnostic attitude towards it). In particular, this definition does not require transition metal coordination to the hydrocarbon in the reaction mechanism. This broader definition includes the narrower definition given above as a subset. However, this definition would also include iron-catalyzed alkane functionalization reaction that proceeds through the oxygen rebound mechanism (e.g. cytochrome P450 enzymes and their synthetic analogues), in which a metal–carbon bond is not believed to be involved. Likewise, the ligand-based reactivity of many metal carbene species with hydrocarbons would also fall under this category, although some cases are mechanistically ambiguous. Some authors similarly define C–H activation broadly as the C–H cleaving step of any mechanism that results in functionalization of a hydrocarbon group (or any observable consequence of C–H bond cleavage, like H/D exchange). Still others maintain the original narrow definition of the term C–H activation, while using C–H functionalization in its broader sense.

Classification

Mechanisms for C-H activations can be classified into three general categories:

- Oxidative addition, in which a low-valent metal center inserts into a carbon-hydrogen bond, which cleaves the bond and oxidizes the metal:

  $L_nM + RH \rightarrow L_nMR(H)$

- Electrophilic activation in which an electrophilic metal attacks the hydrocarbon, displacing a proton:

  $L_nM^+ + RH \rightarrow L_nMR + H^+$

  Substrate undergoes an S_EAr-type mechanism.

- Sigma-bond metathesis, which proceeds through a "four-centered" transition state in which bonds break and form in a single step:

  $L_nMR + R'H \rightarrow L_nMR' + RH$

Directed C-H Activation

Directed-, chelation-assisted-, or "guided" C-H activation involves directing groups that influence regio- and stereochemistry. This is the most useful style of C-H activation in organic synthesis. N,N-dimethylbenzylamine undergoes cyclometalation readily by many transition metals: A semi-practical implementations involve weakly coordinating directing groups, as illustrated by the Murai reaction.

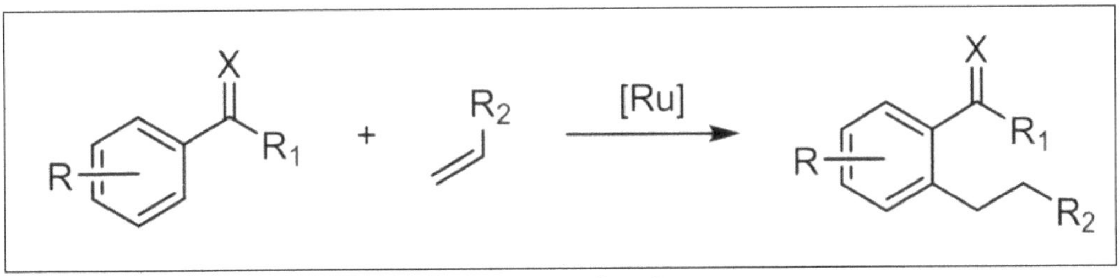

Murai reaction. X = directing group.

The mechanism for the Pd-catalyzed C-H activation reactions of 2-phenylpyridine involves a metallacycle intermediate. The intermediate is oxidized to form a Pd^{IV} species, followed by reductive elimination to form the C-O bond and release the product.

Borylation

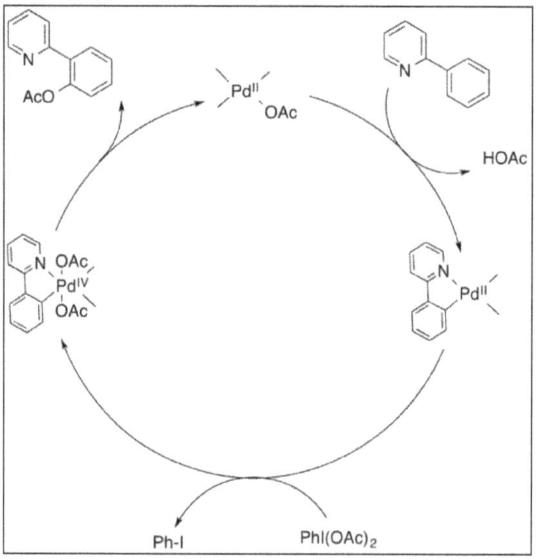

Mechanism for Pd-catalyzed C-H activation.

Transforming C-H bonds into C-B bonds through borylation has been thoroughly investigated due to their utility in synthesis (i.e. for cross-coupling reactions). J. F. Hartwig reported a highly regioselective arene and alkane borylation catalyzed by a rhodium complex. In the case of alkanes, exclusive terminal functionalization was observed.

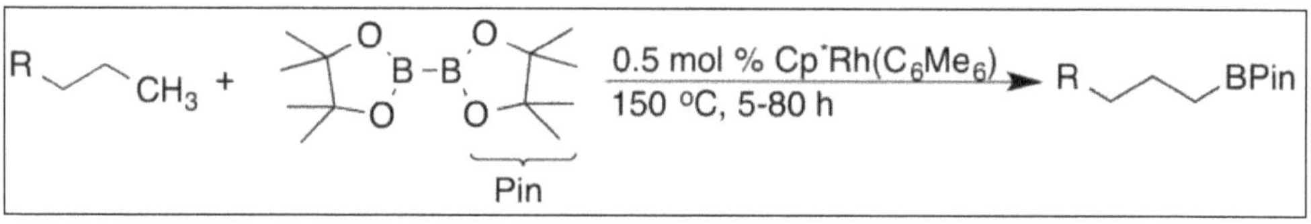

Later, ruthenium catalysts were discovered to have higher activity and functional group compatibility.

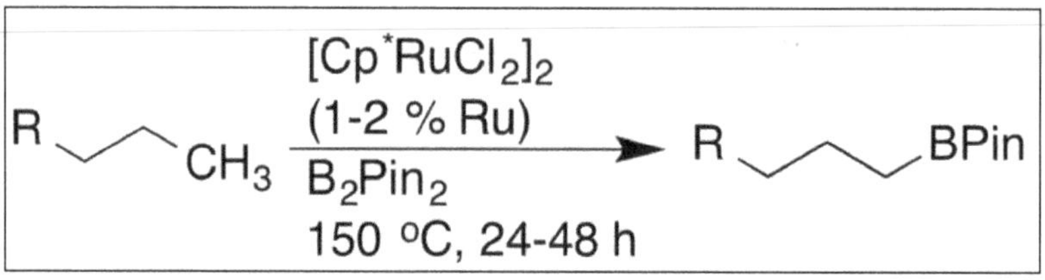

Other borylation catalysts have also been developed, including iridium-based catalysts, which successfully activate C-H bonds with high compatibility.

Natural Gas

Naturally occurring methane is not utilized as a chemical feedstock, despite its abundance and low cost. Current technology makes prodigious use of methane by steam reforming to produce syngas, a mixture of carbon monoxide and hydrogen. This syngas is then used in Fischer-Tropsch reactions to make longer carbon chain products or methanol, one of the most important industrial chemical feedstocks. An intriguing methods to convert these hydrocarbons involves C-H activation. Periana, for example, reported that complexes containing late transition metals, such as Pt, Pd, Au, and Hg, react with methane (CH_4) in H_2SO_4 to yield methyl bisulfate. The process has not however been implemented commercially.

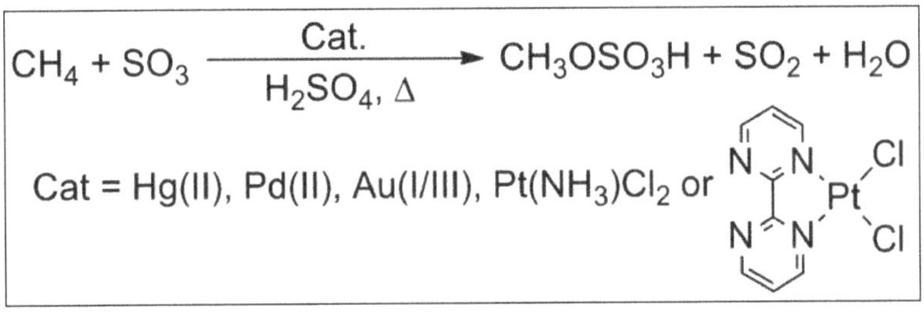

Asymmetric C-H activations

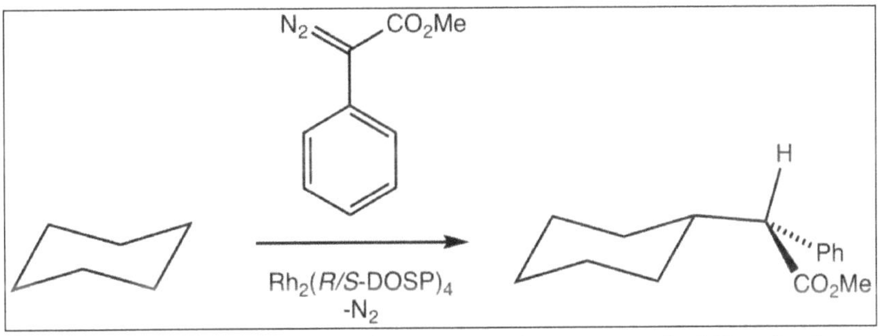

Methyl phenyldiazoacetate is the precursor for asymmetric C-H activation viadonor-acceptor carbene using a chiral dirhodium catalyst.

The total synthesis of lithospermic acid employs guided C-H functionalization late stage to a highly functionalized system. The directing group, a chiral nonracemic imine, is capable of performing an intramolecular alkylation, which allows for the rhodium-catalyzed conversion of imine to the dihydrobenzofuran.

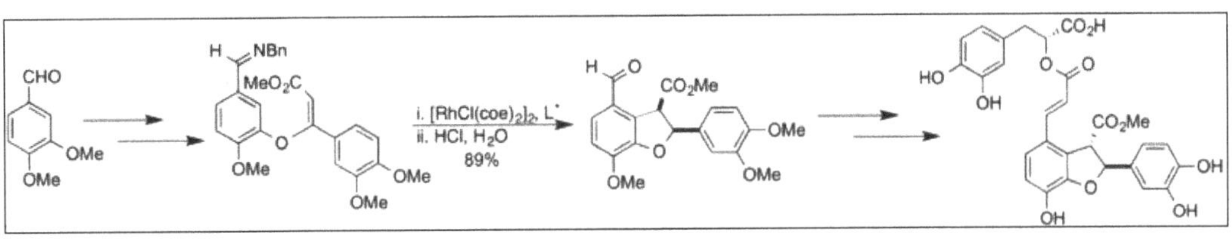

The total synthesis of calothrixin A and B features an intramolecular Pd-catalyzed cross coupling reaction via C-H activation, an example of a guided C-H activation. Cross coupling occurs between aryl C-I and C-H bonds to form a C-C bond. The synthesis of a mescaline analogue employs the rhodium-catalyzed enantioselective annulation of an aryl imine via a C-H activation.

TRANSMETALATION

Transmetalation is a type of organometallic reaction that involves the transfer of ligands from one metal to another. It has the general form:

$$M_1 - R + M_2 - R' \rightarrow M_1 - R' + M_2 - R$$

Where, R and R′ can be, but are not limited to, an alkyl, aryl, alkynyl, allyl, halogen, or pseudo-halogen group. The reaction is usually an irreversible process due to thermodynamic and kinetic reasons. Thermodynamics will favor the reaction based on the electronegativities of the metals and kinetics will favor the reaction if there are empty orbitals on both metals. There are different types of transmetalation including redox-transmetalation and redox-transmetalation/ligand exchange. During transmetalation the metal-carbon bond is activated, leading to the formation of new metal-carbon bonds. Transmetalation is commonly used in catalysis, synthesis of main group complexes, and synthesis of transition metal complexes.

Types of Transmetalation

There are two main types of transmetalation, redox-transmetalation (RT) and redox-transmetalation/ligand-exchange (RTLE). Below, M_1 is usually a 4d or 5d transition metal and M_2 is usually a main group or 3d transition metal. By looking at the electronegativities of the metals and ligands, one can predict whether the RT or RTLE reaction will proceed and what products the reaction will yield. For example, one can predict that the addition of 3 $HgPh_2$ to 2 Al will yield 3 Hg and 2 $AlPh_3$ because Hg is a more electronegative element than Al.

Redox-transmetalation

$$M_1^{n+} - R + M_2 \rightarrow M_1 + M_2^{n+} - R$$

In redox-transmetalation a ligand is transferred from one metal to the other through an intermolecular mechanism. During the reaction one of the metal centers is oxidized and the other is reduced. The electronegativities of the metals and ligands are what causes the reaction to go forward. If M_1 is more electronegative than M_2, it is thermodynamically favorable for the R group to coordinate to the less electronegative M_2.

Redox-transmetalation/Ligand-exchange

$$M_1 - R + M_2 - X \rightarrow M_1 - X + M_2 - R$$

In redox-transmetalation/ligand exchange the ligands of two metal complexes switch places with each other, bonding with the other metal center. The R ligand can be an alkyl, aryl, alkynyl, or allyl group and the X ligand can be a halogen, pseudo-halogen, alkyl, or aryl group. The reaction can proceed by two possible intermediate steps. The first is an associative intermediate, where the R and X ligands bridge the two metals, stabilizing the transition state. The second and less common intermediate is the formation of a cation where R is bridging the two metals and X is anionic. The RTLE reaction proceeds in a concerted manner. Like in RT reactions, the reaction is driven by electronegativity values. The X ligand is attracted to highly electropositive metals. If M_1 is a more electropositive metal than M_2, it is thermodynamically favorable for the exchange of the R and X ligands to occur.

Applications

Cross-coupling Reactions

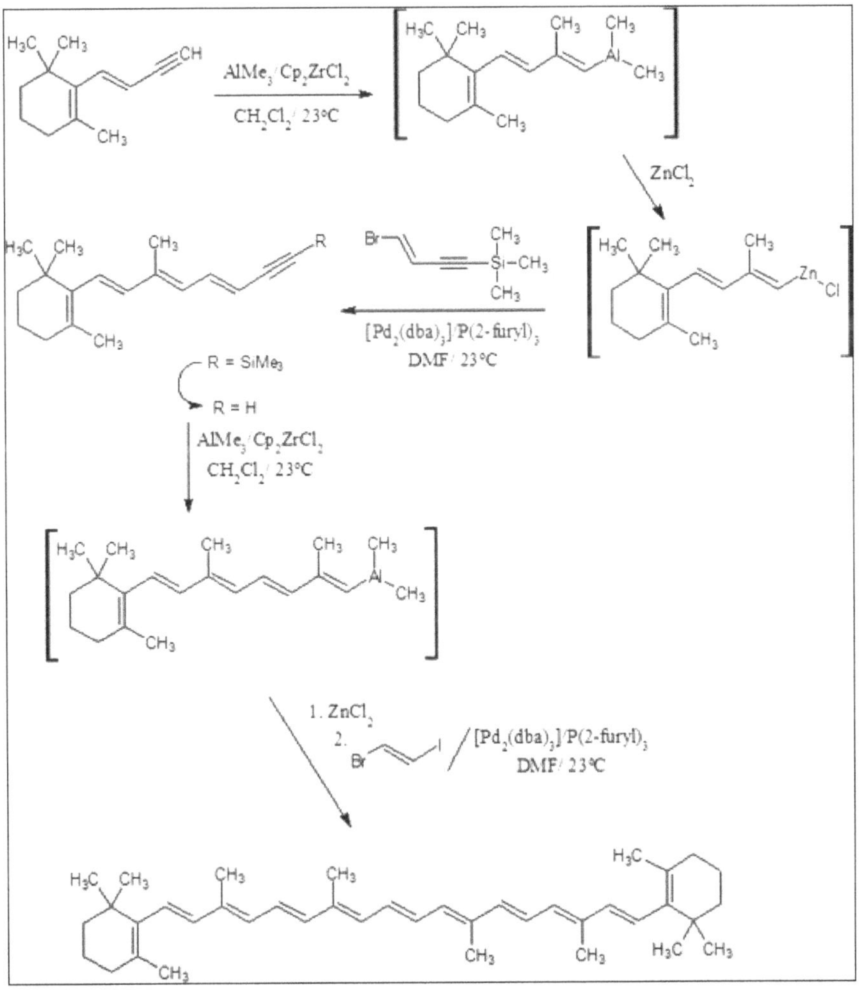

Synthesis of β-carotene by Negishi cross-coupling and transmetalation cascades.

Transmetalation is often used as a step in the catalytic cycles of cross-coupling reactions. Some of the cross-coupling reactions that include a transmetalation step are Stille cross-coupling, Suzuki cross-coupling, Sonogashira cross-coupling, and Negishi cross-coupling. The most useful cross-coupling catalysts tend to be ones that contain palladium. Cross-coupling reactions have the general form of R′–X + M–R → R′–R + M–X and are used to form C–C bonds. R and R′ can be any carbon fragment. The identity of the metal, M, depends on which cross-coupling reaction is being used. Stille reactions use tin, Suzuki reactions use boron, Sonogashira reactions use copper, and Negishi reactions use zinc. The transmetalation step in palladium catalyzed reactions involve the addition of an R–M compound to produce an R′–Pd–R compound. Cross-coupling reactions have a wide range of applications in synthetic chemistry including the area of medicinal chemistry. The Stille reaction has been used to make an antitumor agent, (±)-epi-jatrophone; the Suzuki reaction has been used to make an antitumor agent, oximidine II; the Sonogashira reaction has been used to make an anticancer drug, eniluracil; and the Negishi reaction has been used to make the carotenoid β-carotene via a transmetalation cascade.

Lanthanides

Lanthanide organometallic complexes have been synthesized by RT and RTLE. Lanthanides are very electropositive elements.

Organomercurials, such as $HgPh_2$, are common kinetically inert RT and RTLE reagents that allow functionalized derivatives to be synthesized, unlike organolithiums and Grignard reagents. Diarylmercurials are often used to synthesize lanthanide organometallic complexes. $Hg(C_6F_5)_2$ is a better RT reagent to use with lanthanides than $HgPh_2$ because it does not require a step to activate the metal. However, phenyl-substituted lanthanide complexes are more thermally stable than the pentafluorophenyl complexes. The use of $HgPh_2$ led to the synthesis of a ytterbium complex with different oxidation states on the two Yb atoms:

$$Yb(C_{10}H_8)(THF)_2 + HgPh_2 \rightarrow Yb^{II}Yb^{III}Ph_5(THF)_4$$

In the $Ln(C_6F_5)_2$ complexes, where Ln = Yb, Eu, or Sm, the Ln–C bonds are very reactive, making them useful in RTLE reactions. Protic substrates have been used as a reactant with the $Ln(C_6F_5)_2$ complex as shown: $Ln(C_6F_5)_2 + 2LH \rightarrow Ln(L)_2 + 2C_6F_5H$. It is possible to avoid the challenges of working with the unstable $Ln(C_6F_5)_2$ complex by forming it in situ by the following reaction:

$$Ln + HgR_2 + 2\,LH \rightarrow Ln(L)_2 + Hg + 2\,RH$$

Organotins are also kinetically inert RT and RTLE reagents that have been used in a variety of organometallic reactions. They have applications to the synthesis of lanthanide complexes, such as in the following reaction:

$$Yb + Sn(N(SiMe_3)_2)_2 \rightarrow Yb(N(SiMe_3)_2)_2 + Sn$$

Actinides

RT can be used to synthesize actinide complexes. RT has been used to synthesize uranium halides using uranium metal and mercury halides as shown:

$$U + HgX \rightarrow UX + Hg \quad (X = Cl, Br, I)$$

This actinide RT reaction can be done with multiple mercury compounds to coordinate ligands other than halogens to the metal:

$$2U + 3(C_5H_5)_2Hg + HgCl_2 \rightarrow 2(C_5H_5)_3UCl + 4Hg$$

Alkaline Earth Metals

Alkaline earth metal complexes have been synthesized by RTLE, employing the same methodology used in synthesizing lanthanide complexes. The use of diphenylmercury in alkaline-earth metal reactions leads to the production of elemental mercury. The handling and disposal of elemental mercury is challenging due to its toxicity to humans and the environment. This led to the desire for an alternative RTLE reagent that would be less toxic and still very effective. Triphenylbismuth, $BiPh_3$, was discovered to be a suitable alternative. Mercury and bismuth have similar electronegativity values and behave similarly in RTLE reactions. $BiPh_3$ has been used to synthesize alkaline-earth metal amides and alkaline-earth metal cyclopentadienides. The difference between $HgPh_2$ and $BiPh_3$ in these syntheses was that the reaction time was longer when using $BiPh_3$.

HYDROMETALATION

The addition of a metal hydride to a multiple bond is called hydrometallation, and it leads to the formation of a metal-carbon bond.

$$M-H + H_2C=CH_2 \rightarrow MH_2C-CH_3$$

This reaction is driven mainly by the high C–H bond strength relative to most E–H bond strengths. Two important hydrometallation reactions are hydroboration and hydrosilation, illustrated, respectively, by the following examples.

$$H_3BOR_2 + (H_3C)_2C = CH_2 \xrightarrow{R_2O} (H_3C)_2CH - CH_2BH_2$$
$$SiH_4 + (CH_3)_2C = CH_2 \xrightarrow{H_2PtCl_6} (CH_3)_2CH - CH_2SiH_3$$

In the hydroboration and hydrosilation of an unsymmetrical alkene, the boron or silicon binds to the carbon atom that has less-bulky substituents, and the smaller hydrogen atom binds to the carbon atom that has bulky substituents—$(CH_3)_2C$ in the above equations. Hydroboration was discovered and developed in the United States by Herbert C. Brown, who shared the Nobel Prize for Chemistry in 1979 for this research. Both hydroboration and hydrosilation are widely used in the synthesis of complex organic molecules. In these applications, the B—C or Si—C bond is generally cleaved in a subsequent step to produce a product that is free of boron or silicon.

β-ELIMINATION REACTIONS

Elimination reactions involve the cleavage of a σ bond and formation of a π bond. A nucleophilic pair of electrons (either from another bond or a lone pair) heads into a new π bond as a leaving group departs. This process is called β-elimination because the bond β to the nucleophilic pair of electrons breaks. Transition metal complexes can participate in their own version of β-elimination, and metal alkyl complexes famously do so. Almost by definition, metal alkyls contain a nucleophilic bond—the M–C bond! This bond can be so polarized toward carbon, in fact, that it can promote the elimination of some of the world's worst leaving groups, like –H and $-CH_3$. However, the leaving group is not lost completely in organometallic β-eliminations. As the metal donates electrons, it receives electrons from the departing leaving group. When the reaction is complete, the metal has picked up a new π-bound ligand and exchanged one X-type ligand for another.

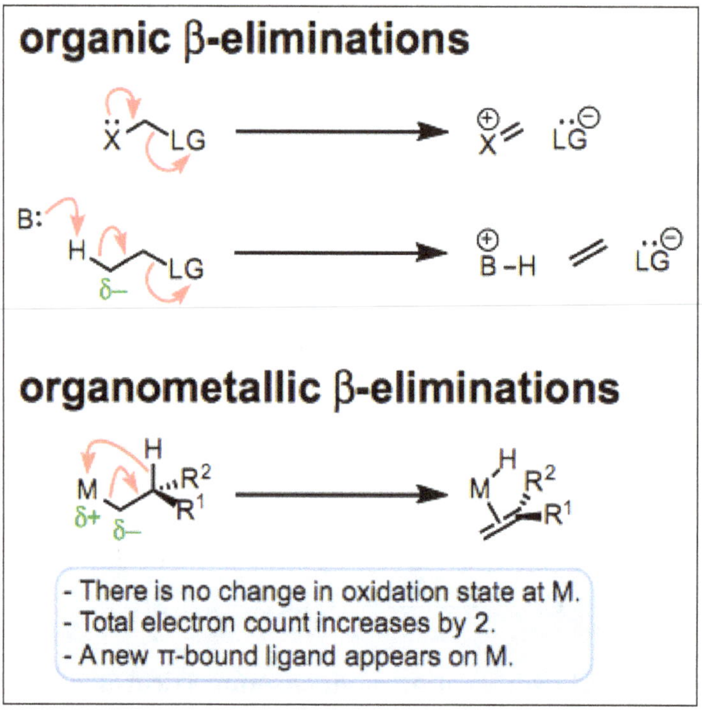

Comparing organic and organometallic β-eliminations. A nucleophilic bond or lone pair promotes loss or migration of a leaving group.

β-Hydride Elimination

The most famous and ubiquitous type of β-elimination is β-hydride elimination, which involves the formation of a π bond and an M–H bond. Metal alkyls that contain β-hydrogens experience rapid elimination of these hydrogens, provided a few other conditions are met.

The complex must have an open coordination site and an accessible, empty orbital on the metal center. The leaving group (–H) needs a place to land. Notice that after β-elimination, the metal has picked up one more ligand—it needs an empty spot for that ligand for elimination to occur. We can envision hydride "attacking" the empty orbital on the metal center as an important orbital interaction in this process.

The M–C_α and C_β–H bonds must have the ability to align in a syn coplanar arrangement. By "syn coplanar" we mean that all four atoms are in a plane and that the M–C_α and C_β–H bonds are on the same side of the C_α–C_β bond (a dihedral angle of 0°). You can see that conformation in the figure above. In the syn coplanar arrangement, the C–H bond departing from the ligand is optimally lined up with the empty orbital on the metal center. Hindered or cyclic complexes that cannot achieve this conformation do not undergo β-hydride elimination. The need for a syn coplanar conformation has important implications for eliminations that may establish diastereomeric olefins: β-elimination is stereospecific. One diastereomer leads to the (E)-olefin, and the other leads to the (Z)-olefin.

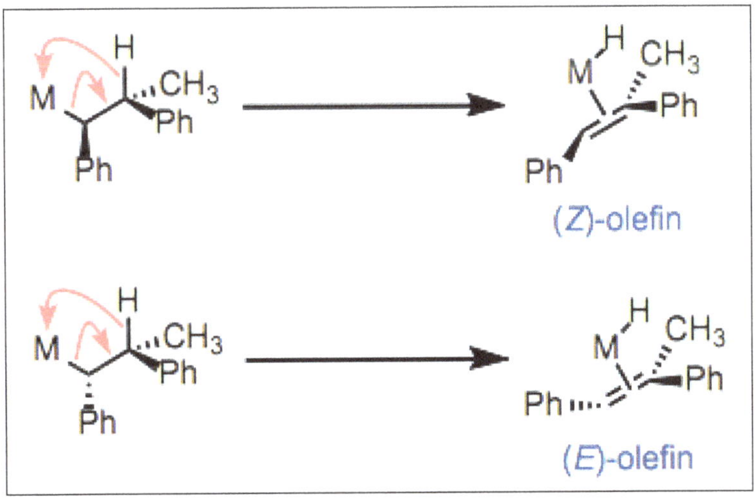

β-elimination is stereospecific. One diastereomer of reactant leads to the (Z)-olefin and the other to the (E)-olefin.

The complex must possess 16 or fewer total electrons. Examine the first figure one more time—notice that the total electron count of the complex increases by 2 during β-hydride elimination. Complexes with 18 total electrons don't undergo β-elimination because the product would end up with 20 total electrons. Of course, dissociation of a loose ligand can produce a 16-electron complex pretty easily, so watch out for ligand dissociation when considering the possibility of β-elimination in a complex. Ligand dissociation may be reversible, but β-Hydride elimination is almost always irreversible.

The metal must bear at least 2 d electrons. Why would the metal center need electrons for β-hydride elimination to occur? The answer lies in an old friend: backbonding. The σ C–H → M orbital interaction is not enough to promote elimination on its own; an M → σ* C–H interaction is also required.

Other β-elimination Reactions

The leaving group does not need to be hydrogen, of course, and a number of more electronegative groups come to mind as better candidates for leaving groups. β-Alkoxy and β-amino eliminations are usually thermodynamically favored thanks to the formation of strong M–O and M–N bonds, respectively. These reactions are so favored in β-alkoxyalkyl "complexes" of alkali and alkaline earth metals (R–Li, R–MgBr, etc.) that using these as σ-nucleophiles at carbon is untenable. Such compounds eliminate immediately upon their formation.

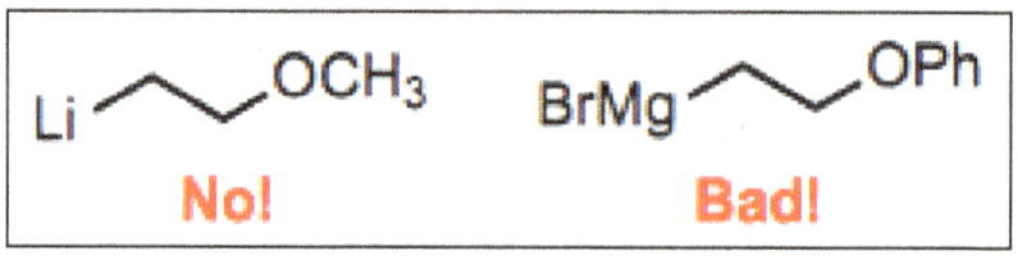

The atom bound to the metal doesn't have to be carbon. β-elimination of alkoxy ligands affords ketones or aldehydes bound at oxygen or through the C=O π bond (this step is important in many transfer hydrogenations, and an analogous process occurs in the Oppenauer oxidation). Amido ligands can undergo β-elimination to afford complexes of imines; however, this process tends to be slower than β-alkoxy elimination.

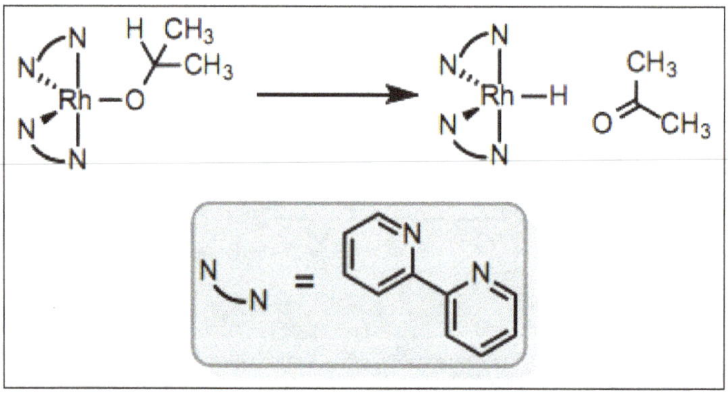

β-Elimination helps transfer the elements of dihydrogen from one organic compound to another.

Applications of β-eliminations

As with many concepts in organometallic chemistry, there are two ways to think about applications of β-elimination. One can take either the "inorganic" perspective, which focuses on the metal center, or the "organic" perspective, which focuses on the ligands.

With the metal center in focus, we can recognize that β-hydride elimination has the wonderful side effect of establishing an M–H bond—a feat generally difficult to achieve in a selective manner via oxidative addition of X–H. If the ligand from which the hydrogen came displaced something more electronegative, the whole process represents reduction at the metal center. For example, imagine rhodium(III) chloride is mixed with sodium isopropoxide, $NaOCH(CH_3)_2$. The isopropoxide easily displaces chloride, and subsequent β-hydride elimination affords a rhodium hydride, formally reduced with respect to the chloride starting material.

With the ligand in focus, the organic ligand is oxidized in the course of β-hydride elimination. Notice that the metal is reduced and the ligand oxidized! A π bond replaces a σ bond in the ligand, and if the conditions are right, this represents a bona fide oxidation (as opposed to a mere elimination). For example, oxidative addition into a C–H bond followed by β-hydride elimination at a C–H bond next door sets up an alkene where two adjacent C–H bonds existed before, an oxidation process. These dehydrogenation reactions are incredibly appealing in a theoretical sense, but still at an early stage when it comes to scope and practicality.

References

- Hydrogenation: courses.lumenlearning.com, Retrieved 02 February, 2020
- Robert Franke, Detlef Selent, Armin Börner (2012). "Applied Hydroformylation". Chem. Rev. 112 (11): 5675–5732. doi:10.1021/cr3001803
- Jay A. Labinger "Tutorial on Oxidative Addition" Organometallics, 2015, volume 34, pp 4784–4795. doi:10.1021/acs.organomet.5b00565
- The-stability-and-reactivity-of-organometallic-compound, science-organometallic-compound-1 001523: britannica.com, Retrieved 16 June, 2020
- Crumpton-Bregel, D. M.; Goldberg, K. I. (2003). "Mechanisms of C-C and C-H Alkane Reductive Eliminations from Octahedral Pt(IV): Reaction via Five-Coordinate Intermediates or Direct Elimination?". J. Am. Chem. Soc. 125: 9442. doi:10.1021/ja029140u. PMID 12889975
- Fekl, U.; Goldberg, K. I. (2003). Homogeneous Hydrocarbon C-H Bond Activation and Functionalization with Platinum. Advances in Inorganic Chemistry. 54. pp. 259–320. doi:10.1016/S0898-8838(03)54005-3. ISBN 9780120236541
- Xu, Shiqing; Negishi, Ei-ichi (2016-10-18). "Zirconium-Catalyzed Asymmetric Carboalumination of Unactivated Terminal Alkenes". Accounts of Chemical Research. 49 (10): 2158–2168. doi:10.1021/acs.accounts.6b00338. ISSN 0001-4842. PMID 27685327

Permissions

All chapters in this book are published with permission under the Creative Commons Attribution Share Alike License or equivalent. Every chapter published in this book has been scrutinized by our experts. Their significance has been extensively debated. The topics covered herein carry significant information for a comprehensive understanding. They may even be implemented as practical applications or may be referred to as a beginning point for further studies.

We would like to thank the editorial team for lending their expertise to make the book truly unique. They have played a crucial role in the development of this book. Without their invaluable contributions this book wouldn't have been possible. They have made vital efforts to compile up to date information on the varied aspects of this subject to make this book a valuable addition to the collection of many professionals and students.

This book was conceptualized with the vision of imparting up-to-date and integrated information in this field. To ensure the same, a matchless editorial board was set up. Every individual on the board went through rigorous rounds of assessment to prove their worth. After which they invested a large part of their time researching and compiling the most relevant data for our readers.

The editorial board has been involved in producing this book since its inception. They have spent rigorous hours researching and exploring the diverse topics which have resulted in the successful publishing of this book. They have passed on their knowledge of decades through this book. To expedite this challenging task, the publisher supported the team at every step. A small team of assistant editors was also appointed to further simplify the editing procedure and attain best results for the readers.

Apart from the editorial board, the designing team has also invested a significant amount of their time in understanding the subject and creating the most relevant covers. They scrutinized every image to scout for the most suitable representation of the subject and create an appropriate cover for the book.

The publishing team has been an ardent support to the editorial, designing and production team. Their endless efforts to recruit the best for this project, has resulted in the accomplishment of this book. They are a veteran in the field of academics and their pool of knowledge is as vast as their experience in printing. Their expertise and guidance has proved useful at every step. Their uncompromising quality standards have made this book an exceptional effort. Their encouragement from time to time has been an inspiration for everyone.

The publisher and the editorial board hope that this book will prove to be a valuable piece of knowledge for students, practitioners and scholars across the globe.

Index

www.ingramcontent.com/pod-product-compliance
Lightning Source LLC
LaVergne TN
LVHW080633210926
848485LV00003B/255
* 9 7 8 1 6 3 9 8 7 4 2 0 0 *